Acta Numerica 1996

Acta Numerica 1996

Volume 5

Published by the Press Syndicate of the University of Cambridge
The Pitt Building, Trumpington Street, Cambridge CB2 1RP
40 West 20th Street, New York, NY 10011-4211, USA
10 Stamford Road, Oakleigh, Melbourne 3166, Australia

© Cambridge University Press 1996

First published 1996

Printed in Great Britain at the University Press, Cambridge

Library of Congress cataloguing in publication data available

A catalogue record for this book is available from the British Library

ISBN 0-521-57234-7 hardback
ISSN 0962-4929

Contents

Hierarchical bases and the finite element method 1
  Randolph E. Bank

Orthogonal polynomials: applications and computation 45
  Walter Gautschi

Automatic grid generation 121
  William D. Henshaw

Eigenvalue optimization 149
  Adrian S. Lewis and Michael L. Overton

On the computation of crystalline microstructure 191
  Mitchell Luskin

Iterative solution of systems of linear differential equations ... 259
  Ulla Miekkala and Olavi Nevanlinna

Theory, algorithms, and applications of level set methods for
propagating interfaces 309
  James A. Sethian

Acta Numerica (1996), *pp.* 1–43

Hierarchical bases and the finite element method

Randolph E. Bank *

Department of Mathematics
University of California at San Diego
La Jolla, CA 92093, USA
E-mail: rbank@ucsd.edu

CONTENTS

1   Introduction   1
2   Preliminaries   3
3   Fundamental two-level estimates   7
4   *A posteriori* error estimates   16
5   Two-level iterative methods   22
6   Multilevel Cauchy inequalities   29
7   Multilevel iterative methods   34
References   41

1. Introduction

In this work we present a brief introduction to hierarchical bases, and the important part they play in contemporary finite element calculations. In particular, we examine their role in *a posteriori* error estimation, and in the formulation of iterative methods for solving the large sparse sets of linear equations arising from finite element discretization.

Our goal is that the development should be largely self-contained, but at the same time accessible and interesting to a broad range of mathematicians and engineers. We focus on the simple model problem of a self-adjoint, positive definite, elliptic equation. For this simple problem, the usefulness of hierarchical bases is already readily apparent, but we are able to avoid some of the more complicated technical hurdles that arise in the analysis of more general situations.

A posteriori error estimates play an important role in two related aspects

* The work of this author was supported by the Office of Naval Research under contract N00014-89J-1440.

of finite element calculations. First, such estimates provide the user of a finite element code with valuable information about the overall accuracy and reliability of the calculation. Second, since most *a posteriori* error estimates are computed locally, they also contain significant information about the distribution of error among individual elements, which can form the basis of adaptive procedures such as local mesh refinement. Space considerations prevent us from exploring these two topics in depth, and we will limit our discussion here to the error estimation procedure itself.

Hierarchical basis iterative methods have enjoyed a fair degree of popularity as elliptic solvers. These methods are closely related to the classical multigrid V-cycle and the BPX methods. Hierarchical basis methods typically have a growth in condition number of order k^2, where k is the number of levels*. This is in contrast to multigrid and BPX methods, where the generalized condition number is usually bounded independent of the number of unknowns. Although the rate of convergence is less than optimal, hierarchical basis methods offer several important advantages. First, classical multigrid methods require a sequence of subspaces of geometrically increasing dimension, having work estimates per cycle proportional to the number of unknowns. Such a sequence is sometimes difficult to achieve if adaptive local mesh refinement is used. Hierarchical basis methods, on the other hand, require work per cycle proportional to the number of unknowns for any distribution of unknowns among levels. Second, the analysis of classical multigrid methods often relies on global properties of the mesh and solution (e.g. quasi-uniformity of the meshes, $\mathcal{H}^2$ regularity of the solution), whereas analysis of hierarchical basis methods relies mainly on local properties of the mesh (e.g. shape regularity of the triangulation). This yields a method which is very robust over a broad range of problems.

Our analysis of *a posteriori* error estimates and hierarchical basis iterative methods is based on so-called strengthened Cauchy–Schwartz inequalities. The basic inequality for two levels, along with some other important properties of the hierarchical basis decomposition, is presented in Section 3. In Section 4 we use these results to analyse *a posteriori* error estimates, while in Section 5 we analyse basic two-level iterative methods. In Section 6, we develop a suite of strengthened Cauchy–Schwartz inequalities for k-level hierarchical decompositions, which are then used in Section 7 to analyse multilevel hierarchical basis iterations.

Notation is often a matter of personal preference and provokes considerable debate. We have chosen to use a mixture of the function space notation typical in the mathematical analysis of finite element methods, and matrix-vector notation, which is often most useful when considering questions of practical

---

* This result is for two space dimensions. For three space dimensions the growth is much faster, like $N^{1/3}$, where N is the number of unknowns.

implementation. We switch freely and frequently between these two types of notation, using that which we believe affords the clearest statement of a particular result. Some important results are presented using both types of notation.

2. Preliminaries

For background on finite element discretizations, we refer the reader to Aziz and Babuška (1972), Brenner and Scott (1994), and Ciarlet (1980). For simplicity, we will consider the solution of the self-adjoint elliptic partial differential equation

$$-\nabla(a\nabla u) + bu = f \tag{2.1}$$

in a polygonal region $\Omega \subset \mathbb{R}^2$, with the homogeneous Neumann boundary conditions

$$\nabla u \cdot n = 0 \tag{2.2}$$

on $\partial\Omega$, where n is the outward pointing unit normal. Most of our results apply with small modification to the case of Dirichlet boundary conditions $u = 0$ on $\partial\Omega$. We assume that $a(x)$, $b(x)$ are smooth functions satisfying $0 < \underline{a} \leq a(x) \leq \bar{a}$ and $0 < \underline{b} \leq b(x) \leq \bar{b}$ for $x \in \Omega$. The requirement that $\underline{b} > 0$ rather than $\underline{b} \geq 0$ is mainly for convenience.

The $\mathcal{L}^2(\Omega)$ inner product $(\cdot, \cdot)$ is defined by

$$(u, v) = \int_\Omega uv \, dx$$

and the corresponding norm

$$\|u\|^2 = (u, u) = \int_\Omega u^2 \, dx.$$

Let $\mathcal{H} = \mathcal{H}^1(\Omega)$ denote the usual Sobolev space equipped with the norm

$$\|u\|_1^2 = \|\nabla u\|^2 + \|u\|^2 = \int_\Omega |\nabla u|^2 + u^2 \, dx,$$

where $|\cdot|$ denotes the Euclidean norm on $\mathbb{R}^2$. The energy inner product $a(\cdot, \cdot)$ is defined by

$$a(u, v) = \int_\Omega a\nabla u^t \nabla v + buv \, dx, \tag{2.3}$$

for $u, v \in \mathcal{H}$. For $u \in \mathcal{H}$, we define the energy norm $\|u\|$ by

$$\|u\|^2 = a(u, u).$$

This norm is comparable to the $\mathcal{H}^1$ norm in the sense that there exist positive constants c_1 and c_2, depending on a and b, such that

$$c_1\|u\| \leq \|u\|_1 \leq c_2\|u\|.$$

The weak form of the elliptic boundary value problem (2.1)–(2.2) is as follows: find $u \in \mathcal{H}$ such that

$$a(u, v) = (f, v) \tag{2.4}$$

for all $v \in \mathcal{H}$.

Let $\mathcal{T}$ be a triangulation of the region Ω. While the results presented here do not depend on the uniformity or quasiuniformity of the triangulation, many of the constants depend on the shape regularity of the mesh. Let h_t denote the diameter of triangle $t \in \mathcal{T}$, and let d_t denote the diameter of the inscribed circle for t. We assume there exists a positive constant δ_0 such that

$$h_t \delta_0 \leq d_t \tag{2.5}$$

for all $t \in \mathcal{T}$. Later, when we consider sequences or families of triangulations, the constant δ_0 will be assumed to be uniform over all triangulations considered. While a shape regularity condition like (2.5) does not imply a *globally* quasiuniform triangulation, it does imply a *local* quasiuniformity for the mesh.

Many of the constants in our estimates depend only on the local variation of the functions a and b; thus we define

$$\alpha_0 = \max_{t \in \mathcal{T}} \frac{\max_{x \in t} a(x)}{\min_{x \in t} a(x)} \quad \text{and} \quad \beta_0 = \max_{t \in \mathcal{T}} \frac{\max_{x \in t} b(x)}{\min_{x \in t} b(x)}.$$

The fact that our estimates have only a local dependence on the coefficients can be very important in practice. For example, suppose a is piecewise constant, varying by orders of magnitude over the region Ω. If the jumps in a are aligned with edges of the triangulation, then our estimates will be independent of a ($\alpha_0 = 1$), irrespective of the magnitudes of the jumps.

Let $\mathcal{M}$ be an N-dimensional finite element subspace of $\mathcal{H}$, consisting of continuous piecewise polynomials with respect to the triangulation $\mathcal{T}$. We will be more specific about requirements for $\mathcal{M}$ later. The finite element approximation $u_h \in \mathcal{M}$ satisfies

$$a(u_h, v) = (f, v) \tag{2.6}$$

for all $v \in \mathcal{M}$. From (2.4) and (2.6), it is easy to see that the finite element solution is the *best approximation* of u with respect to the energy norm

$$\|u - u_h\| = \inf_{v \in \mathcal{M}} \|u - v\|.$$

Let ϕ_i $1 \leq i \leq N$ be a basis for $\mathcal{M}$. Then (2.6) can be transformed to the linear system of equations

$$AU = F \tag{2.7}$$

where

$$A_{ij} = a(\phi_j, \phi_i), \quad F_i = (f, \phi_i), \quad \text{and} \quad u_h = \sum_{i=1}^{N} U_i \phi_i.$$

The matrix A is typically large, sparse, symmetric, and positive definite. We note that

$$\|x\|_A^2 \equiv x^t A x = \|\chi\|^2,$$

where

$$\chi = \sum_{i=1}^{N} x_i \phi_i.$$

Thus the A-norm of a vector in $\mathbb{R}^N$ is equivalent to the energy norm of the corresponding finite element function.

At the computational level, many aspects of implementation of the finite element method are carried out on an elementwise basis. For example, the *stiffness matrix* A is typically computed as the sum of element stiffness matrices, in which integration is restricted to a single element $t \in \mathcal{T}$. The element stiffness matrix is usually computed by first mapping t to a fixed *reference element* t_r, and then computing the relevant integrals on the reference element. Because such mappings play an important role in our analysis, we begin by considering them in some detail.

Let $\mathcal{S}$ denote the set of triangles t satisfying $h_t = 1$, $\delta_0 \leq d_t/h_t$ and one vertex at the origin. Roughly speaking, the set $\mathcal{S}$ characterizes all shape regular triangles of diameter one. We will denote a particular triangle $t_r \in \mathcal{S}$ as the reference triangle. The reference triangle t_r can be mapped to any other triangle $t \in \mathcal{S}$ using a simple linear transformation (which can be represented as a 2×2 matrix). Shape regularity of the triangles in $\mathcal{S}$ implies that such transformations are well conditioned, with condition numbers depending only on the constant δ_0.

Let $\mathcal{A}$ denote the set of linear transformations mapping the reference triangle t_r to $t \in \mathcal{S}$. Since the triangles in the triangulation $\mathcal{T}$ are all shape regular, any triangle $l \in \mathcal{T}$ can be generated by a simple scaling and translation of an element $\hat{t} \in \mathcal{S}$. Thus the reference element t_r can be mapped to t using a linear transformation from the set $\mathcal{A}$ followed by a simple scaling and translation.

We now suppose that the finite element space $\mathcal{M}$ has the direct sum hierarchical decomposition $\mathcal{M} = \mathcal{V} \oplus \mathcal{W}$. Thus for $u \in \mathcal{M}$ we have the unique decomposition $u = v + w$, where $v \in \mathcal{V}$ and $w \in \mathcal{W}$. Let $\mathcal{V}_t$ and $\mathcal{W}_t$ denote the restrictions of $\mathcal{V}$ and $\mathcal{W}$ to each triangle $t \in \mathcal{T}$, and write $u_t = v_t + w_t$. Often, $\mathcal{V}_t$ and $\mathcal{W}_t$ will be polynomial spaces (as opposed to *piecewise* polynomial spaces), being restricted to a single element. Let $\mathcal{V}_r$ and $\mathcal{W}_r$ denote reference spaces of polynomials defined with respect to the reference triangle

t_r. We require that the finite element space $\mathcal{M} = \mathcal{V} \oplus \mathcal{W}$ satisfy the following assumptions for all $t \in \mathcal{T}$:

A1. If $u_t = c$ is constant then $w_t = 0$ and $v_t = c$.

A2. The mapping from t_r to t, consisting of a linear mapping from $\mathcal{A}$ followed by simple scaling and translation, induces maps from $\mathcal{V}_r$ onto $\mathcal{V}_t$ and $\mathcal{W}_r$ onto $\mathcal{W}_t$.

These conditions are very weak and are satisfied by many common finite element spaces, although sometimes with a nonstandard choice of basis functions. For example, consider the spaces of continuous piecewise polynomials of degree $p > 1$. For this choice, we let $\mathcal{V}$ be the space of continuous piecewise *linear* polynomials and $\mathcal{W}$ be the space of piecewise polynomials of degree p which are zero at the vertices of the triangulation $\mathcal{T}$. A basis for $\mathcal{V}$ is just the usual nodal basis for the space of continuous piecewise linear polynomials. A basis for $\mathcal{W}$ consists of all the nodal basis functions for the continuous piecewise polynomials of degree p *except* those associated with the triangle vertices. For example, for $p = 2$, $\mathcal{W}$ consists of the span of the quadratic 'bump functions' associated with edge midpoints in the triangulation. This is called the hierarchical basis for the piecewise quadratic polynomial space, in contrast to the usual nodal basis, and is often employed in practice in the p-version of the finite element method. It is typically the case that the dimension of the space $\mathcal{W}$ is larger than that of $\mathcal{V}$. In this example, the space $\mathcal{V}$ has a dimension of approximately N/p^2, or about $\dim \mathcal{M}/4$ for the case $p = 2$, and an increasingly smaller fraction as p increases.

We now consider a decomposition of the form $\mathcal{M} = \mathcal{V} \oplus \mathcal{W}$ for the case of continuous piecewise linear polynomials. In this case, we imagine that the triangulation $\mathcal{T} \equiv \mathcal{T}_f$, which we will call the fine grid, arose from the refinement of a coarse grid triangulation $\mathcal{T}_c$. For example, we can consider the case of uniform refinement, in which each triangle $t \in \mathcal{T}_c$ is refined into four similar triangles in $\mathcal{T}$ by pairwise connecting the midpoints of the edges of t. In this case the space $\mathcal{V} \equiv \mathcal{M}_c$ is just the space of continuous piecewise linear polynomials associated with the coarse mesh, while $\mathcal{W}$ consists of the span of the fine grid nodal basis functions associated with vertices in $\mathcal{T}$ which are not in $\mathcal{T}_c$. If uniform refinement is used, then the space $\mathcal{V}$ has a dimension of approximately $N/4$ while the dimension of $\mathcal{W}$ will be approximately $3N/4$. For iterative methods, it is important in practice that the dimension of the space $\mathcal{V}$ be as small as conveniently possible. In this vein, we note that the hierarchical decomposition of $\mathcal{M}$ can be recursively applied to the space $\mathcal{V}$, assuming that $\mathcal{T}_c$ arose from the refinement of an even coarser triangulation. This anticipates the k-level iterations discussed in later sections.

Let $\mathcal{M} = \mathcal{V} \oplus \mathcal{W}$. Let $\dim \mathcal{V} = N_\mathcal{V}$ and $\dim \mathcal{W} = N_\mathcal{W} = N - N_\mathcal{V}$, and let $\{\phi_i\}_{i=1}^{N_\mathcal{V}}$ be a basis for $\mathcal{V}$ and $\{\phi_i\}_{N_\mathcal{V}+1}^{N}$ be a basis for $\mathcal{W}$. This induces a natural block 2×2 partitioning of the linear system of (2.7) as

$$\begin{bmatrix} A_{11} & A_{12} \\ A_{21} & A_{22} \end{bmatrix} \begin{bmatrix} U_1 \\ U_2 \end{bmatrix} = \begin{bmatrix} F_1 \\ F_2 \end{bmatrix} \qquad (2.8)$$

where A_{11} is of order $N_{\mathcal{V}}$, and A_{22} is of order $N_{\mathcal{W}}$.

We note that if the vector $U \in \mathbb{R}^N$ corresponds to the finite element function $u = v + w \in \mathcal{M}$, then

$$U_1^t A_{11} U_1 = \|v\|^2, \quad U_2^t A_{22} U_2 = \|w\|^2, \quad \text{and} \quad U_1^t A_{12} U_2 = a(v, w).$$

3. Fundamental two-level estimates

In this section we develop some of the mathematical properties of the hierarchical basis. Chief among these properties is the so-called strengthened Cauchy inequality. One interesting feature of this strengthened Cauchy inequality is that it is a local property of the hierarchical basis: that is, it is true for the hierarchical decomposition corresponding to individual elements in the mesh as well as on the space as a whole. As a result, the constant in the strengthened Cauchy inequality does not depend strongly on such things as global regularity of solutions, the shape of the domain, quasiuniformity of the mesh, global variation of coefficient functions, and other properties that typically appear in the mathematical analysis of finite element methods. By the same reasoning, it is not surprising that the constant in the strengthened Cauchy inequality does depend on local properties like the shape regularity of the elements.

Our analysis of the strengthened Cauchy inequality in this section is taken from Bank and Dupont (1980), but see also Eijkhout and Vassilevski (1991). We begin our analysis with a preliminary technical lemma.

Lemma 1 Let $(\cdot, \cdot)$ and $\langle \cdot, \cdot \rangle$ denote two inner products defined on a vector space X. Let $\| \cdot \|$ and $| \cdot |$ denote the corresponding norms. Suppose that there exist positive constants $\underline{\lambda}$ and $\bar{\lambda}$ such that

$$0 < \underline{\lambda} < \frac{(z, z)}{\langle z, z \rangle} < \bar{\lambda}, \qquad (3.1)$$

for all nonzero $z \in X$. For any nonzero $x, y \in X$, let

$$\beta = \frac{(x, y)}{\|x\| \, \|y\|} \quad \text{and} \quad \gamma = \frac{\langle x, y \rangle}{|x| \, |y|}. \qquad (3.2)$$

Then

$$1 - \beta^2 \geq \mathcal{K}^{-2}(1 - \gamma^2) \qquad (3.3)$$

where $\mathcal{K} = \bar{\lambda}/\underline{\lambda}$.

Proof. Lemma 1 states that if two inner products give rise to norms that are comparable as in (3.1), then the angles measured by those inner products must also be comparable. Without loss of generality, we can assume $|x| = |y| = 1$. Then from (3.1), we have

$$
\begin{aligned}
1 - \beta^2 \ &= \ (1 - \beta)(1 + \beta) \\
&= \ \frac{1}{4} \left\| \frac{x}{\|x\|} + \frac{y}{\|y\|} \right\|^2 \left\| \frac{x}{\|x\|} - \frac{y}{\|y\|} \right\|^2 \\
&= \ \frac{1}{4\|x\|^4} \|x + \theta y\|^2 \|x - \theta y\|^2 \\
&\geq \ \frac{\lambda^2}{4\|x\|^4} |x + \theta y|^2 |x - \theta y|^2,
\end{aligned}
$$

where $\theta = \|x\|/\|y\|$. Since

$$
|x \pm \theta y|^2 = 1 + \theta^2 \pm 2\theta\gamma,
$$

we have

$$
\begin{aligned}
1 - \beta^2 \ &\geq \ \frac{\lambda^2}{\|x\|^4} \left\{ \theta^2(1 - \gamma^2) + \frac{1}{4}(1 - \theta^2)^2 \right\} \\
&\geq \ \frac{\lambda^2 \theta^2}{\|x\|^4}(1 - \gamma^2) \\
&= \ \frac{\lambda^2}{\|x\|^2 \|y\|^2}(1 - \gamma^2) \\
&\geq \ \mathcal{K}^{-2}(1 - \gamma^2).
\end{aligned}
$$

$\square$

We now state the main Lemma of this section, the strengthened Cauchy inequality.

Lemma 2 Let $\mathcal{M} = \mathcal{V} \oplus \mathcal{W}$ satisfy the assumptions A1 and A2 above. Then there exists a number $\gamma = \gamma(\alpha_0, \beta_0, \delta_0, \mathcal{V}_r, \mathcal{W}_r) \in [0, 1)$, such that

$$
|a(v, w)| \leq \gamma \|v\| \|w\| \tag{3.4}
$$

for all $v \in \mathcal{V}$ and all $w \in \mathcal{W}$.

Proof. This proof is done in detail, as many later proofs follow a similar pattern. The first step is to reduce (3.4) to an element-by-element estimate. In particular, suppose that for each $t \in \mathcal{T}$,

$$
|a(v, w)_t| \leq \gamma_t \|v\|_t \|w\|_t, \tag{3.5}
$$

where

$$
a(v, w)_t = \int_t a \nabla v^t \nabla w + b v w \ \mathrm{d}x
$$

is the restriction of $a(\cdot, \cdot)$ to t, and $\|\!|\cdot|\!\|_t$ is the corresponding norm. Then

$$
\begin{aligned}
|a(v, w)| &= \left| \sum_t a(v, w)_t \right| \\
&\leq \sum_t |a(v, w)_t| \\
&\leq \sum_t \gamma_t \|\!|v|\!\|_t \|\!|w|\!\|_t \\
&\leq \gamma \left(\sum_t \|\!|v|\!\|_t^2 \right)^{1/2} \left(\sum_t \|\!|w|\!\|_t^2 \right)^{1/2} \\
&= \gamma \|\!|v|\!\| \|\!|w|\!\|,
\end{aligned}
$$

where

$$
\gamma = \max_{t \in \mathcal{T}} \gamma_t.
$$

Thus, if we can show (3.5), then (3.4) follows.

To prove (3.5), we derive the pair of inequalities

$$
\begin{aligned}
|a(v, w)_{1,t}| &\leq \gamma_{1,t} \|v\|_{1,t} \|w\|_{1,t}, & (3.6) \\
|a(v, w)_{0,t}| &\leq \gamma_{0,t} \|v\|_{0,t} \|w\|_{0,t}, & (3.7)
\end{aligned}
$$

where

$$
a(v, w)_{1,t} = \int_t a \nabla v^t \nabla w \, dx, \qquad a(v, w)_{0,t} = \int_t bvw \, dx,
$$

and $\|\!|\cdot|\!\|_{i,t}$, $i = 0, 1$, are the corresponding seminorms. If (3.6)–(3.7) hold, then for

$$
\gamma_t = \max(\gamma_{0,t}, \gamma_{1,t}),
$$

we have

$$
\begin{aligned}
a(v, w)_t^2 &= (a(v, w)_{0,t} + a(v, w)_{1,t})^2 \\
&\leq \gamma_t^2 \left(\|v\|_{0,t} \|w\|_{0,t} + \|v\|_{1,t} \|w\|_{1,t} \right)^2 \\
&\leq \gamma_t^2 \left(\|v\|_{0,t}^2 + \|v\|_{1,t}^2 \right) \left(\|w\|_{0,t}^2 + \|w\|_{1,t}^2 \right) \\
&= \gamma_t^2 \|\!|v|\!\|_t^2 \|\!|w|\!\|_t^2.
\end{aligned}
$$

We now restrict attention to (3.6); the proof of (3.7) follows a similar pattern. We note that $\|\!|\cdot|\!\|_{1,t}$ defines a strong norm of $\mathcal{W}_t$, but only a seminorm on $\mathcal{V}_t$, since $\mathcal{V}_t$ contains the constant function, and $\|c\|_{1,t} = 0$ for any constant c. It is sufficient to show (3.6) only for the subspace $\tilde{\mathcal{V}}_t = \{v \in \mathcal{V}_t | \int_t v \, dx = 0\}$, whose elements have average value zero. For any $v \in \mathcal{V}_t$ let $c = \int_t v \, dx$, and note $v - c \in \tilde{\mathcal{V}}_t$. Then

$$
a(v, w)_{1,t} = a(v - c, w)_{1,t} \quad \text{and} \quad a(v, v)_{1,t} = a(v - c, v - c)_{1,t}
$$

for any $w \in \mathcal{W}_t$. Thus we need show (3.6) only for $v \in \tilde{\mathcal{V}}_t$ and $w \in \mathcal{W}_t$ and note that $\|| \cdot \||_{1,t}$ is a strong norm on the space $\tilde{\mathcal{V}}_t \oplus \mathcal{W}_t$.

A simple homogeneity argument now shows that $\gamma_{1,t}$ does not depend on the size of the element h_t. Making the change of variable

$$\hat{x} = \frac{x - x_0}{h_t},$$

where x_0 is any vertex of t, (3.6) becomes

$$\left| \int_{\hat{t}} \hat{a} \nabla \hat{v}^t \nabla \hat{w} \, \mathrm{d}\hat{x} \right| \leq \gamma_{1,t} \left(\int_{\hat{t}} \hat{a} |\nabla \hat{v}|^2 \, \mathrm{d}\hat{x} \right)^{1/2} \left(\int_{\hat{t}} \hat{a} |\nabla \hat{w}|^2 \, \mathrm{d}\hat{x} \right)^{1/2}, \qquad (3.8)$$

where $\hat{t} \in \mathcal{S}$ is the image of t under the change of variables, $\hat{v}(\hat{x}) = v(x)$, $\hat{w}(\hat{x}) = w(x)$, and $\hat{a}(\hat{x}) = a(x)$. In view of (3.8), we can restrict our attention to the set of triangles $\mathcal{S}$, the set of linear mappings $\mathcal{A}$, and the reference spaces $\mathcal{V}_r$ and $\mathcal{W}_r$.

Let $J \in \mathcal{A}$ be the linear mapping that takes the reference triangle t_r to $\hat{t}$. Then we have

$$\int_{\hat{t}} \hat{a} \nabla \hat{v}^t \nabla \hat{w} \, \mathrm{d}\hat{x} = |\det J| \int_{t_r} \tilde{a} (J^{-t} \nabla \tilde{v})^t (J^{-t} \nabla \tilde{w}) \, \mathrm{d}\tilde{x}. \qquad (3.9)$$

The right-hand side of (3.9) defines an inner product on the reference triangle t_r. A second inner product is given by the right-hand side of (3.9) with $\tilde{a} \equiv 1$ and $J = I$

$$\langle v, w \rangle = \int_{t_r} \nabla v^t \nabla w \, \mathrm{d}\tilde{x}.$$

Since $\hat{t} \in \mathcal{S}$, there is a positive constant $C = C(\delta_0)$ such that, for all $z \in \tilde{\mathcal{V}}_r \oplus \mathcal{W}_r$,

$$\underline{a}_t C^{-1} \leq \frac{|\det J| \int_{t_r} \tilde{a} |J^{-t} \nabla z|^2 \, \mathrm{d}\tilde{x}}{\int_{t_r} |\nabla \tilde{z}|^2 \, \mathrm{d}\tilde{x}} \leq C \bar{a}_t. \qquad (3.10)$$

Here $\underline{a}_t \leq a \leq \bar{a}_t$ for $x \in t$, and $\tilde{\mathcal{V}}_r = \left\{ v \in \mathcal{V}_r | \int_{t_r} v \, \mathrm{d}\tilde{x} = 0 \right\}$. Lemma 1 now tells us that angles measured by these two inner products are comparable.

The last step of the proof is to note that for $v \in \tilde{\mathcal{V}}_r$ and $w \in \mathcal{W}_r$, there exists $\gamma_r = \gamma_r(\mathcal{V}_r, \mathcal{W}_r)$, $0 \leq \gamma_r < 1$ for which

$$\left| \int_{t_r} \nabla v^t \nabla w \, \mathrm{d}\tilde{x} \right| \leq \gamma_r \left(\int_{t_r} |\nabla v|^2 \, \mathrm{d}\tilde{x} \right)^{1/2} \left(\int_{t_r} |\nabla w|^2 \, \mathrm{d}\tilde{x} \right)^{1/2}. \qquad (3.11)$$

Estimate (3.11) is true because $\tilde{\mathcal{V}}_r$ and $\mathcal{W}_r$ are linearly independent subspaces, so there must be a nonzero angle between them. Through the use of Lemma 1, it follows that $0 \leq \gamma_{1,t}(\alpha_0, \delta_0, \mathcal{V}_r, \mathcal{W}_r) < 1$. The estimate $0 \leq \gamma_{0,t}(\beta_0, \delta_0, \mathcal{V}_r, \mathcal{W}_r) < 1$ follows by similar reasoning, except that the reduction to $\tilde{\mathcal{V}}_t$ is unnecessary. $\square$

Analysis of methods employing hierarchical bases is often framed in terms of bounds of certain interpolation operators between fine and coarse spaces. See for example Borneman and Yserentant (1993), Bramble (1993) Oswald (1994), Xu (1989) and (1992), and Yserentant (1992). In the present context, the fine space is $\mathcal{M}$ while the coarse space is $\mathcal{V}$. The following lemma shows that this approach is entirely equivalent to the use of strengthened Cauchy inequalities.

Lemma 3 Suppose $\mathcal{M} = \mathcal{V} \oplus \mathcal{W}$, and let $\mathcal{I}$ denote the interpolation operator defined as follows: if $u = v + w \in \mathcal{M}$, $v \in \mathcal{V}$, and $w \in \mathcal{W}$, then $\mathcal{I}(u) = v$. Then

$$\|\mathcal{I}(u)\| \leq C\|u\| \qquad (3.12)$$

if and only if

$$|a(v, w)| \leq \gamma \|v\| \, \|w\| \qquad (3.13)$$

for $\gamma < 1$ and for all $v \in \mathcal{V}$ and $w \in \mathcal{W}$.

Proof. First, we assume (3.13) in order to prove (3.12). Let $u = v + w$, $v \in \mathcal{V}$, $w \in \mathcal{W}$. Then

$$
\begin{aligned}
\|u\|^2 &= a(v + w, v + w) \\
&= \|v\|^2 + \|w\|^2 + 2a(v, w) \\
&\geq \|v\|^2 + \|w\|^2 - 2\gamma\|v\| \, \|w\| \\
&\geq (1 - \gamma^2)\|v\|^2.
\end{aligned}
$$

Therefore

$$\|\mathcal{I}(u)\| \leq (1 - \gamma^2)^{-1/2}\|u\|.$$

Now we assume (3.12) to show (3.13). It suffices to take $\|v\| = \|w\| = 1$. Then, from (3.12)

$$\|v - w\| \geq \frac{1}{C}\|v\| - \frac{1}{C}.$$

Thus,

$$a(v, w) = \frac{1}{2}\left(\|v\|^2 + \|w\|^2 - \|v - w\|^2\right) \leq 1 - \frac{1}{2C^2}.$$

$\square$

The last result in this section is related to the space $\mathcal{W}$. The functions in $\mathcal{W}$ are necessarily quite oscillatory, since by assumption $\mathcal{V}$ contains local constants. Indeed, typically $\mathcal{V}$ contains the larger space of local linear functions, although it has not been necessary to assume this. The solution of equations using the space $\mathcal{W}$ should be quite simple, because on such an oscillatory space, an elliptic differential operator behaves very much like a large multiple of the identity.

To make this more precise, suppose that there is a basis for the reference space $\mathcal{W}_r$ whose elements are mapped onto the computational basis functions $\{\phi_j\}_{j=1}^{N_r}$ for $\mathcal{W}_t$ by the affine mapping of t_r onto t. This is a very natural assumption for the case of nodal finite elements, and is typically exploited in practical computations in algorithms for the assembly of the stiffness matrix and the right-hand side. With this additional assumption, we have the following lemma.

Lemma 4   Suppose $\{\phi_j\}_{j=1}^{N_{\mathcal{W}}}$ is the basis for $\mathcal{W}$ and let

$$w = \sum_{j=1}^{N_{\mathcal{W}}} w_j \phi_j(x, y).$$

Then there exist finite positive constants $\underline{\mu}$ and $\bar{\mu}$, depending only on α_0, β_0, and δ_0, such that

$$\underline{\mu}\, \|w\|^2 \leq \sum_{j=1}^{N_{\mathcal{W}}} w_j^2 \|\phi_j\|^2 \leq \bar{\mu}\, \|w\|^2. \tag{3.14}$$

Proof.   The proof follows the pattern of Lemma 2, so we will provide only a short sketch here. One first shows it is sufficient to prove

$$\underline{\mu}_t\, \|w\|_t^2 \leq \sum_{j=1}^{N_r} w_j^2 \|\phi_j\|_t^2 \leq \bar{\mu}_t\, \|w\|_t^2,$$

and set $\underline{\mu} = \min_t \underline{\mu}_t$ and $\bar{\mu} = \max_t \bar{\mu}_t$. (We have been a bit sloppy in our use of subscripts on w_j and ϕ_j in order to avoid more complicated notation.) We then reduce this to showing the pair of inequalities

$$\underline{\mu}_{0,t}\, \|w\|_{0,t}^2 \leq \sum_{j=1}^{N_r} w_j^2 \|\phi_j\|_{0,t}^2 \leq \bar{\mu}_{0,t}\, \|w\|_{0,t}^2,$$

and

$$\underline{\mu}_{1,t}\, \|w\|_{1,t}^2 \leq \sum_{j=1}^{N_r} w_j^2 \|\phi_j\|_{1,t}^2 \leq \bar{\mu}_{1,t}\, \|w\|_{1,t}^2,$$

with $\underline{\mu}_t = \min\{\underline{\mu}_{0,t}, \underline{\mu}_{1,t}\}$ and $\bar{\mu}_t = \max\{\bar{\mu}_{0,t}, \bar{\mu}_{1,t}\}$.

A change of variable as in (3.8), mapping $t \in \mathcal{T}$ to an element $\hat{t} \in \mathcal{S}$, proves that $\underline{\mu}$ and $\bar{\mu}$ are independent of h_t. Finally, changing variables as in (3.9) and using equivalence of norms as in (3.10)–(3.11) yields the result. $\square$

We now apply Lemmas 2 and 4 to several finite element spaces having hierarchical decompositions. Much of our analysis of these examples comes from the work of Maitre and Musy (1982); see also Braess (1981). In these examples, we will compute the constants $\gamma_{1,t}$, $\underline{\mu}_{1,t}$, and $\bar{\mu}_{1,t}$ for the case $a = 1$, illustrating the effect of shape regularity on the estimates. Let t be a triangle with vertices ν_i, edges ϵ_i, and angles θ_i, $1 \leq i \leq 3$.

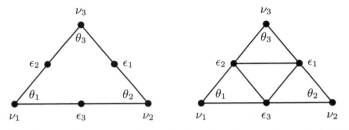

Fig. 1. Quadratic element (left) and piecewise linear element (right).

In our first example, we consider the space of continuous piecewise quadratic finite elements, illustrated on the left in Figure 1. Let ϕ_i $1 \leq i \leq 3$ denote the linear basis functions for t. Then $\mathcal{V}_t = \langle \phi_i \rangle_{i=1}^3$. The space $\mathcal{W}_t$ is composed of the quadratic bump functions $\mathcal{W}_t = \langle \psi_i \rangle_{i=1}^3$, where $\psi_i = 4\phi_j\phi_k$, and (i, j, k) is a cyclic permutation of $(1, 2, 3)$.

In the second example, we consider the space of continuous piecewise linear polynomials on a refined mesh, illustrated in Figure 1 on the right. Here $\mathcal{V}_t$ contains the linear polynomials on the coarse mesh element t; $\mathcal{V}_t = \langle \phi_i \rangle_{i=1}^3$, with ϕ_i defined as in the first example. The space $\mathcal{W}_t$ contains the continuous piecewise polynomials on the fine grid that are zero at the vertices of t. Thus $\mathcal{W}_t = \langle \hat{\phi}_i \rangle_{i=1}^3$, where $\hat{\phi}_i$ is the standard nodal piecewise linear basis function associated with the midpoint of edge ϵ_i of triangle t.

By direct computation, we establish the relation

$$-|t|\nabla\phi_j^t\nabla\phi_k = \frac{1}{2}\cot\theta_i = \frac{1}{2}L_i.$$

Let

$$A = \begin{bmatrix} L_2 + L_3 & -L_3 & -L_2 \\ -L_3 & L_3 + L_1 & -L_1 \\ -L_2 & -L_1 & L_1 + L_2 \end{bmatrix}, \tag{3.15}$$

and

$$D = \begin{bmatrix} L_1 & 0 & 0 \\ 0 & L_2 & 0 \\ 0 & 0 & L_3 \end{bmatrix}. \tag{3.16}$$

Then the element stiffness matrix for the quadratic hierarchical basis can be shown to be

$$M_Q = \begin{bmatrix} A/2 & -2A/3 \\ -2A/3 & 4(A+D)/3 \end{bmatrix}. \tag{3.17}$$

We know that

$$
\begin{aligned}
\gamma_{1,t} &= \max\{a(v,w) : \|v\| = \|w\| = 1\} \\
&= \max\{2x^t Ay/3 : x^t Ax = 2,\ y^t(A+D)y = 3/4\}.
\end{aligned}
$$

Standard algebraic manipulations yield

$$
\gamma_{1,t}^2 = \frac{2}{3}(1 - \lambda_{\min}),
$$

where $\lambda_{\min}$ is the smallest eigenvalue of the generalized eigenvalue problem

$$
Dx = \lambda(A+D)x. \tag{3.18}
$$

By direct computation and the use of various trigonometric identities, in particular $L_1 L_2 + L_2 L_3 + L_3 L_1 = 1$, we can compute

$$
\det\{D - \lambda(A+D)\} = 2(p-s)\lambda^3 + 3(s-p)\lambda^2 - s\lambda + p = 0,
$$

where

$$
\begin{aligned}
p &= L_1 L_2 L_3, \\
s &= L_1 + L_2 + L_3.
\end{aligned}
$$

The corresponding eigenvalues are $\lambda = 1$ and $\lambda = (1 \pm \sqrt{4c-3})/4$, where

$$
c = \cos^2 \theta_1 + \cos^2 \theta_2 + \cos^2 \theta_3,
$$

and

$$
\frac{p}{s} = \frac{1-c}{3-c}.
$$

Thus

$$
\gamma_{1,t}^2 = \frac{3 + \sqrt{4c-3}}{6}. \tag{3.19}
$$

For the second example, the element stiffness matrix for the piecewise linear hierarchical basis is given by

$$
M_L = \begin{bmatrix} A/2 & -A \\ -A & A+D \end{bmatrix}. \tag{3.20}
$$

We see that repeating the arguments for the first example leads to the same values for $\gamma_{1,t}$ but scaled by $\sqrt{3}/2$, that is

$$
\gamma_{1,t}^2 = \frac{3 + \sqrt{4c-3}}{8}. \tag{3.21}
$$

We now turn to the bounds for μ and $\bar{\mu}$ of Lemma 4. These may be expressed in terms of the largest and smallest eigenvalues in the generalized eigenvalue problem

$$
(A+D)x = s\lambda x, \tag{3.22}
$$

so that

$$\det\{A + D - s\lambda I\} = s^3(1 - \lambda)^3 - s(s^2 - 2)(1 - \lambda) - 2p = 0.$$

One can easily write down the analytic solutions of this cubic equation in terms of p and s, but there is no major simplification as in the case of $\gamma_{1,t}$. The bounds for the case of the piecewise linear hierarchical basis are given by $\underline{\mu}_{1,t} = \lambda_{\min}$ and $\bar{\mu}_{1,t} = \lambda_{\max}$. Those for the quadratic case are a simple scaling by $4/3$; $\underline{\mu}_{1,t} = 4\lambda_{\min}/3$ and $\bar{\mu}_{1,t} = 4\lambda_{\max}/3$.

Fig. 2. The contour map for $\gamma_{1,t}$ (left) and for $\underline{\mu}_{1,t}/\bar{\mu}_{1,t}$ (right).

In Figure 2, we have plotted $\gamma_{1,t}$ and the ratio $\kappa_t^{-1} = \underline{\mu}_{1,t}/\bar{\mu}_{1,t}$ as a function of $0 \leq \theta_1 \leq \pi$ and $0 \leq \theta_2 \leq \pi - \theta_1$, with $\theta_3 = \pi - \theta_1 - \theta_2$. For the case of quadratic elements, the smallest value $\gamma_{1,t} = 1/\sqrt{2}$ occurs for an equilateral triangle, while the largest value $\gamma_{1,t} = 1$ occurs for the degenerate cases $\theta_i = \theta_j = 0, \theta_k = \pi$. For the case of piecewise linear elements, one should scale all values of $\gamma_{1,t}$ by $\sqrt{3}/2$; for this case $\gamma_{1,t} < 1$, even in the degenerate cases.

It is the ratio $\kappa = \bar{\mu}/\underline{\mu}$ that plays a central role in our later analysis. However, we plot the reciprocal to confine the ratio to the interval $[0, 1]$. Here the largest value occurs again for the equilateral triangle, where $\kappa_t^{-1} \quad 1/4$, while $\kappa_t = 0$ whenever $\theta_i = 0, 1 \leq i \leq 3$. A special case occurs in the corners of the domain where the function κ_t^{-1} is discontinuous. For example, if one approaches the origin along the edge $\theta_1 = 0$, then the limiting cubic equation is $(1 - \lambda)^3 - (1 - \lambda) = 0$, with a corresponding $\kappa_t^{-1} = 0$. However, if we approach along, say, the line $\theta_1 = \theta_2 = \delta$, then the limiting cubic is $(2/3 - \lambda)(\lambda^2 - 7\lambda/3 + 4/9) = 0$, and $\kappa_{1,t}^{-1} = (7 - \sqrt{33})/(7 + \sqrt{33}) > 0$.

4. *A posteriori* error estimates

A posteriori error estimates are now widely used in the solution of partial differential equations. A recent survey of the field is given by Verfürth (1995), which contains a good bibliography on the subject. See also Ainsworth and Oden (1992 and 1993), Babuška and Gui (1986), Babuška and Rheinboldt (1978a) and (1978b), Bank and Weiser (1985), Weiser (1981), Zienkiewicz et al. (1982), and the book edited by Babuška et al. (1986). Our discussion here is motivated by Bank and Smith (1993).

A posteriori error estimates provide useful indications of the accuracy of a calculation and also provide the basis of adaptive local mesh refinement or local order refinement schemes. For example, if one has solved a problem for a given order p, corresponding to a finite element space $\mathcal{M}$, one can enrich the space to, say, order $p+1$ by adding certain hierarchical basis functions to the set of basis functions already used for $\mathcal{M}$. If $\bar{\mathcal{M}}$ is the new space, then we have the hierarchical decomposition

$$\bar{\mathcal{M}} = \mathcal{M} \oplus \mathcal{W},$$

where $\mathcal{W}$ is the subspace spanned by the additional basis functions.

If we resolve the problem with the space $\bar{\mathcal{M}}$ using the hierarchical basis, one expects intuitively that the component of the new solution lying in $\mathcal{M}$ will change very little from the previous calculation. Therefore, the component lying in $\mathcal{W}$ should be a good approximation to the error for the solution on the original space $\mathcal{M}$.

In fact, for our error estimate, we simply solve an (approximate) problem in the space $\mathcal{W}$ rather than $\bar{\mathcal{M}}$ to estimate the error. Let $\bar{u}_h \in \bar{\mathcal{M}}$ be the finite element solution on the enriched space satisfying

$$a(\bar{u}_h, v) = (f, v) \tag{4.1}$$

for all $v \in \bar{\mathcal{M}}$, and

$$\|u - \bar{u}_h\| = \inf_{v \in \bar{\mathcal{M}}} \|u - v\|. \tag{4.2}$$

Although we don't explicitly compute $\bar{u}_h$, it enters into our theoretical analysis of the *a posteriori* error estimate for $u - u_h$. In particular, we assume that the approximate solutions $\bar{u}_h$ converge to u more rapidly than u_h. This is expressed in terms of the *saturation assumption*

$$\|u - \bar{u}_h\| \leq \beta \|u - u_h\|, \tag{4.3}$$

where $\beta < 1$ independent of h. (We note that since $\mathcal{M} \subset \bar{\mathcal{M}}$, $\beta \leq 1$ is insured by the best approximation property.) In a typical situation, due to the higher degree of approximation for the space $\bar{\mathcal{M}}$, one can anticipate that $\beta = O(h^r)$, for some $r > 0$. In this case, $\beta \to 0$ as $h \to 0$, which is stronger than required by our theorems.

We seek to approximate the error $u - u_h$ in the space $\mathcal{W}$. Our first *a posteriori* error estimator $e_h \in \mathcal{W}$ is defined by

$$a(e_h, v) = (f, v) - a(u_h, v) \tag{4.4}$$

for all $v \in \mathcal{W}$.

To express this using matrix notation, we consider the linear system of equations corresponding to (4.1), expressed in terms of the hierarchical basis

$$\begin{bmatrix} A_{11} & A_{12} \\ A_{21} & A_{22} \end{bmatrix} \begin{bmatrix} \bar{U}_1 \\ \bar{U}_2 \end{bmatrix} = \begin{bmatrix} F_1 \\ F_2 \end{bmatrix}. \tag{4.5}$$

The vector $(\bar{U}_1^t, \bar{U}_2^t)$ corresponds to the function $\bar{u}_h = v + w \in \bar{\mathcal{M}}$ expanded in terms of the hierarchical basis, with $\bar{U}_1$ corresponding to $v \in \mathcal{M}$ and $\bar{U}_2$ corresponding to $w \in \mathcal{W}$. In this notation, the linear system solved to compute $u_h \in \mathcal{M}$ is given by $A_{11}U_1 = F_1$. If we combine this with the linear system for e_h corresponding to (4.4), we have

$$\begin{bmatrix} A_{11} & 0 \\ A_{21} & A_{22} \end{bmatrix} \begin{bmatrix} U_1 \\ E_2 \end{bmatrix} = \begin{bmatrix} F_1 \\ F_2 \end{bmatrix}, \tag{4.6}$$

where the vector E_2 corresponds to $e_h \in \mathcal{W}$.

We begin our analysis by noting the orthogonality relations

$$\begin{array}{llll} a(u - u_h, v) &=& 0 & \text{for all } v \in \mathcal{M}, & (4.7) \\ a(u - \bar{u}_h, v) &=& 0 & \text{for all } v \in \bar{\mathcal{M}}, & (4.8) \\ a(\bar{u}_h - u_h, v) &=& 0 & \text{for all } v \in \mathcal{M}, & (4.9) \\ a(u - u_h - e_h, v) &=& 0 & \text{for all } v \in \mathcal{W}, & (4.10) \\ a(\bar{u}_h - u_h - e_h, v) &=& 0 & \text{for all } v \in \mathcal{W}. & (4.11) \end{array}$$

Equations (4.7)–(4.11) are proved using various combinations of (2.4), (2.6), (4.1), and (4.4), restricted to the indicated subspaces. We can use the orthogonality relationships (4.7)–(4.9) to show

$$\|u - u_h\|^2 = \|u - \bar{u}_h\|^2 + \|\bar{u}_h - u_h\|^2. \tag{4.12}$$

Using (4.12) in conjunction with the saturation assumption (4.3) shows

$$(1 - \beta^2)\|u - u_h\|^2 \le \|\bar{u}_h - u_h\|^2 \le \|u - u_h\|^2, \tag{4.13}$$

demonstrating $\bar{u}_h - u_h$ to be a good approximation to the error. However, our goal is to show the easily computed function e_h also yields a good approximation of the error. This is shown next.

Theorem 1 Let $\bar{\mathcal{M}} = \mathcal{M} \oplus \mathcal{W}$ as above and assume (4.3) and Lemma 2 hold. Then

$$(1 - \beta^2)(1 - \gamma^2)\|u - u_h\|^2 \le \|e_h\|^2 \le \|u - u_h\|^2. \tag{4.14}$$

Proof. The right inequality in (4.14) is a simple consequence of (4.10) for the choice $v = e_h$. Now let $\bar{u}_h = \hat{u}_h + \hat{e}_h$, where $\hat{u}_h \in \mathcal{M}$, and $\hat{e}_h \in \mathcal{W}$. Then, using (4.9) with $v = \hat{u}_h - u_h$ and (4.11) with $v = \hat{e}_h$, we obtain

$$\|\bar{u}_h - u_h\|^2 = a(\bar{u}_h - u_h, \hat{e}_h) = a(e_h, \hat{e}_h). \tag{4.15}$$

Combining this with (4.12), we get

$$\|u - u_h\|^2 = \|u - \bar{u}_h\|^2 + a(\hat{e}_h, e_h). \tag{4.16}$$

To complete the proof, we must estimate $\|\hat{e}_h\|$ in terms of $\|e_h\|$. We apply the strengthened Cauchy inequality (3.4) to obtain

$$\begin{aligned} \|\bar{u}_h - u_h\|^2 &\geq \|\hat{u}_h - u_h\|^2 + \|\hat{e}_h\|^2 - 2\gamma \|\hat{u}_h - u_h\| \, \|\hat{e}_h\| \\ &\geq (1 - \gamma^2)\|\hat{e}_h\|^2. \end{aligned} \tag{4.17}$$

Combine this with (4.15) to obtain

$$(1 - \gamma^2)\|\hat{e}_h\| \leq \|e_h\|. \tag{4.18}$$

Using (4.16) and (4.18), we have

$$\|u - u_h\|^2 \leq \beta^2 \|u - u_h\|^2 + \frac{1}{1 - \gamma^2}\|e_h\|^2.$$

Rearranging this inequality leads directly to the left-hand inequality in (4.14). $\square$

We note that computing e_h in (4.4) requires the solution of a linear system involving the matrix A_{22} in (4.6). This is a rather an expensive calculation, given that typically the dimension of the space $\mathcal{W}$ is much larger than that of $\mathcal{M}$. Therefore it is of great interest to explore ways in which this calculation can be made more efficient. In situations where Lemma 4 can be applied, one possibility is to replace A_{22} by its diagonal $D_{22} = \text{diag} A_{22}$. In finite element notation, let $d(\cdot, \cdot)$ be the bilinear form corresponding to D_{22}. If $w = \sum_j w_j \phi_j \in \mathcal{W}$, and $z = \sum_j z_j \phi_j \in \mathcal{W}$, and $\{\phi_j\}$ are the basis functions used in Lemma 4, then

$$d(z, w) = \sum_j z_j w_j a(\phi_j, \phi_j).$$

We compute an approximation $\tilde{e}_h \in \mathcal{W}$ satisfying

$$d(\tilde{e}_h, v) = (f, v) - a(u_h, v). \tag{4.19}$$

In our proof of Theorem 1, we replace the orthogonality relations (4.10)–(4.11) with

$$\begin{aligned} a(u - u_h, v) &= d(\tilde{e}_h, v) & \text{for all } v \in \mathcal{W}, & \tag{4.20} \\ a(\bar{u}_h - u_h, v) &= d(\tilde{e}_h, v) & \text{for all } v \in \mathcal{W}. & \tag{4.21} \end{aligned}$$

Theorem 2 Let $d(\cdot,\cdot)$ be defined as above, and assume Theorem 1 and Lemma 4 hold. Then

$$\bar{\mu}^{-1}(1-\beta^2)(1-\gamma^2)\,\|u-u_h\|^2 \le \|\tilde{e}_h\|^2 \le \underline{\mu}^{-1}\|u-u_h\|^2. \qquad (4.22)$$

Proof. One can follow the proof of Theorem 1 with small modifications to show (4.22). However, we will take a more direct approach. From (4.10) and (4.20), we have

$$d(\tilde{e}_h, v) = a(e_h, v)$$

for all $v \in \mathcal{W}$. Taking $v = \tilde{e}_h$ and $v = e_h$, and applying Lemma 4, we have

$$\underline{\mu}\|\tilde{e}_h\|^2 \le \|e_h\|^2 \le \bar{\mu}\|\tilde{e}_h\|^2.$$

Combining this with Theorem 1 proves (4.22). □

A second possibility for improving the efficiency of the computation of the *a posteriori* error estimate is to use a nonconforming space $\bar{\mathcal{W}}$ of *discontinuous* piecewise polynomials to approximate the error. We assume that $\mathcal{W} \subset \bar{\mathcal{W}}$, but $\bar{\mathcal{W}} \not\subset \mathcal{H}$. The advantage of this approach is that the resulting stiffness matrix $\bar{A}_{22}$ is block diagonal, with each diagonal block corresponding to a single element. Thus the error can be computed element by element, by solving a small linear system for each triangle.

To analyse such an error estimator, we need to consider the effect of using nonconforming elements. First, we consider the continuous problem. Let $\mathcal{E}$ denote the set of interior edges of $\mathcal{T}$. For each edge $e \in \mathcal{E}$, we denote a fixed unit normal n_e, chosen arbitrarily from the two possibilities. For w discontinuous along e, let w_A and w_J denote the average and jump of w on e, the sign of w_J being chosen consistently with the choice of n_e. Let $v \in \mathcal{H} \cup \bar{\mathcal{W}}$ and u be the solution of (2.4). Then a straightforward calculation shows that

$$a(u, v) = (f, v) + g(u, v), \qquad (4.23)$$

where

$$g(u, v) = \sum_{e \in \mathcal{E}} \int_e \{a\nabla u^t n_e\}_A v_J \; dx, \qquad (4.24)$$

and

$$a(u, v) = \sum_{t \in \mathcal{T}} a(u, v)_t.$$

The error estimator $\bar{e}_h \in \bar{\mathcal{W}}$ based on this formulation is given by

$$a(\bar{e}_h, v) = (f, v) + g(u_h, v) - a(u_h, v) \qquad (4.25)$$

for all $v \in \bar{\mathcal{W}}$. Note that (4.25) consists of a collection of decoupled problems having the appearance of local Neumann problems on each element; since the space $\bar{\mathcal{W}}$ cannot contain local constants, all problems must be nonsingular and have unique solutions.

To analyse this process, we note that the orthogonality conditions (4.10)–(4.11) are now replaced by

$$a(u - u_h - \bar{e}_h, v) \;=\; g(u - u_h, v) \qquad \text{for all } v \in \bar{\mathcal{W}}, \qquad (4.26)$$

$$a(\bar{u}_h - u_h - \bar{e}_h, v) \;=\; 0 \qquad\qquad \text{for all } v \in \mathcal{W}. \qquad (4.27)$$

Here $\bar{u}_h \in \bar{\mathcal{M}}$ is still the conforming finite element approximation defined in (4.1). The bilinear form $g(\cdot, \cdot)$ does not appear in (4.27) since $v_J = 0$ for $v \in \mathcal{W}$.

In examining the proof of Theorem 1, we note that the argument used in proving the left inequality in (4.14) remains unchanged when applied to $\|\bar{e}_h\|$. The difficulty arises only in the upper bound, where the choice $v = \bar{e}_h$ in (4.26) leads to

$$\|\bar{e}_h\|^2 \leq \|u - u_h\| \, \|\bar{e}_h\| + |g(u - u_h, \bar{e}_h)|.$$

Obtaining a bound for the nonconforming term is fairly technical and lengthy, and we will only sketch the arguments here. The interested reader is referred to Bank and Weiser (1985) for a more complete discussion. First note that the presence of the nonconforming term demands more (local) regularity of the solution since line integrals of $\nabla(u - u_h)^t n_e$ appear. Here we will make the simplifying assumption

$$\sum_{t \in \mathcal{T}} h_t^2 \|\nabla^2(u - u_h)\|_t^2 \leq \alpha^2 \|u - u_h\|^2, \qquad (4.28)$$

which essentially states that a standard *a priori* estimate for $\|u - u_h\|$ is sharp. A more complicated form of the saturation assumption could be used in place of (4.28).

Using standard trace inequalities edge by edge for $e \in \mathcal{E}$, we are led to the estimate

$$|g(u - u_h, \bar{e}_h)|^2 \;\leq\; C \left(\sum_{t \in \mathcal{T}} \|\sqrt{a}\nabla(u - u_h)\|_t^2 + h_t^2 \|\sqrt{a}\nabla^2(u - u_h)\|_t^2 \right)$$

$$\left(\sum_{t \in \mathcal{T}} h_t^{-2} \|\sqrt{a}\bar{e}_h\|_t^2 + \|\sqrt{a}\nabla \bar{e}_h\|_t^2 \right).$$

See Brenner and Scott (1994) for a discussion of trace inequalities.

Now, using (4.28), and a slight generalization of Lemma 4,

$$\underline{\nu}\|w\|_t^2 \leq h_t^{-2}\|w\|_t^2 \leq \bar{\nu}\|w\|_t^2,$$

for all $w \in \mathcal{W}_t$, we obtain the bound

$$|g(u - u_h, \bar{e}_h)| \leq \delta \|u - u_h\| \, \|\bar{e}_h\|,$$

which yields our next result.

Theorem 3 Let $\bar{e}_h \in \bar{\mathcal{W}}$ satisfy (4.25). Assume (4.3), (4.28), and Lemmas 2 and 4. Then

$$(1 - \beta^2)(1 - \gamma^2) \|u - u_h\|^2 \le \|\bar{e}_h\|^2 \le (1 + \delta)^2 \|u - u_h\|^2, \qquad (4.29)$$

where β and γ are as in Theorem 1 and $\delta = \delta(\alpha_0, \beta_0, \delta_0, \mathcal{V}_r, \mathcal{W}_r)$.

We remark that one could make the diagonal approximation to the systems of linear equations to be solved in computing $\bar{e}_h$. One would then have an estimate modified as in Theorem 2. However, there is less advantage to be gained in the current situation because $\bar{A}_{22}$ is already block diagonal with diagonal blocks of small order. Another possibility is to use a different bilinear form $b(\cdot, \cdot)$ in place of $a(\cdot, \cdot)$ on the left-hand side of (4.25). Such an algorithm would calculate $\check{e}_h \in \bar{\mathcal{W}}$ such that

$$b(\check{e}_h, v) = (f, v) + g(u_h, v) - a(u_h, v). \qquad (4.30)$$

One choice, suggested by Ainsworth and Oden (1992 and 1993), is to let $b(\cdot, \cdot)$ correspond to the Laplace operator $-\Delta$. If there exist finite, positive constants $\underline{\mu}$ and $\bar{\mu}$ such that

$$\underline{\mu} \|w\|^2 \le b(w, w) \le \bar{\mu} \|w\|^2$$

in analogy to (3.14), then the analysis of such approximations may be carried out in a fashion similar to Theorem 2. Durán and Rodríguez (1992) and Durán, Muschietti and Rodríguez (1991) analyse the asymptotic exactness of error estimates of the type developed here, a topic we will not consider in detail.

We now develop some examples of *a posteriori* error estimates for the space of continuous piecewise linear polynomials. We let $\bar{\mathcal{M}}$ be the space of continuous piecewise quadratic polynomials, and $\mathcal{W}$ the space of quadratic bump functions. The basis functions, denoted $\{\psi_i\}$, will be the standard quadratic nodal basis functions associated with edge midpoints for all edges of the triangulation $\mathcal{T}$. We first consider the estimate $\mathring{e}_h$ defined in (4.19). Let

$$\mathring{e}_h = \sum_i \tilde{E}_i \psi_i.$$

Let ψ_i be associated with an interior edge e of the triangulation and have support in triangles t_1 and t_2, the two triangles sharing edge e. Then

$$\tilde{E}_i = \frac{(f, \psi_i)_{t_1} - a(u_h, \psi_i)_{t_1} + (f, \psi_i)_{t_2} - a(u_h, \psi_i)_{t_2}}{a(\psi_i, \psi_i)_{t_1} + a(\psi_i, \psi_i)_{t_2}}.$$

Here we see that the calculation of $\tilde{E}_i$ involves only local computations. Standard element-by-element assembly techniques can be used to compute all the relevant quantities.

We next consider the computation of $\bar{e}_h$ in (4.25). Let $\bar{\mathcal{W}}$ be the space of discontinuous piecewise quadratic bump functions. There are now two basis functions associated with each interior edge, one with support in each element sharing that edge, so the dimension of $\bar{W}$ is approximately twice that of W. However, at the level of a single element t, we have $W_t = \bar{W}_t$. Let $\{\bar{\psi}_i\}$ be the basis for $\bar{W}$. Then the function $\bar{e}_h$ of (4.25) can be expressed as

$$\bar{e}_h = \sum_i \bar{E}_i \bar{\psi}_i.$$

Suppose $\bar{\psi}_i$, $\bar{\psi}_j$, and $\bar{\psi}_k$ are the three discontinuous quadratic bump functions having support in the element $t \in \mathcal{T}$. Then we must assemble and solve the 3×3 linear system

$$\left[\begin{array}{ccc} a(\bar{\psi}_i, \bar{\psi}_i)_t & a(\bar{\psi}_j, \bar{\psi}_i)_t & a(\bar{\psi}_k, \bar{\psi}_i)_t \\ a(\bar{\psi}_i, \bar{\psi}_j)_t & a(\bar{\psi}_j, \bar{\psi}_j)_t & a(\bar{\psi}_k, \bar{\psi}_j)_t \\ a(\bar{\psi}_i, \bar{\psi}_k)_t & a(\bar{\psi}_j, \bar{\psi}_k)_t & a(\bar{\psi}_k, \bar{\psi}_k)_t \end{array} \right] \left[\begin{array}{c} \bar{E}_i \\ \bar{E}_j \\ \bar{E}_k \end{array} \right] =$$

$$\left[\begin{array}{c} (f, \bar{\psi}_i)_t - a(u_h, \bar{\psi}_i)_t \\ (f, \bar{\psi}_j)_t - a(u_h, \bar{\psi}_j)_t \\ (f, \bar{\psi}_k)_t - a(u_h, \bar{\psi}_k)_t \end{array} \right] + \left[\begin{array}{c} g(u_h, \bar{\psi}_i)_t \\ g(u_h, \bar{\psi}_j)_t \\ g(u_h, \bar{\psi}_k)_t \end{array} \right].$$

As in the case of $\tilde{e}_h$, only local computations are involved. All are completely standard except for the evaluation of the nonconforming terms. For example, to evaluate $g(u_h, \bar{\psi}_i)_t$, we first note that $\bar{\psi}_i$ is nonzero on only one edge of t, say edge e. Thus

$$g(u_h, \bar{\psi}_i)_t = \int_e \{a\nabla u_h^t n\}_A \bar{\psi}_i \, dx,$$

where n is the outward normal for t. To evaluate the average, we must compute $a\nabla u_h$ for both t and the adjacent triangle sharing edge e.

5. Two-level iterative methods

In this section we analyse several two-level iterations for solving (2.6) (in finite element notation) or, equivalently, (2.7) (in matrix notation). Much of our development is based on Bank and Dupont (1980) and Bank, Dupont, and Yserentant (1988). See also the books of Hackbusch (1985) and Bramble (1993).

Let $\mathcal{M} = \mathcal{V} \oplus \mathcal{W}$, let A be the stiffness matrix computed using the hierarchical basis, and partitioned according to (2.8), and let

$$A = L + D + L^t, \tag{5.1}$$

where

$$D = \left[\begin{array}{cc} A_{11} & 0 \\ 0 & A_{22} \end{array} \right] \quad \text{and} \quad L = \left[\begin{array}{cc} 0 & 0 \\ A_{21} & 0 \end{array} \right].$$

We consider the following iteration for solving (2.6). Let $u_0 \in \mathcal{M}$ be given.

We define the sequence $u_k = v_k + w_k$, with $v_k \in \mathcal{V}$ and $w_k \in \mathcal{W}$ by

$$a(v_{k+1} - v_k, \chi) = \omega\{(f, \chi) - a(u_k, \chi)\}, \qquad \chi \in \mathcal{V}, \tag{5.2}$$

and

$$a(w_{k+1} - w_k, \chi) = \omega\{(f, \chi) - a(u_k, \chi)\} \qquad \chi \in \mathcal{W}. \tag{5.3}$$

The iteration (5.2)–(5.3) can be written in matrix notation as

$$D(x_{k+1} - x_k) = \omega\{F - Ax_k\}, \tag{5.4}$$

where the vector $x_k \in \mathbb{R}^N$ corresponds to the finite element function $u_k \in \mathcal{M}$. Equations (5.2)–(5.4) represent a standard block Jacobi iteration for solving (2.6)–(2.7). Although we have written (5.4) as a stationary iteration, practically we expect to use D as a preconditioner in the conjugate gradient procedure. We refer the interested reader to Golub and Van Loan (1983) or Golub and O'Leary (1989) for a complete discussion of the preconditioned conjugate gradient algorithm. Here we analyse the generalized condition number of the preconditioned system.

Theorem 4 Let $A = L + D + L^t$ as defined above. Then for all $x \neq 0$,

$$\frac{1}{1+\gamma} \leq \frac{x^t D x}{x^t A x} \leq \frac{1}{1-\gamma}, \tag{5.5}$$

where $0 \leq \gamma < 1$ is given in Lemma 2.

Proof. It is easiest to analyse (5.5) using finite element notation. Let $u = v + w$, with $v \in \mathcal{V}$ and $w \in \mathcal{W}$, correspond to $x \in \mathbb{R}^N$. Then

$$x^t A x = \|u\|^2 \quad \text{and} \quad x^t D x = \|v\|^2 + \|w\|^2.$$

Now

$$\|u\|^2 = \|v\|^t + \|w\|^t + 2a(v, w).$$

Applying Lemma 2, we have

$$(1 - \gamma)(\|v\|^2 + \|w\|^2) \leq \|u\|^2 \leq (1 + \gamma)(\|v\|^2 + \|w\|^2),$$

proving (5.5). $\square$

The generalized condition number $\mathcal{K}$ is given by

$$\mathcal{K} = \frac{1+\gamma}{1-\gamma}.$$

The optimum value for ω for the stationary iteration (5.4) is $\omega = 1$, and the rate of convergence is given by

$$\frac{\mathcal{K} - 1}{\mathcal{K} + 1} = \gamma.$$

See Dupont, Kendall and Rachford (1968) for an analysis of the stationary method.

If conjugate gradient acceleration is used, the estimate for the rate of convergence is bounded by

$$\frac{\sqrt{\mathcal{K}} - 1}{\sqrt{\mathcal{K}} + 1} = \frac{\gamma}{1 + \sqrt{1 - \gamma^2}}.$$

We note that (5.4) requires the solution of linear systems involving the diagonal blocks A_{11} and A_{22} in each iteration. We next show that the systems involving A_{22} can be effectively solved using an inner iteration. Those involving A_{11} should either be solved directly, or solved recursively, using a multilevel iteration.

Let $\hat{A}_{22}$ be a symmetric, positive definite preconditioner for A_{22}, and suppose we approximately solve the linear system $A_{22}x = b$, using $m \geq 1$ steps of the iterative process

$$\hat{A}_{22}(x_{k+1} - x_k) = b - A_{22}x_k. \tag{5.6}$$

The iteration (5.6) should not be accelerated, but should be implemented as a stationary iteration to allow the use of conjugate gradient acceleration for the overall (outer) iteration. We assume that any fixed parameters for (5.6) have been already incorporated in the definition of $\hat{A}_{22}$. Let

$$G = I - A_{22}^{1/2}\hat{A}_{22}^{-1}A_{22}^{1/2}.$$

We assume G is symmetric with

$$\|G\|_{\ell_2} = \rho < 1. \tag{5.7}$$

Let

$$R_m = G^m(I - G^m)^{-1}. \tag{5.8}$$

The eigenvalues of R_m lie on the interval

$$0 \leq \lambda \leq \frac{\rho^m}{1 - \rho^m} \tag{5.9}$$

when m is even or if all eigenvalues of G are nonnegative. In the latter case, G is sometimes called a *smoother*. If G is not a smoother and m is odd, we must use the weaker bound

$$-\frac{\rho^m}{1 + \rho^m} \leq \lambda \leq \frac{\rho^m}{1 - \rho^m}. \tag{5.10}$$

An induction argument shows the m-step process in (5.6) is mathematically equivalent to the solution of

$$A_{22}^{1/2}(I + R_m)A_{22}^{1/2}x_m = b + A_{22}^{1/2}R_m A_{22}^{1/2}x_0. \tag{5.11}$$

In our current situation, the initial guess $x_0 = 0$, simplifying the right-hand side of (5.11). Our overall preconditioner, using m inner iterations, is thus

$$\hat{D} = D + \begin{bmatrix} 0 & 0 \\ 0 & A_{22}^{1/2} R_m A_{22}^{1/2} \end{bmatrix}. \tag{5.12}$$

Theorem 5   Let $A = L + D + L^t$ and $\hat{D}$ be defined as above. Then for all $x \neq 0$,

$$\frac{1}{(1+\gamma)(1+\rho^m)} \leq \frac{x^t \hat{D} x}{x^t A x} \leq \frac{1}{(1-\gamma)(1-\rho^m)}. \tag{5.13}$$

Proof.   As in the proof of Theorem 4, we let $u = v + w \in \mathcal{M}$ correspond to $x \in \mathbb{R}^N$. Then

$$\|v\|^2 + (1+\rho^m)^{-1}\|w\|^2 \leq x^t \hat{D} x \leq \|v\|^2 + (1-\rho^m)^{-1}\|w\|^2.$$

Thus

$$\frac{1}{1+\rho^m} \leq \frac{x^t \hat{D} x}{x^t D x} \leq \frac{1}{1-\rho^m},$$

and the theorem follows from Theorem 4 and

$$\frac{x^t \hat{D} x}{x^t A x} = \left(\frac{x^t \hat{D} x}{x^t D x}\right)\left(\frac{x^t D x}{x^t A x}\right).$$

□

The generalized condition number $\mathcal{K}$ is bounded by

$$\mathcal{K} \leq \left(\frac{1+\gamma}{1-\gamma}\right)\left(\frac{1+\rho^m}{1-\rho^m}\right).$$

Here we see that the use of inner iterations has only a modest effect on the generalized condition number, provided that ρ is small or m is large. We remark that by bounding $x^t \hat{D} x / x^t A x$ directly, instead of bounding $x^t \hat{D} x / x^t D x$ and $x^t D x / x^t A x$ separately, one can achieve a somewhat smaller but more complicated bound for $\mathcal{K}$. If G is a smoother, then the bound on $\mathcal{K}$ can be improved to

$$\mathcal{K} \leq \left(\frac{1+\gamma}{1-\gamma}\right)\left(\frac{1}{1-\rho^m}\right).$$

We now consider the symmetric block Gauss–Seidel iteration

$$(D + L)(x_{k+1/2} - x_k) = F - Ax_k \tag{5.14}$$
$$(D + L^t)(x_{k+1} - x_{k+1/2}) = F - Ax_{k+1/2}.$$

In finite element notation, we may write (5.14) as

$$a(v_{k+1/2} + w_k, \chi) = (f, \chi) \tag{5.15}$$

for $\chi \in \mathcal{V}$,

$$a(v_{k+1/2} + w_{k+1}, \chi) = (f, \chi) \qquad (5.16)$$

for $\chi \in \mathcal{W}$, and

$$a(v_{k+1} + w_{k+1}, \chi) = (f, \chi) \qquad (5.17)$$

for $\chi \in \mathcal{V}$. A careful analysis of (5.15)–(5.17) will show that block Gauss–Seidel and block symmetric Gauss–Seidel are equivalent as stationary iterative methods (that is, $v_{k+1/2} = v_k$), but this is no longer true when symmetric Gauss–Seidel is used as a preconditioner for the conjugate gradient algorithm.

Let $e_k = x - x_k$. Then from (5.14),

$$e_{k+1/2} = \{I - (D+L)^{-1}A\}e_k,$$
$$e_{k+1} = \{I - (D+L)^{-t}A\}e_{k+1/2},$$

from which it follows that

$$
\begin{aligned}
e_{k+1} &= \{I - (D+L)^{-t}A\}\{I - (D+L)^{-1}A\}e_k \\
&= \{I - [(D+L)^{-t} + (D+L)^{-1}]A + (D+L)^{-t}A(D+L)^{-1}A\}e_k \\
&= \{I - (D+L)^{-t}(L + 2D + L^t - A)(D+L)^{-1}A\}e_k \\
&= \{I - (D+L)^{-t}D(D+L)^{-1}A\}e_k \\
&= \{I - B^{-1}A\}e_k, \qquad (5.18)
\end{aligned}
$$

where

$$B = (D+L)D^{-1}(D+L^t) = A + LD^{-1}L^t. \qquad (5.19)$$

Once again, our task is to determine the generalized condition number by estimating the Rayleigh quotient.

Theorem 6   Let $A = L + D + L^t$ as defined above, and let B be given by (5.19). Then

$$1 \leq \frac{x^t B x}{x^t A x} \leq \frac{1}{1 - \gamma^2}, \qquad (5.20)$$

where $0 \leq \gamma < 1$ is given in Lemma 2.

Proof.   Since $LD^{-1}L^t$ is symmetric, positive semidefinite, it is clear from (5.19) that the lower bound is one. The upper bound is given by $1 + \mu$ where

$$\mu = \max_{x \neq 0} \frac{x^t L D^{-1} L^t x}{x^t A x}. \qquad (5.21)$$

This can be written as

$$\mu = \max_{x \neq 0} \frac{y^t D y}{x^t A x},$$

where

$$Dy = L^t x.$$

In finite element notation, this becomes

$$\mu = \max_{u \neq 0} \frac{\|\hat{v}\|^2}{\|u\|^2}, \tag{5.22}$$

where $u = v + w$, $v \in \mathcal{V}$, $w \in \mathcal{W}$ and $\hat{v} \in \mathcal{V}$ satisfies

$$a(\hat{v}, \chi) = a(w, \chi) \tag{5.23}$$

for all $\chi \in \mathcal{V}$. Written in finite element language, (5.22)–(5.23) is easy to analyse in terms of the strengthened Cauchy inequality. We take $\chi = \hat{v}$ in (5.23) to see

$$\|\hat{v}\| \leq \gamma \|w\|.$$

On the other hand

$$\begin{aligned}
\|u\|^2 &= \|v\|^2 + \|w\|^2 + 2a(v, w) \\
&\geq \|v\|^2 + \|w\|^2 - 2\gamma \|v\| \|w\| \\
&\geq (1 - \gamma^2)\|w\|^2 \\
&\geq (1 - \gamma^2)\gamma^{-2}\|\hat{v}\|^2.
\end{aligned}$$

The theorem now follows from combining this estimate and (5.22). $\square$

The analysis of the block symmetric Gauss–Seidel scheme with inner iterations is a little more complicated. We formally consider the iteration

$$\begin{aligned}
(\hat{D} + L)(x_{k+1/2} - x_k) &= F - Ax_k, \\
(\hat{D} + L^t)(x_{k+1} - x_{k+1/2}) &= F - Ax_{k+1/2},
\end{aligned} \tag{5.24}$$

where $\hat{D}$ is given in (5.12). A calculation similar to (5.18) shows that

$$\begin{aligned}
e_{k+1} &= \{I - (\hat{D} + L)^{-t}A\}\{I - (\hat{D} + L)^{-1}A\}e_k \\
&= \{I - [(\hat{D} + L)^{-t} + (\hat{D} + L)^{-1}]A + (\hat{D} + L)^{-t}A(\hat{D} + L)^{-1}A\}e_k \\
&= \{I - (\hat{D} + L)^{-t}(L + 2\hat{D} + L^t - A)(\hat{D} + L)^{-1}A\}e_k \\
&= \{I - (\hat{D} + L)^{-t}(2\hat{D} - D)(\hat{D} + L)^{-1}A\}e_k \\
&= \{I - \hat{B}^{-1}A\}e_k,
\end{aligned} \tag{5.25}$$

where

$$\begin{aligned}
\hat{B} &= (\hat{D} + L)(2\hat{D} - D)^{-1}(\hat{D} + L)^t \\
&= A + (D - \hat{D} + L)(2\hat{D} - D)^{-1}(D - \hat{D} + L)^t \\
&= A + LD^{-1}L^t + \Delta.
\end{aligned} \tag{5.26}$$

Here

$$\Delta = \begin{bmatrix} 0 & 0 \\ 0 & A_{22}^{1/2} R_m^2 (I + 2R_m)^{-1} A_{22}^{1/2} \end{bmatrix} = \begin{bmatrix} 0 & 0 \\ 0 & A_{22}^{1/2} R_{2m} A_{22}^{1/2} \end{bmatrix},$$

and R_m is defined in (5.8).

Theorem 7   Let $A = L + D + L^t$ as defined above, and let $\hat{B}$ be given by (5.26). Then

$$1 \le \frac{x^t \hat{B} x}{x^t A x} \le \frac{1}{(1 - \gamma^2)(1 - \rho^{2m})}, \tag{5.27}$$

where $0 \le \gamma < 1$ is given in Lemma 2, and ρ is given in (5.7).

Proof.   Since $LD^{-1}L^t + \Delta$ is symmetric, positive semidefinite, the lower bound is one. For the upper bound, $x^t LD^{-1}L^t x / x^t A x$ was estimated in the proof of Theorem 6. Let $u = v + w \in \mathcal{M}$ correspond to $x \in \mathbb{R}^N$. Then, using (5.7)–(5.8) and Lemma 2, we have

$$\frac{x^t \Delta x}{x^t A x} \le \left(\frac{\rho^{2m}}{1 - \rho^{2m}} \right) \frac{\|w\|^2}{\|u\|^2} \le \left(\frac{\rho^{2m}}{1 - \rho^{2m}} \right) \left(\frac{1}{1 - \gamma^2} \right).$$

Combining these estimates, we have

$$\frac{x^t \hat{B} x}{x^t A x} \le 1 + \left(\frac{\gamma^2}{1 - \gamma^2} \right) + \left(\frac{\rho^{2m}}{1 - \rho^{2m}} \right) \left(\frac{1}{1 - \gamma^2} \right) = \frac{1}{(1 - \gamma^2)(1 - \rho^{2m})}.$$

$\square$

We now consider some possibilities for the inner iterations. One obvious choice is a Jacobi method based on the diagonal matrix $D_{22} = \mathrm{diag}\, A_{22}$ with $\hat{A}_{22} = D_{22}/\omega$. Using Lemma 4, for the choice $\omega = 2/(\mu + \bar{\mu})$, we have

$$\rho \le \frac{\kappa - 1}{\kappa + 1},$$

where $\kappa = \bar{\mu}/\mu$.

A second possibility is to use a symmetric Gauss–Seidel iteration. Let $A_{22} = L_{22} + D_{22} + L_{22}^t$, where L_{22} is lower triangular. We then take

$$\hat{A}_{22} = (D_{22} + L_{22}) D_{22}^{-1} (D_{22} + L_{22})^t. \tag{5.28}$$

Lemma 5   Suppose the hypotheses of Lemma 4 hold, and let $\hat{A}_{22}$ be given by (5.28). Then there exists a finite positive constant η depending only on α_0, β_0, and δ_0, such that

$$1 \le \frac{x^t \hat{A}_{22} x}{x^t A_{22} x} \le 1 + \eta. \tag{5.29}$$

Proof.   As usual, the lower bound is one, since $\hat{A}_{22} = A_{22} + L_{22}D_{22}^{-1}L_{22}^t$, and $L_{22}D_{22}^{-1}L_{22}^t$ is symmetric and positive semidefinite. Now

$$\eta = \max_x \frac{y^t D_{22} y}{x^t A_{22} x},$$

where

$$D_{22}y = L_{22}^t x.$$

In finite element notation, this is

$$\eta = \max_w \frac{\sum_j \hat{w}_j^2 \|\phi_j\|^2}{\|w\|^2},$$

where $\hat{w} \in \mathcal{W}$ corresponds to y, $w \in \mathcal{W}$ corresponds to x, and $\{\phi_j\}$ are the basis functions for $\mathcal{W}$. Since the basis functions for $\mathcal{W}$ are developed from a fixed set of functions defined on the reference element, the support of a given basis function can intersect that of only a small number of other basis functions (there are at most a fixed number of nonzeros in any row of L_{22}^t, independent of the number of elements in the mesh). Therefore we must have

$$\sum_j \hat{w}_j^2 \|\phi_j\|^2 \le C \sum_j w_j^2 \|\phi_j\|^2,$$

where $C = C(\delta_0)$. The result now follows directly from Lemma 4. $\square$

Using Lemma 5, we can estimate

$$\rho = \|I - A_{22}^{1/2} \hat{A}_{22}^{-1} A_{22}^{1/2}\|_{\ell_2} \le \frac{\eta}{1 + \eta}.$$

Thus we see that although these inner iterations perturb the rate of convergence, they do not affect the essential feature that the rate depends only on local properties of the finite element spaces, and is independent of such things as the dimension of the space, uniformity or nonuniformity of the mesh, and regularity of the solution.

6. Multilevel Cauchy inequalities

In this section we will develop several strengthened Cauchy inequalities of use in analysing hierarchical basis iterations with more than two levels. These estimates are developed for the special case of continuous piecewise linear finite elements; they can be combined with the two-level analysis of Section 5 to develop multilevel algorithms for higher-degree polynomial spaces. We will return to this point in Section 7. Much of the material here is based on Bank and Dupont (1979), Yserentant (1986), and Bank, Dupont and Yserentant (1988). See also the books of Hackbusch (1985), Bramble (1993), and Oswald (1994).

Let $\mathcal{T}_1$ be a coarse, shape regular triangulation of Ω. We will inductively construct a sequence of uniformly refined triangulations $\mathcal{T}_j$, $2 \leq j \leq k$, as follows. For each triangle $t \in \mathcal{T}_{j-1}$, we will construct 4 triangles in $\mathcal{T}_j$ by pairwise connecting the midpoints of t. All triangulations will be shape regular, as every triangle $t \in \mathcal{T}_j$ will be geometrically similar to the triangle in $\mathcal{T}_0$ which contains it. We could also allow nonuniform refinements that control shape regularity, for example those of the type used in the adaptive finite element program $PLTMG$ (Bank 1994). See also Rüde (1993) and Deuflhard, Leinen and Yserentant (1989).

With this definition, it is easy to introduce the notion of the level of a given vertex in the triangulation $\mathcal{T}_j$. All vertices in the original triangulation $\mathcal{T}_1$ are called level-1 vertices. The new vertices created in forming $\mathcal{T}_j$ from $\mathcal{T}_{j-1}$ are called level-j vertices. Notice that all vertices in $\mathcal{T}_j$ have a level less than or equal to j. Also note that each vertex has a unique level, and this unique level is the same in all triangulations that contain it.

Let $\mathcal{M}_j$ be the space of continuous piecewise linear polynomials associated with $\mathcal{T}_j$. Functions in $\mathcal{M}_j$ will be represented using the hierarchical basis, which is easily constructed in an inductive fashion. Let $\{\phi_i\}_{i=1}^{N_1}$ denote the usual nodal basis functions for the space $\mathcal{M}_1$; this is also the hierarchical basis for $\mathcal{M}_1$. To construct the hierarchical basis for $\mathcal{M}_j$, $j > 1$, we take the union of the hierarchical basis for $\mathcal{M}_{j-1}$, $\{\phi_i\}_{i=1}^{N_{j-1}}$, with the nodal basis functions associated with the newly introduced level j vertices, $\{\phi_i\}_{i=N_{j-1}+1}^{N_j}$.

Let $\mathcal{V}_j$ be the subspace spanned by the basis functions associated with the level-j vertices, $\{\phi_i\}_{i=N_{j-1}+1}^{N_j}$, where $N_0 = 0$. Note that $\mathcal{V}_1 = \mathcal{M}_1$. Then we can write for $j > 1$,

$$\mathcal{M}_j = \mathcal{M}_{j-1} \oplus \mathcal{V}_j = \mathcal{V}_1 \oplus \mathcal{V}_2 \oplus \ldots \oplus \mathcal{V}_j.$$

Let $\mathcal{N}_j$, $1 \leq j \leq k-1$ be defined by

$$\mathcal{N}_j = \mathcal{V}_{j+1} \oplus \mathcal{V}_{j+2} \oplus \ldots \oplus \mathcal{V}_k$$

with $\mathcal{N}_k = \emptyset$. Then we have the decompositions

$$\mathcal{M}_k = \mathcal{M}_j \oplus \mathcal{N}_j$$

for $1 \leq j \leq k$.

Before proceeding to the Cauchy inequalities, we need a preliminary technical result.

Lemma 6 Let $t \in \mathcal{S}$, where $\mathcal{S}$ is defined as in Section 3. Let $\mathcal{T}'$ be a shape regular triangulation of t, whose elements have a minimum diameter of h. Let $\mathcal{M}'$ be the space of continuous piecewise linear polynomials associated with $\mathcal{T}'$. Then there exists a constant $c = c(\delta_0)$, independent of h, such that, for all $v \in \mathcal{M}'$,

$$\|v\|_{\infty,t} \le c|\log h|^{1/2}\|v\|_{1,t}. \tag{6.1}$$

Proof.   Here we will only sketch a proof, following ideas in Bank and Scott (1989), but see Yserentant (1986) for a more detailed, but also more elementary proof. We remark that estimate (6.1) is restricted to two space dimensions.

Our proof is based on an inverse inequality, and the Sobolev inequality; see Brenner and Scott (1994) or Ciarlet (1980) for a general discussion of these topics. Let t' be a shape regular triangle of size $h_{t'}$, and let v be a linear polynomial. The inverse inequality we require states

$$\|v\|_{\mathcal{L}^\infty(t')} \le C_0 h_{t'}^{-2/p}\|v\|_{\mathcal{L}^p(t')}$$

for $1 \le p \le \infty$. Let D be a closed bounded region with a piecewise smooth boundary; then the Sobolev inequality we need states

$$\|v\|_{\mathcal{L}^p(D)} \le C_1\sqrt{p}\|v\|_{\mathcal{H}^1(D)}$$

for all $v \in \mathcal{H}^1(D)$ and all $p < \infty$. Now let $t \in \mathcal{S}$ and $v \in \mathcal{M}'$; then

$$
\begin{aligned}
\|v\|_{\mathcal{L}^\infty(t)} &= \max_{t' \in \mathcal{T}'} \|v\|_{\mathcal{L}^\infty(t')} \\
&\le C_0 h^{-2/p} \max_{t' \in \mathcal{T}'} \|v\|_{\mathcal{L}^p(t')} \\
&\le C_0 h^{-2/p}\|v\|_{\mathcal{L}^p(t)} \\
&\le C_0 C_1 h^{-2/p}\sqrt{p}\|v\|_{\mathcal{H}^1(t)}.
\end{aligned}
$$

The proof is now completed by taking $p \approx -4\log h$. $\square$

Lemma 7 Let $\mathcal{M}_k = \mathcal{M}_j \oplus \mathcal{N}_j$ as above. Then there exist positive constants γ_j, $1 \le j \le k - 1$ such that

$$\gamma_j \le 1 - \frac{c}{k-j}, \tag{6.2}$$

and the strengthened Cauchy inequality

$$|a(v, w)| \le \gamma_j \|v\| \, \|w\| \tag{6.3}$$

holds for $v \in \mathcal{M}_j$ and $w \in \mathcal{N}_j$. The positive constant c in (6.2) is independent of j and k.

Proof.   Our proof is based on that of Bank and Dupont (1979). Following the pattern used in proving Lemma 2, we first reduce the estimate (6.3) to an elementwise estimate for $t \in \mathcal{T}_j$. If we show

$$|a(v, w)_t| \le \gamma_{j,t}\|v\|_t\|w\|_t, \tag{6.4}$$

then

$$\gamma_j = \max_{t \in \mathcal{T}_j} \gamma_{j,t}.$$

Let $t \in \mathcal{T}_j$, and let x_i, $1 \le i \le 3$ denote the three vertices of t. We map t to a triangle $\hat{t} \in \mathcal{S}$ using the change of variable

$$\hat{x} = \frac{x - x_1}{h_t}.$$

As in the proof of Lemma 2, this verifies that $\gamma_{j,t}$ is independent of h_t. Notice that $\mathcal{M}_{j,t}$, the restriction of $\mathcal{M}_j$ to t, is just the space of linear polynomials on t and has dimension three. In the case of uniform refinement, the space $\mathcal{N}_{j,t}$ is the space of piecewise linear polynomials on a uniform grid of 4^{k-j} congruent triangles, which are zero at the three vertices of t. The (local) constant function is thus contained in $\mathcal{M}_{j,t}$, and $\mathcal{M}_{j,t} \oplus \mathcal{N}_{j,t}$ is just the space of continuous piecewise linear polynomials on t.

Let $v \in \mathcal{M}_{j,t}$ and $w \in \mathcal{N}_{j,t}$. Then

$$
\begin{aligned}
\gamma_{j,t} &= \max_{\|v\|_t = \|w\|_t = 1} a(v,w)_t \\
&= \max_{\|v\|_t = \|w\|_t = 1} 1 - \frac{\|v - w\|_t^2}{2} \\
&\le \max_{\|v\|_t = \|w\|_t = 1} 1 - c\|v - w\|_{1,t}^2,
\end{aligned}
$$

where $c = c(\alpha_0, \beta_0)$.

We now apply Lemma 6, noting that $h \approx 2^{k-j}$ for the triangulation of $\hat{t}$.

$$\gamma_{j,t} \le \max_{\|v\|_t = \|w\|_t = 1} 1 - \frac{C\|v - w\|_{\infty,t}^2}{\log 2^{k-j}},$$

where $C = C(\alpha_0, \beta_0, \delta_0)$.

Next we note that, since v is just a linear polynomial on t with $\|v\|_t = 1$, and $w(x_i) = 0$, $1 \le i \le 3$, we have a fixed constant $c' > 0$, independent of j and k, such that

$$c' < \max_{x_i} |v(x_i)| = \max_{x_i} |v(x_i) - w(x_i)| \le \|v - w\|_{\infty,t}.$$

Thus it follows that

$$\gamma_{j,t} \le 1 - \frac{Cc'}{\log 2^{k-j}},$$

and the lemma follows. $\square$

We next describe the result of Lemma 7 in terms of interpolation operators.

Lemma 8 Let $u = v_j + w_j \in \mathcal{M}_k$, $v_j \in \mathcal{M}_j$ and $w_j \in \mathcal{N}_j$. Define the interpolation operator $\mathcal{I}_j$, mapping $\mathcal{M}_k$ to $\mathcal{M}_j$, by $\mathcal{I}_j(u) = v_j$. Then

$$\|\mathcal{I}_j(u)\| \leq C\sqrt{k-j}\|u\|. \tag{6.5}$$

The positive constant C is independent of j and k.

Proof. Apply Lemmas 3 and 7. See also Yserentant (1986), and Bank, Dupont and Yserentant (1988). $\square$

We finish this section with

Lemma 9 Let $\mathcal{V}_i$ and $\mathcal{V}_j$ for $1 \leq i, j \leq k$ be defined as above. Then there exist positive constants $\Gamma_{i,j}$ satisfying

$$\Gamma_{i,j} \leq c2^{-|i-j|/2}, \tag{6.6}$$

such that

$$|a(v,w)| \leq \Gamma_{i,j}\|v\|\,\|w\| \tag{6.7}$$

for all $v \in \mathcal{V}_i$ and $w \in \mathcal{V}_j$. The constant c in (6.6) is independent of i and j.

Proof. Our proof is similar to that given by Yserentant (1986). Without loss of generality, suppose $i < j$. We need consider no triangulation finer than $\mathcal{T}_j$, since subsequent refinements do not affect either v or w. As in the other Cauchy inequalities, one first reduces the estimate to a single element $t \in \mathcal{T}_i$, that is

$$|a(v,w)_t| \leq \Gamma_{i,j,t}\|v\|_t\|w\|_t. \tag{6.8}$$

We then consider the gradient terms and the lower order terms separately as in (3.6)–(3.7). For the highest order term, we must again consider the special importance of the (local) constant function, which in this case belongs to $\mathcal{V}_{i,t}$. Following the pattern in the proof of Lemma 2, we next map $t \in \mathcal{T}_i$ to an element $\hat{t} \in \mathcal{S}$ by scaling and translation, showing that the estimate must be independent of h_t. Also note that under this mapping, triangles in $\mathcal{T}_j$ become triangles with size $\hat{h} \approx 2^{i-j}$.

The central estimate is to show that

$$|a(\hat{v},\hat{w})_{1,\hat{t}}| \leq \Gamma_{i,j,1,t}\|\hat{v}\|_{1,\hat{t}}\|\hat{w}\|_{1,\hat{t}}, \tag{6.9}$$

where

$$a(\hat{v},\hat{w})_{1,\hat{t}} = \int_{\hat{t}} \hat{a}\nabla\hat{v}^t\nabla\hat{w}\,\mathrm{d}\hat{x}$$

$$\|\hat{v}\|^2_{1,\hat{t}} = a(\hat{v},\hat{v})_{1,\hat{t}}.$$

We will also use the norms

$$\|\hat{v}\|_{\hat{t}}^2 = \int_{\hat{t}} \hat{v}^2 \, d\hat{x} \quad \text{and} \quad \|\hat{v}\|_{\partial\hat{t}}^2 = \int_{\partial\hat{t}} \hat{v}^2 \, d\hat{x}.$$

The function $\hat{v}$ is just a linear polynomial on $\hat{t}$, while $\hat{w}$ is a piecewise linear polynomial vanishing at all the vertices with level smaller than j. Such a function is necessarily very oscillatory, and for such a function the differential operator behaves very much like $\hat{h}^{-1}$ times the identity operator. In particular, we have the estimates

$$\|\hat{w}\|_{\hat{t}} \le C\hat{h}\|\hat{w}\|_{1,\hat{t}} \le \hat{C}2^{i-j}\|\hat{w}\|_{1,\hat{t}} \tag{6.10}$$

and

$$\|\hat{w}\|_{\partial\hat{t}} \le C\hat{h}^{1/2}\|\hat{w}\|_{1,\hat{t}} \le \hat{C}2^{(i-j)/2}\|\hat{w}\|_{1,\hat{t}}, \tag{6.11}$$

where $\hat{C} = \hat{C}(\alpha_0, \delta_0)$.

Now, using integration by parts, the fact that $\Delta v = 0$ in $\hat{t}$, and (6.10)–(6.11) we have

$$\begin{aligned}
a(\hat{v}, \hat{w})_{1,\hat{t}} &= \int_{\hat{t}} -\nabla\hat{a}^t\nabla\hat{v}\hat{w} \, d\hat{x} + \int_{\partial\hat{t}} \hat{a}\nabla\hat{v}^t n\hat{w} \, d\hat{s} \\
&\le C\{\|\nabla\hat{v}\|_{\hat{t}}\|w\|_{\hat{t}} + \|\nabla\hat{v}\|_{\partial\hat{t}}\|\hat{w}\|_{\partial\hat{t}}\} \\
&\le C'2^{(i-j)/2}\|\hat{v}\|_{1,\hat{t}}\|\hat{w}\|_{1,\hat{t}}.
\end{aligned}$$

The lower order term is easy to treat in this case because of (6.10). □

7. Multilevel iterative methods

In this section, we will analyse block Jacobi and block symmetric Gauss–Seidel iterations using the hierarchical decomposition

$$\mathcal{M}_k = \mathcal{V}_1 \oplus \mathcal{V}_2 \oplus \ldots \oplus \mathcal{V}_k$$

defined in Section 6. Much of this material comes from Bank, Dupont and Yserentant (1988), but see also Bramble (1993), Bramble, Pasciak, and Xu (1990), Bramble, Pasciak, Wang, and Xu (1991), Griebel (1994), Hackbusch (1985), Ong (1989), Xu (1989) and (1992), and Yserentant (1986) and (1992).

As before, we let $\{\phi_i\}_{i=N_{j-1}+1}^{N_j}$ denote piecewise linear nodal basis functions for the level-j vertices in $\mathcal{T}_k$. Then the stiffness matrix A can be expressed as the symmetric, positive definite block $k \times k$ matrix

$$A = \begin{bmatrix} A_{11} & A_{12} & \cdots & A_{1k} \\ A_{21} & A_{22} & & A_{2k} \\ \vdots & & \ddots & \vdots \\ A_{k1} & A_{k2} & \cdots & A_{kk} \end{bmatrix}, \tag{7.1}$$

where A_{jj} is the $(N_j - N_{j-1}) \times (N_j - N_{j-1})$ matrix of energy inner products involving just the level-j basis functions. In similar fashion to the analysis in Section 5, we set

$$A = L + D + L^t, \tag{7.2}$$

where

$$D = \begin{bmatrix} A_{11} & & & \\ & A_{22} & & \\ & & \ddots & \\ & & & A_{kk} \end{bmatrix} \quad \text{and} \quad L = \begin{bmatrix} 0 & & & \\ A_{21} & 0 & & \\ \vdots & & \ddots & \\ A_{k1} & A_{k2} & \cdots & 0 \end{bmatrix}.$$

We first consider the block Jacobi iteration. Let $u_0 \in \mathcal{M}_k$ be given. We define the sequence

$$u_i = v_{1,i} + v_{2,i} + \ldots + v_{k,i},$$

where $v_{j,i} \in \mathcal{V}_j$, $1 \le j \le k$. In finite element notation, the block Jacobi iteration is written

$$a(v_{j,i+1} - v_{j,i}, \chi) = \omega\{(f, \chi) - a(u_i, \chi)\} \tag{7.3}$$

for $\chi \in \mathcal{V}_j$, $1 \le j \le k$. The iteration (7.3) can be written in matrix notation as

$$D(x_{i+1} - x_i) = \omega\{F - Ax_i\}, \tag{7.4}$$

where the vector $x_i \in \mathbb{R}^{N_k}$ corresponds to the finite element function $u_i \in \mathcal{M}_k$. To estimate the rate of convergence, we must bound the Rayleigh quotient

$$0 < \underline{\lambda} \le \frac{x^t D x}{x^t A x} \le \bar{\lambda} \tag{7.5}$$

for $x \ne 0$. In finite element notation, this is written

$$0 < \underline{\lambda} \le \frac{\sum_{i=1}^{k} \|v_i\|^2}{\|v\|^2} \le \bar{\lambda}, \tag{7.6}$$

where $v_i \in \mathcal{V}_i$ and $v = \sum_{i=1}^{k} v_i \ne 0$.

For any $v = v_1 + v_2 + \ldots + v_k$, we define

$$z_j = v_1 + v_2 + \ldots + v_j, \tag{7.7}$$

for $1 \le j \le k$, with $z_0 = 0$,

$$w_j = v_{j+1} + v_{j+2} + \ldots + v_k, \tag{7.8}$$

for $0 \le j \le k - 1$, with $w_k = 0$. Thus we have $v = z_j + w_j$, $0 \le j \le k$. Note $z_j \in \mathcal{M}_j$, while $w_j \in \mathcal{N}_j$.

We begin our analysis with an upper bound for (7.6). First note that the angle between the spaces $\mathcal{V}_1 \oplus \mathcal{V}_2 \oplus \ldots \oplus \mathcal{V}_{j-1} = \mathcal{M}_{j-1}$ and $\mathcal{V}_j$ is just the angle between the spaces $\mathcal{V}$ and $\mathcal{W}$ of Lemma 2. Therefore the constant in the strengthened Cauchy inequality for these spaces, which we will denote by $\tilde{\gamma}$, does not depend on j. Now

$$
\begin{aligned}
\|z_j\|^2 &= \|z_{j-1} + v_j\|^2 \\
&= \|z_{j-1}\|^2 + \|v_j\|^2 + 2a(z_{j-1}, v_j) \\
&\geq \|z_{j-1}\|^2 + \|v_j\|^2 - 2\tilde{\gamma}\|z_{j-1}\|\,\|v_j\| \\
&\geq (1 - \tilde{\gamma}^2)\|v_j\|^2.
\end{aligned}
$$

We now use Lemma 7 to deduce

$$
\begin{aligned}
\|v\|^2 &= \|z_j + w_j\|^2 \\
&= \|z_j\|^2 + \|w_j\|^2 + 2a(z_j, w_j) \\
&\geq \|z_j\|^2 + \|w_j\|^2 - 2\gamma_j\|z_j\|\,\|w_j\| \\
&\geq (1 - \gamma_j^2)\|z_j\|^2 \\
&\geq (1 - \gamma_j^2)(1 - \tilde{\gamma}^2)\|v_j\|^2.
\end{aligned}
$$

Thus we have

$$
\sum_{i=1}^{k} \|v_i\|^2 \leq \frac{\|v\|^2}{1 - \tilde{\gamma}^2} \sum_{i=1}^{k} \frac{1}{1 - \gamma_i^2} \leq Ck^2\|v\|^2.
$$

To find a lower bound, we note that

$$
\|\sum_{i=1}^{k} v_i\|^2 = \sum_{i=1}^{k}\sum_{j=1}^{k} a(v_i, v_j) \leq \sum_{i=1}^{k}\sum_{j=1}^{k} \Gamma_{i,j}\|v_i\|\,\|v_j\| = E^t \Gamma E,
$$

where $E_i = \|v_i\|$, and Γ is the $k \times k$ matrix introduced in Lemma 9. One can easily see that $\|\Gamma\|_{\ell^2} < C$, so that

$$
\|v\|^2 = \|\sum_{i=1}^{k} v_i\|^2 \leq C\sum_{i=1}^{k} \|v_i\|^2.
$$

Thus we have proved the following result.

Theorem 8 Let $A = L + D + L^t$ as defined above. Then

$$
C_1 \leq \frac{x^t D x}{x^t A x} \leq C_2 k^2, \tag{7.9}
$$

where $C_i = C_i(\alpha_0, \beta_0, \delta_0)$, $i = 1, 2$.

Note that the generalized condition number $\mathcal{K} \leq ck^2$ now depends on the number of levels. For the case of uniform refinement, $k = O(\log N_k)$, so this introduces a logarithmic-like term into the convergence rate. Note that $\sqrt{\mathcal{K}} \leq \tilde{c}k$, so that conjugate gradient acceleration can be expected to have a more significant impact on the k-level iteration than on the two-level method.

As in the case of the two-level iteration, we may solve linear systems of the form $A_{ii}x = b$ by an inner iteration for all $i > 1$. Following the development given in Section 5, let $\hat{A}_{ii}$ be the preconditioner for A_{ii} and let $G_i = I - A_{ii}^{1/2}\hat{A}_{ii}^{-1}A_{ii}^{1/2}$. Suppose

$$\max_{i>1} \|G_i\|_{\ell_2} = \rho < 1,$$

and assume for simplicity that $m \geq 1$ inner iterations are used for all $i > 1$. Let $R_{i,m} = G_i^m(I - G_i^m)^{-1}$. Then, using reasoning similar to that of (5.12), we replace (7.4) with

$$\hat{D}(x_{i+1} - x_i) = \omega\{F - Ax_i\} \tag{7.10}$$

where

$$\hat{D} = D + D^{1/2} \begin{bmatrix} 0 & & & \\ & R_{2,m} & & \\ & & \ddots & \\ & & & R_{k,m} \end{bmatrix} D^{1/2} = D + Z.$$

Theorem 9   Let $A = L + D + L^t$ and $\hat{D}$ be defined as above. Then

$$\frac{C_1}{1 + \rho^m} \leq \frac{x^t\hat{D}x}{x^tAx} \leq \frac{C_2k^2}{1 - \rho^m}, \tag{7.11}$$

where C_i, $i = 1, 2$ are given in Theorem 8.

Proof.   Following the proof of Theorem 5, we see for all $x \neq 0$,

$$\frac{1}{1 + \rho^m} \leq \frac{x^t\hat{D}x}{x^tDx} \leq \frac{1}{1 - \rho^m}.$$

The theorem then follows easily from this estimate and Theorem 8. $\square$

We next consider the symmetric block Gauss–Seidel iteration. In finite element notation, we may write this as

$$a(v_{j,i+1/2} - v_{j,i}, \chi) = (f, \chi) - a(z_{j-1,i+1/2} + w_{j-1,i}, \chi) \tag{7.12}$$

for $\chi \subset V_j$, $j = 1, 2, \ldots, k$, and

$$a(v_{j,i+1} - v_{j,i+1/2}, \chi) = (f, \chi) - a(z_{j,i+1/2} + w_{j,i+1}, \chi) \tag{7.13}$$

for $\chi \in V_j$, $j = k, k - 1, \ldots, 1$. Here $z_{j,i}$ and $w_{j,i}$ are defined analogously to v_j and w_j in (7.7)–(7.8). In matrix notation the iteration is written

$$\begin{aligned}(D + L)(x_{i+1/2} - x_i) &= F - Ax_i, \tag{7.14}\\ (D + L^t)(x_{i+1} - x_{i+1/2}) &= F - Ax_{i+1/2}.\end{aligned}$$

As in the two-level scheme, the preconditioner B is given by

$$B = (D + L)D^{-1}(D + L^t) = A + LD^{-1}L^t. \tag{7.15}$$

Theorem 10 Let $A = L + D + L^t$ and B be defined as above. Then

$$1 \le \frac{x^t B x}{x^t A x} \le 1 + \mu, \tag{7.16}$$

where

$$\mu \le C_3 k^2, \qquad C_3 = C_3(\alpha_0, \beta_0, \delta_0). \tag{7.17}$$

Proof. The lower bound is clear since $LD^{-1}L^t$ is symmetric and positive semidefinite. For the upper bound, we estimate

$$\mu = \max_{x \neq 0} \frac{y^t D y}{x^t A x}$$

where

$$D y = L^t x.$$

Let $v = v_1 + v_2 + \ldots + v_k = z_j + w_j$, with $v_i \in \mathcal{V}_i$ and $z_j \in \mathcal{M}_j$ and $w_j \in \mathcal{N}_j$ as in (7.7)–(7.8). Then in finite element notation, we have

$$\mu = \max_{v \neq 0} \frac{\sum_{i=1}^{k-1} \|\tilde{v}_i\|^2}{\|v\|^2}, \tag{7.18}$$

where

$$a(\tilde{v}_i, \chi) = a(w_i, \chi) \tag{7.19}$$

for all $\chi \in \mathcal{V}_i$.

Taking $\chi = \tilde{v}_i$ in (7.19) and applying Lemma 7, we have

$$\|\tilde{v}_i\| \le \gamma_i \|w_i\|,$$

and

$$\begin{aligned}
\|v\|^2 &= \|z_i + w_i\|^2 \\
&= \|z_i\|^2 + \|w_i\|^2 + 2a(z_i, w_i) \\
&\ge \|z_i\|^2 + \|w_i\|^2 - 2\gamma_i \|z_i\| \|w_i\| \\
&\ge (1 - \gamma_i^2) \|w_i\|^2 \\
&\ge (1 - \gamma_i^2) \gamma_i^{-2} \|\tilde{v}_i\|^2.
\end{aligned}$$

Thus we have

$$\mu \le \sum_{i=1}^{k-1} \frac{\gamma_i^2}{1 - \gamma_i^2} < C_3 k^2.$$

$\square$

We next analyse the effect of inner iterations on the symmetric block Gauss–Seidel iteration. Thus we replace D with $\hat{D}$ in (7.14) and obtain the iteration

$$\begin{aligned}
(\hat{D} + L)(x_{i+1/2} - x_i) &= F - Ax_i \tag{7.20} \\
(\hat{D} + L^t)(x_{i+1} - x_{i+1/2}) &= F - Ax_{i+1/2}
\end{aligned}$$

Following arguments similar to (5.26), we have

$$
\begin{aligned}
\hat{B} &= (\hat{D} + L)(2\hat{D} - D)^{-1}(\hat{D} + L^t) \\
&= A + (D - \hat{D} + L)(2\hat{D} - D)^{-1}(D - \hat{D} + L^t) \qquad (7.21) \\
&= A + (L - Z)(D + 2Z)^{-1}(L^t - Z).
\end{aligned}
$$

As usual, we need to estimate the Rayleigh quotient $x^t \hat{B} x / x^t A x$. Since $(L - Z)(D + 2Z)^{-1}(L^t - Z)$ is symmetric, positive semidefinite, the lower bound is just 1. To obtain an upper bound, the essential estimate we must make is

$$
\begin{aligned}
\hat{\mu} &= \max_{x \neq 0} \frac{x^t (L - Z)(D + 2Z)^{-1}(L^t - Z)x}{x^t A x} \\
&= \|(D + 2Z)^{-1/2}(L^t - Z)A^{-1/2}\|_{\ell^2}^2 \\
&\leq \Big(\|(D + 2Z)^{-1/2}D^{1/2}\|_{\ell^2} \, \|D^{-1/2}L^t A^{-1/2}\|_{\ell^2} \\
&\qquad + \|(D + 2Z)^{-1/2}ZD^{-1/2}\|_{\ell^2} \, \|D^{1/2}A^{-1/2}\|_{\ell^2} \Big)^2 .
\end{aligned}
$$

Now

$$
\|(D + 2Z)^{-1/2}D^{1/2}\|_{\ell^2} \leq \frac{1 + \rho^m}{1 - \rho^m}
$$

and

$$
\|(D + 2Z)^{-1/2}ZD^{-1/2}\|_{\ell^2} \leq \frac{\rho^{2m}}{1 - \rho^{2m}}.
$$

The norms $\|D^{-1/2}L^t A^{-1/2}\|_{\ell^2}$ and $\|D^{1/2}A^{-1/2}\|_{\ell^2}$ are estimated using Theorems 10 and 8, respectively, and we now combine these estimates.

Theorem 11 Let $A = L + D + L^t$ and $\hat{B}$ be defined as above. Then

$$
1 \leq \frac{x^t \hat{B} x}{x^t A x} \leq 1 + \hat{\mu}, \qquad (7.22)
$$

where

$$
\hat{\mu} \leq \left(\sqrt{\frac{1 + \rho^m}{1 - \rho^m}} C_3 + \sqrt{\frac{\rho^{2m}}{1 - \rho^{2m}}} C_2 \right)^2 k^2 \leq C_4 k^2, \qquad (7.23)
$$

and C_2 and C_3 are given in Theorems 8 and 10, respectively.

If G is a smoother, then using (5.9) we have $\|(D + 2Z)^{-1/2}D^{1/2}\|_{\ell^2} \leq 1$, and the improved estimate

$$
\hat{\mu} \leq \left(\sqrt{C_3} + \sqrt{\frac{\rho^{2m}}{1 - \rho^{2m}}} C_2 \right)^2 k^2 \leq C_4 k^2.
$$

We conclude with several remarks about the two-level and k-level methods.

Although the k-level method was developed for only the case of continuous piecewise linear polynomials, this is sufficient to construct efficient methods for higher-degree spaces. For example, we consider the case of continuous piecewise quadratic polynomials on a sequence of meshes $\mathcal{T}_j$, $1 \leq j \leq k$. At first glance, one might be tempted to try to develop a method in which one used piecewise quadratic spaces on all levels. Further reflection would lead one to the conclusion that such a method could potentially be very complicated, as it is not clear that there is a simple way to develop a hierarchical basis. It is also not clear that the analysis of such a method could be based on the results in this work.

On the other hand, we could begin by making the usual two-level decomposition $\mathcal{M} = \mathcal{V} \oplus \mathcal{W}$, where $\mathcal{V}$ is the space of piecewise linear polynomials on $\mathcal{T}_k$ and $\mathcal{W}$ is the space of piecewise quadratic bump functions that are zero at the vertices of $\mathcal{T}_k$. The dimension of $\mathcal{W}$ is then approximately $3N/4$ where N is the dimension of $\mathcal{M}$. For the space $\mathcal{V}$, which is just the space of piecewise linear polynomials on $\mathcal{T}_k$, we can make the hierarchical decomposition

$$\mathcal{V} = \mathcal{V}_1 \oplus \mathcal{V}_2 \oplus \ldots \oplus \mathcal{V}_k$$

as described here. Overall, we have the hierarchical decomposition

$$\mathcal{M} = \mathcal{V}_1 \oplus \mathcal{V}_2 \oplus \ldots \oplus \mathcal{V}_k \oplus \mathcal{W}.$$

Based on this decomposition, there is an obvious multilevel hierarchical basis iteration that can be developed. This iteration could be viewed as a two-level iteration, with an elaborate k-level inner iteration used to solve the linear systems associated with the space $\mathcal{V}$. Alternatively, this iteration could be viewed as a $k+1$-level iteration, in which the the first k levels are the standard ones, but level $k+1$ is special, in that the degree of approximation is increased instead of the mesh being refined. For either viewpoint, the algorithm is the same, and its analysis is straightforward using the results in Sections 3–7.

Another possibility along these lines is to make some further hierarchical decomposition of the space $\mathcal{W}$. For example, suppose now that $\mathcal{M}$ is the space of continuous piecewise quartic polynomials on $\mathcal{T}_k$. We can begin by making a decomposition $\mathcal{M} = \mathcal{V} \oplus \mathcal{W}$, where $\mathcal{V}$ is the space of continuous piecewise linear polynomials and $\mathcal{W}$ is the space of quartic polynomials that are zero at the vertices of $\mathcal{T}_k$. We make a further decomposition of $\mathcal{V}$ as in the previous example. We can also conveniently make the further decomposition $\mathcal{W} = \mathcal{W}_2 \oplus \mathcal{W}_4$, where $\mathcal{W}_2$ is the space of continuous piecewise quadratic polynomials that are zero at the vertices of $\mathcal{T}_k$. This is the same as the space $\mathcal{W}$ in our last example. The space $\mathcal{W}_4$ is now the space of continuous piecewise quartic polynomials that are zero at the vertices and edge midpoints of $\mathcal{T}_k$ (that is, all the nodes associated with the piecewise linear and piecewise quadratic spaces). This space can be characterized in terms of a subset of the standard nodal basis functions for the piecewise quartic space, the bump functions associated with the 1/4 and 3/4 points on each edge, and the bubble functions associated with the barycentric coordinates $(1/4, 1/4, 1/2)$, $(1/4, 1/2, 1/4)$, and $(1/2, 1/4, 1/4)$ in each element. This leads to an overall decomposition

$$\mathcal{M} = \mathcal{V}_1 \oplus \mathcal{V}_2 \oplus \ldots \oplus \mathcal{V}_k \oplus \mathcal{W}_2 \oplus \mathcal{W}_4.$$

The resulting hierarchical basis iteration could then be viewed as a basic two-level iteration in which elaborate inner iterations are used for solving linear systems associated with both the $\mathcal{V}$ and $\mathcal{W}$ spaces, or as a $k + 2$-level scheme in which the last two levels involve an increase in degree of approximation rather than a refinement of the mesh.

REFERENCES

M. Ainsworth and J. T. Oden (1992), 'A procedure for *a posteriori* error estimation for h-p finite element methods', *Comp. Meth. Appl. Mech. Engrg.* **101**, 73–96.

M. Ainsworth and J. T. Oden (1993), 'A unified approach to *a posteriori* error estimation using element residual methods', *Numer. Math.* **65**, 23–50.

A. K. Aziz and I. Babuška (1972), 'Survey lectures on the mathematical foundations of the finite element method', in *The Mathematical Foundations of the Finite Element Method with Applications to Partial Differential Equations* (A. K. Aziz, ed.), Academic Press, New York, 1–362.

I. Babuška and W. Gui (1986), 'Basic principles of feedback and adaptive approaches in the finite element method', *Comp. Meth. Appl. Mech. Engrg.* **55**, 27–42.

I. Babuška, O. C. Zienkiewicz, J. P. de S.R. Gago, and E. R. de Arantes e Oliveira, eds (1986) *Accuracy Estimates and Adaptive Refinements in Finite Element Computations*, Wiley, New York.

I. Babuška and W. C. Rheinboldt (1978), 'Error estimates for adaptive finite element computations', *SIAM J. Numer. Anal.* **15**, 736–754.

I. Babuška and W. C. Rheinboldt (1978), '*A posteriori* error estimates for the finite element method', *Internat. J. Numer. Methods Engrg.* **12**, 1597–1615.

R. E. Bank (1994), *PLTMG: A Software Package for Solving Elliptic Partial Differential Equations, Users' Guide 7.0.* Frontiers in Applied Mathematics **15**, SIAM, Philadelphia.

R. E. Bank and T. F. Dupont (1979), 'Notes on the k-level iteration', unpublished notes.

R. E. Bank and T. F. Dupont (1980), 'Analysis of a two level scheme for solving finite element equations', Technical Report CNA-159, Center for Numerical Analysis, University of Texas at Austin.

R. E. Bank, T. F. Dupont, and H. Yserentant (1988), 'The hierarchical basis multigrid method', *Numer. Math.* **52**, 427–458.

R. E. Bank and L. R. Scott (1989), 'On the conditioning of finite element equations with highly refined meshes', *SIAM J. Numer. Anal.* **26**, 1383–1394.

R. E. Bank and R. K. Smith (1993), '*A posteriori* error estimates based on hierarchical bases, *SIAM J. Numer. Anal.* **30**, 921–935.

R. E. Bank and A. Weiser (1985), 'Some *a posteriori* error estimates for elliptic partial differential equations', *Math. Comp.* **44**, 283–301.

F. Bornemann and H. Yserentant (1993), 'A basic norm equivalence for the theory of multilevel methods', *Numer. Math.* **64**, 455–476.

D. Braess (1981), 'The contraction number of a multigrid method for solving the Poisson equation', *Numer. Math.* **37**, 387–404.

J. H. Bramble (1993), *Multigrid Methods*, Pitman Research Notes in Mathematical Sciences **294**, Longman Sci. & Techn., Harlow, UK.

J. H. Bramble, J. E. Pasciak, J. Wang, and J. Xu (1991), 'Convergence estimate for product iterative methods with application to domain decomposition and multigrid', *Math. Comp.* **57**, 1–21.

J. H. Bramble, J. E. Pasciak, and J. Xu (1990), 'Parallel multilevel preconditioners', *Math. Comp.* **55**, 1–22.

S. C. Brenner and L. R. Scott (1994), *The Mathematical Theory of Finite Element Methods*, Springer, Heidelberg.

P. G. Ciarlet (1980), *The Finite Element Method for Elliptic Problems*, North-Holland, Amsterdam.

P. Deuflhard, P. Leinen, and H. Yserentant (1989), 'Concepts of an adaptive hierarchical finite element code', *IMPACT of Comput. in Sci. and Eng.* **1**, 3–35.

T. F. Dupont, R. P. Kendall, and H. H. Rachford (1968), 'An approximate factorization procedure for self-adjoint elliptic difference equations', *SIAM J. Numer. Anal.* **5**, 559–573.

R. Durán, M. A. Muschietti, and R. Rodríguez (1991), 'On the asymptotic exactness of error estimators for linear triangular finite elements', *Numer. Math.* **59**, 107–127.

R. Durán and R. Rodríguez (1992), 'On the asymptotic exactness of Bank–Weiser's estimator', *Numer. Math.* **62**, 297–303.

V. Eijkhout and P. Vassilevski (1991), 'The role of the strengthened Cauchy–Buniakowskii–Schwarz inequality in multilevel methods', *SIAM Review* **33**, 405–419.

G. H. Golub and C. F. Van Loan (1983), *Matrix Computations*, Johns Hopkins University Press, Baltimore.

G. H. Golub and D. P. O'Leary (1989), 'Some history of the conjugate gradient and Lanczos algorithms: 1948–1976', *SIAM Review* **31**, 50–102.

M. Griebel (1994), 'Multilevel algorithms considered as iterative methods on semi-definite systems', *SIAM J. Sci. Comput.* **15**, 547–565.

W. Hackbusch (1985), *Multigrid Methods and Applications*, Springer, Berlin.

J. F. Maitre and F. Musy (1982), 'The contraction number of a class of two level methods; an exact evaluation for some finite element subspaces and model problems', in *Multigrid Methods: Proceedings, Cologne 1981*, Lecture Notes in Mathematics **960**, Springer, Heidelberg, 535–544.

E. Ong (1989), 'Hierarchical basis preconditioners for second order elliptic problems in three dimensions', PhD thesis, University of Washington.

P. Oswald (1994), *Multilevel Finite Element Approximation: Theory and Applications*, Teubner Skripten zur Numerik, B. G. Teubner, Stuttgart.

U. Rüde (1993), *Mathematical and Computational Techniques for Multilevel Adaptive Methods*, Frontiers in Applied Mathematics **13**, SIAM, Philadelphia.

R. Verfürth (1995), *A Posteriori Error Estimation and Adaptive Mesh Refinement Techniques*, Teubner Skripten zur Numerik, B. G. Teubner, Stuttgart.

A. Weiser (1981), 'Local-mesh, local-order, adaptive finite element methods with *a posteriori* error estimators for elliptic partial differential equations', PhD thesis, Yale University.

J. Xu (1989), 'Theory of multilevel methods', PhD thesis, Cornell University.

J. Xu (1992), 'Iterative methods by space decomposition and subspace correction', *SIAM Review* **34**, 581–613.

H. Yserentant (1986), 'On the multi-level splitting of finite element spaces', *Numer. Math.* **49**, 379–412.

H. Yserentant (1992), 'Old and new convergence proofs for multigrid methods', in *Acta Numerica*, Cambridge University Press.

O. C. Zienkiewicz, D. W. Kelley, J. P. de S. R. Gago, and I. Babuška (1982), 'Hierarchical finite element approaches, adaptive refinement, and error estimates', in *The Mathematics of Finite Elements and Applications*, Academic Press, New York, 313–346.

Acta Numerica (1996), *pp.* 45–119

Orthogonal polynomials: applications and computation

Walter Gautschi *

Department of Computer Sciences

Purdue University

West Lafayette, IN 47907–1398, USA

E-mail: wxg@cs.purdue.edu

We give examples of problem areas in interpolation, approximation, and quadrature, that call for orthogonal polynomials not of the classical kind. We then discuss numerical methods of computing the respective Gauss-type quadrature rules and orthogonal polynomials. The basic task is to compute the coefficients in the three-term recurrence relation for the orthogonal polynomials. This can be done by methods relying either on moment information or on discretization procedures. The effect on the recurrence coefficients of multiplying the weight function by a rational function is also discussed. Similar methods are applicable to computing Sobolev orthogonal polynomials, although their recurrence relations are more complicated. The paper concludes with a brief account of available software.

* Work supported in part by the National Science Foundation under grant DMS-9305430. A preliminary account of this material was presented at the 10th Summer School in Computational Mathematics, Maratea, Italy, in September of 1992. The material of Section 1.2 is taken, with permission, from the author's article in Bowers and Lund, eds (1989, pp. 63–95).

CONTENTS

0   Introduction   46

PART I: APPLICATIONS

1   Interpolation   52
2   Approximation   56
3   Quadrature   64

PART II: COMPUTATION

4   Computation of Gauss-type quadrature rules   71
5   Moment-based methods   79
6   Discretization methods   91
7   Modification algorithms   98
8   Orthogonal polynomials of Sobolev type   107
9   Software   111
References   112

0. Introduction

The subject of orthogonal polynomials, if not in name then in substance, is quite old, having its origin in the 19th-century theories of continued fractions and the moment problem. Classical orthogonal polynomials, such as those of Legendre, Laguerre and Hermite, but also discrete ones, due to Chebyshev, Krawtchouk and others, have found widespread use in all areas of science and engineering. Typically, they are used as basis functions in which to expand other more complicated functions. In contrast, polynomials orthogonal with respect to general, nonstandard, weight functions and measures have received much less attention in applications, in part because of the considerable difficulties attending their numerical generation. Some progress, nevertheless, has been made in the last fifteen years or so, both in novel applications of non-classical orthogonal polynomials and in methods of their computation. The purpose of this article is to review some of these recent developments.

In Part I, we outline a number of (somewhat disconnected) problem areas that have given rise to unconventional orthogonal polynomials. These include problems in interpolation and least squares approximation, Gauss quadrature of rational functions, slowly convergent series, and moment-preserving spline approximation. Part II then takes up the problem of actually generating the respective orthogonal polynomials. Since most applications involve Gauss quadrature in one way or another, the computation of these quadrature rules is discussed first. Constructive methods for generating orthogonal polynomials, including those of Sobolev type, then follow, among them moment-based methods, discretization methods, and modification algorithms. We conclude by giving a brief account of available software.

The choice of topics treated here reflects the author's past interest and involvement in orthogonal polynomials. There are other applications and computational aspects that would deserve equal treatment. Foremost among these are applications to iterative methods of solving large (and usually sparse) systems of linear algebraic equations and eigenvalue problems. The pioneering work on this was done in the 1950s by Stiefel (1958) and Lanczos (1950); modern accounts can be found, for instance in Hageman and Young (1981), Golub and Van Loan (1989) and Freund, Golub and Nachtigal (1991). Among additional computational issues there is the problem of constructing the measure underlying a set of orthogonal polynomials, given their recursion coefficients. Some discussion of this can be found in Askey and Ismail (1984), and Dombrowski and Nevai (1986).

Before we start, we recall two items of particular importance in the constructive theory of orthogonal polynomials: the Gaussian quadrature formula, and the basic three-term recurrence relation. This will also provide us with an opportunity to introduce relevant notation.

0.1. Gauss-type quadrature rules

The concept of orthogonality arises naturally in the context of quadrature formulae, when one tries to maximize, or nearly maximize, their degree of exactness. Thus suppose we are given a positive measure[1] $\mathrm{d}\lambda$ on the real line $\mathbb{R}$ with respect to which polynomials can be integrated, that is, for which $\int_{\mathbb{R}} t^k \, \mathrm{d}\lambda(t)$ exists for each nonnegative integer $k \in \mathbb{N}_0$. A quadrature formula

$$\int_{\mathbb{R}} f(t) \, \mathrm{d}\lambda(t) = \sum_{\nu=1}^{n} \lambda_\nu f(\tau_\nu) + R_n(f), \qquad (0.1)$$

---

[1] For our purposes it suffices to assume that $\mathrm{d}\lambda$ is either a discrete measure, $\mathrm{d}\lambda(t) = \mathrm{d}\lambda_N(t)$, concentrated on a finite number N of points $t_1 < t_2 < \cdots < t_N$, that is, $\lambda(t)$ is constant on each open interval (t_i, t_{i+1}), $i = 0, 1, \ldots, N$ (where $t_0 = -\infty$, $t_{N+1} = +\infty$), and has a positive jump $w_i = \lambda(t_i+0) - \lambda(t_i-0)$ at t_i, $i = 1, 2, \ldots, N$, or $\mathrm{d}\lambda(t) = w(t) \, \mathrm{d}t$ is an absolutely continuous measure, where $w \geq 0$ is integrable on $\mathbb{R}$ and $\int_{\mathbb{R}} w(t) \mathrm{d}t > 0$, or a combination of both. Then for suitable functions f,

$$\int_{\mathbb{R}} f(t) \, \mathrm{d}\lambda(t) = \begin{cases} \sum_{i=1}^{N} w_i f(t_i), & \mathrm{d}\lambda \text{ discrete}, \\ \int_{\mathrm{supp}(\mathrm{d}\lambda)} f(t) w(t) \, \mathrm{d}t, & \mathrm{d}\lambda \text{ absolutely continuous}, \end{cases}$$

where $\mathrm{supp}(\mathrm{d}\lambda)$ denotes the support of $\mathrm{d}\lambda$, typically an interval or a union of disjoint intervals.

with distinct nodes $\tau_\nu \in \mathbb{R}$ and real weights λ_ν, is said to have *degree of exactness d* if

$$R_n(p) = 0, \qquad \text{all } p \in \mathbb{P}_d, \tag{0.2}$$

where $\mathbb{P}_d$ is the set of polynomials of degree $\leq d$. It is well known that for given τ_ν we can always achieve degree of exactness $n - 1$ by interpolating at the points τ_ν and integrating the interpolation polynomial instead of f. The resulting quadrature rule (0.1) is called the *Newton–Cotes formula* (relative to the points τ_ν and the measure $d\lambda$). Indeed, any quadrature formula having degree of exactness $d = n - 1$ can be so obtained, and is therefore called *interpolatory*. A natural question to ask is: what conditions must the nodes τ_ν and weights λ_ν satisfy in order for (0.1) to have degree of exactness larger than $n - 1$, say $d = n - 1 + m$, where $m > 0$ is a given integer? The complete answer is given by the following theorem, essentially due to Jacobi (1826).

Theorem 1 Given an integer $m > 0$, the quadrature rule (0.1) has degree of exactness $d = n - 1 + m$ if and only if the following two conditions are satisfied:

(i) The formula (0.1) is interpolatory.
(ii) The node polynomial $\omega_n(t) = \Pi_{\nu=1}^n (t - \tau_\nu)$ satisfies

$$\int_{\mathbb{R}} \omega_n(t)p(t)\,d\lambda(t) = 0 \qquad \text{for each } p \in \mathbb{P}_{m-1}. \tag{0.3}$$

Condition (ii) is clearly a condition involving only the nodes τ_ν of (0.1); it says that the node polynomial must be *orthogonal* to all polynomials of degree $\leq m - 1$. Here, orthogonality is in the sense of the inner product

$$(u, v)_{d\lambda} = \int_{\mathbb{R}} u(t)v(t)\,d\lambda(t), \qquad u, v \in \mathbb{P}, \tag{0.4}$$

in terms of which (0.3) can be stated as $(\omega_n, p)_{d\lambda} = 0$ for every $p \in \mathbb{P}_{m-1}$. Once a set of distinct nodes τ_ν has been found that satisfies this orthogonality constraint, condition (i) then determines uniquely the weights λ_ν, for example by requiring that (0.1) be exact for each power $f(t) = t^k$, $k = 0, 1, \ldots, n - 1$. This is a system of linear equations for the weights λ_ν whose matrix is a Vandermonde matrix in the nodes τ_ν, hence nonsingular, since they are assumed distinct.

It is clear that $m \leq n$; otherwise, we could take $p = \omega_n$ in (ii) and get $\int_{\mathbb{R}} \omega_n^2(t)\,d\lambda(t) = 0$, which is impossible if $d\lambda$ has more than n points of increase. (In the context of quadrature rules, $d\lambda$ indeed is usually assumed to be absolutely continuous and thus to have infinitely many points of increase.) Thus, $m = n$ is optimal and gives rise to the condition

$$(\omega_n, p)_{d\lambda} = 0, \qquad \text{all } p \in \mathbb{P}_{n-1}. \tag{0.5}$$

This means that ω_n must be orthogonal to all polynomials of lower degree, hence (see Section 0.2 below) is the unique (monic) *orthogonal polynomial* of degree n relative to the measure $\mathrm{d}\lambda$. We will denote this polynomial by $\pi_n(\cdot) = \pi_n(\cdot\,; \mathrm{d}\lambda)$. The formula (0.1) then becomes the n-point *Gaussian quadrature formula* (with respect to the measure $\mathrm{d}\lambda$), that is, the interpolatory quadrature rule of maximum degree of exactness $d = 2n - 1$ whose nodes are the zeros of $\pi_n(\cdot\,; \mathrm{d}\lambda)$. It is known from the theory of orthogonal polynomials (Szegő 1975) that these zeros are all simple and contained in the smallest interval containing the support of $\mathrm{d}\lambda$.

There are other interesting special cases of Theorem 1. We mention four:

(1) Assume that the infimum $a = \inf \mathrm{supp}\,(\mathrm{d}\lambda)$ is a finite number. We choose one of the nodes τ_ν to be equal to a, say $\tau_1 = a$. Then $\omega_n(t) = (t-a)\omega_{n-1}(t)$, where $\omega_{n-1}(t) = \Pi_{\nu=2}^{n}(t - \tau_\nu)$, and condition (ii) requires that

$$\int_a^\infty \omega_{n-1}(t)p(t)(t - a)\,\mathrm{d}\lambda(t) = 0, \qquad \text{all } p \in \mathbb{P}_{m-1}. \tag{0.6}$$

The optimal value of m is now clearly $m = n - 1$, in which case ω_{n-1} is the unique (monic) polynomial of degree $n - 1$ orthogonal with respect to the modified measure $\mathrm{d}\lambda_a(t) = (t - a)\,\mathrm{d}\lambda(t)$ – also a positive measure – that is, $\omega_{n-1}(t) = \pi_{n-1}(\cdot\,; \mathrm{d}\lambda_a)$. Again, all zeros of ω_{n-1} are distinct and larger than a; the resulting formula (0.1) is called the n-point *Gauss–Radau formula* (with respect to the measure $\mathrm{d}\lambda$).

(2) Similarly, if both $a = \inf \mathrm{supp}\,(\mathrm{d}\lambda)$ and $b = \sup \mathrm{supp}\,(\mathrm{d}\lambda)$ are finite numbers, and $n \geq 2$, and if we want $t_1 = a$ and (say) $t_n = b$, then $\omega_n(t) = -(t-a)(b-t)\omega_{n-2}(t)$, and $\omega_{n-2}(\cdot) = \pi_{n-2}(\cdot\,; \mathrm{d}\lambda_{a,b})$ for optimal $m = n - 2$, where $\mathrm{d}\lambda_{a,b}(t) = (t-a)(b-t)\,\mathrm{d}\lambda(t)$ is again a positive measure. The formula (0.1) with the interior nodes being the (distinct) zeros of $\pi_{n-2}(\cdot\,; \mathrm{d}\lambda_{a,b})$ then becomes the n-point *Gauss–Lobatto quadrature rule* (for the measure $\mathrm{d}\lambda$).

(3) Replace n in (0.1) by $2n + 1$, let $\tau_\nu = \tau_\nu^{(n)}$ be the zeros of $\pi_n(\cdot\,; \mathrm{d}\lambda)$ for some positive measure $\mathrm{d}\lambda$, and choose $n + 1$ additional nodes $\hat{\tau}_\mu$ such that the $(2n + 1)$-point formula (0.1) with nodes τ_ν and $\hat{\tau}_\mu$ has maximum degree of exactness $d > 3n + 1$. By Theorem 1 (with n replaced by $2n + 1$), the $n + 1$ nodes $\hat{\tau}_\mu$ to be inserted must be the zeros of the (monic) polynomial $\hat{\pi}_{n+1}$ satisfying

$$\int_{\mathbb{R}} \hat{\pi}_{n+1}(t)p(t)\pi_n(t; \mathrm{d}\lambda)\,\mathrm{d}\lambda(t) = 0, \qquad \text{all } p \in \mathbb{P}_n. \tag{0.7}$$

Here, the measure of orthogonality is $\mathrm{d}\hat{\lambda}(t) = \pi_n(t; \mathrm{d}\lambda)\,\mathrm{d}\lambda(t)$, which is no longer positive, but oscillatory. This calls for special techniques of computation; see, for instance, Monegato (1982), Kautsky and Elhay (1984), Caliò, Gautschi and Marchetti (1986, Section 2) and Laurie (1996). While $\hat{\pi}_{n+1}$ can be shown to exist uniquely, its zeros are not necessarily contained in the support of $d\lambda$ and may even be complex. The resulting $(2n + 1)$-point quadrature formula is called the *Gauss–Kronrod rule*. It has an interesting history and has received considerable attention in recent years. For surveys, see Monegato (1982), Gautschi (1988) and Notaris (1994).

(4) Consider $s > 1$ different measures $d\lambda_\sigma$, $\sigma = 1, 2, \ldots, s$, with common support, and for each an n-point quadrature rule (0.1) with a common set of nodes $\{\tau_\nu\}$ but individual weights $\{\lambda_{\nu,\sigma}\}, \sigma = 1, 2, \ldots, s$. Assume $n = ms$ to be an integer multiple of s. Find s such quadrature rules, each having degree of exactness $n-1+m$. (This is expected to be optimal since there are $n(s+1)$ unknowns and $(n+m)s = ns+s$ conditions imposed.) According to Theorem 1, each quadrature rule has to be interpolatory, and the node polynomial ω_n must be orthogonal to polynomials of degree $m - 1$ with respect to each measure,

$$\int_{\mathbb{R}} \omega_n(t)p(t)\, d\lambda_\sigma(t) = 0, \qquad \text{all } p \in \mathbb{P}_{m-1}, \quad \sigma = 1, 2, \ldots, s. \qquad (0.8)$$

One obtains the *shared-nodes quadrature rules* recently introduced by Borges (1994) in connection with computer graphics illumination models, where the models $d\lambda_\sigma$ are colour matching functions. Instead of assuming $n = ms$, one could require (0.8) to hold for $p \in \mathbb{P}_{m_\sigma-1}$, where $\sum_{\sigma=1}^{s} m_\sigma = n$, and thus 'distribute' the degrees of exactness differently among the s measures $d\lambda_\sigma$. The construction of such quadrature rules calls for *quasi-orthogonal polynomials*, that is, polynomials that are only partially orthogonal, as in (0.8), and not fully orthogonal, as in (0.5).

0.2. The three-term recurrence relation

Next to the Gauss formula, another important fact about orthogonal polynomials is that they always satisfy a three-term recurrence relation. The reason for this is the basic property

$$(tu, v)_{d\lambda} = (u, tv)_{d\lambda} \qquad (0.9)$$

satisfied by the inner product (0.4). Indeed, assume that $d\lambda$ has at least N points of increase. Then the system of orthogonal polynomials $\pi_k(\,\cdot\,; d\lambda)$, $k = 0, 1, \ldots, N - 1$, is easily seen to form a basis of $\mathbb{P}_{N-1}$. For any integer $k \leq N - 1$, therefore, since the polynomial

$$\pi_{k+1}(t) - t\pi_k(t)$$

is a polynomial of degree $\leq k$ (both π_{k+1} and $t\pi_k$ being monic of degree $k + 1$), there exist constants α_k, β_k and γ_{kj} such that

$$\pi_{k+1}(t) - t\pi_k(t) = -\alpha_k \pi_k(t) - \beta_k \pi_{k-1}(t) + \sum_{j=0}^{k-2} \gamma_{kj}\pi_j(t),$$
$$k = 0, 1, \ldots, N - 1,$$
$$(0.10)$$

where it is understood that $\pi_{-1}(t) \equiv 0$ and empty sums are zero. To determine α_k, take the inner product of both sides of (0.10) with π_k; this yields, by orthogonality,

$$-(t\pi_k, \pi_k) = -\alpha_k(\pi_k, \pi_k),$$

hence

$$\alpha_k = \frac{(t\pi_k, \pi_k)}{(\pi_k, \pi_k)}.$$

Similarly, forming the inner product with π_{k-1} $(k \geq 1)$ gives

$$-(t\pi_k, \pi_{k-1}) = -\beta_k(\pi_{k-1}, \pi_{k-1}).$$

This can be simplified by noting $(t\pi_k, \pi_{k-1}) = (\pi_k, t\pi_{k-1}) = (\pi_k, \pi_k + \cdots)$, where dots stand for a polynomial of degree $< k$. By orthogonality, then, $(t\pi_k, \pi_{k-1}) = (\pi_k, \pi_k)$, and we get

$$\beta_k = \frac{(\pi_k, \pi_k)}{(\pi_{k-1}, \pi_{k-1})}.$$

Finally, taking the inner product with π_i, $i < k - 1$, in (0.10), we find

$$-(t\pi_k, \pi_i) - \gamma_{ki}(\pi_i, \pi_i).$$

It is here where (0.9) is crucially used to obtain $\gamma_{ki} = 0$, since $(\pi_i, \pi_i) \neq 0$ and $(t\pi_k, \pi_i) = (\pi_k, t\pi_i) = 0$ because of $t\pi_i \in \mathbb{P}_{k-1}$. Thus, we have shown that

$$\pi_{k+1}(t) = (t - \alpha_k)\pi_k(t) - \beta_k\pi_{k-1}(t), \qquad k = 0, 1, \ldots, N - 1,$$
$$\pi_{-1}(t) = 0, \qquad \pi_0(t) = 1, \tag{0.11}$$

where

$$\alpha_k = \frac{(t\pi_k, \pi_k)}{(\pi_k, \pi_k)}, \qquad k = 0, 1, \ldots, N - 1,$$
$$\beta_k = \frac{(\pi_k, \pi_k)}{(\pi_{k-1}, \pi_{k-1})}, \qquad k = 1, 2, \ldots, N - 1. \tag{0.12}$$

This is the basic *three-term recurrence relation* satisfied by orthogonal polynomials. Since $\pi_{-1} = 0$, the coefficient β_0 in (0.11) can be arbitrary. It is convenient, however, to define it by

$$\beta_0 = (\pi_0, \pi_0) = \int_{\mathbb{R}} d\lambda(t). \tag{0.13}$$

Note that by construction, π_N is orthogonal to all polynomials of degree $< N$. If $d\lambda = d\lambda_N$ is a discrete measure with exactly N points of increase, there can be at most N orthogonal polynomials, $\pi_0, \pi_1, \ldots, \pi_{N-1}$, which implies that $(\pi_N, \pi_N) = 0$, that is, π_N vanishes at all the support points of $d\lambda_N$. On the other hand, if $N = \infty$, then (0.11) holds for all $k \in \mathbb{N}_0$. Vice versa, if (0.11) holds for all $k \in \mathbb{N}_0$, with $\beta_k > 0$, then by a well-known theorem of Favard (see, for instance, Natanson 1964/65, Volume II, Chapter VIII, Section 6) the system of polynomials $\{\pi_k\}$ is orthogonal relative to some positive measure $\mathrm{d}\lambda$ having infinitely many support points.

The recurrence relation (0.11) is generally quite stable, numerically, and indeed provides an excellent means of computing the orthogonal polynomials $\pi_k(\,\cdot\,;\mathrm{d}\lambda)$, both inside and outside the interval of orthogonality. For discrete measures $\mathrm{d}\lambda_N$, however, there is a good chance that the recurrence relation exhibits a phenomenon of 'pseudostability' (*cf.* Gautschi 1993a; Gautschi 1996b, Section 3.4.2), particularly if the support points of $\mathrm{d}\lambda_N$ are equally spaced. As a consequence, the accuracy of the $\pi_k(\,\cdot\,;\mathrm{d}\lambda_N)$, if computed by (0.11), may severely deteriorate as k approaches N.

PART I: APPLICATIONS

1. Interpolation

1.1. *Extended Lagrange interpolation*

Our interest here is in the convergence of Lagrange interpolation and quadrature processes on a finite interval $[-1,1]$, assuming only that the function to be interpolated is continuous on $[-1,1]$. A well-known negative result of Faber (see, for instance, Natanson 1965, Volume III, Chapter II, Theorem 2) tells us that there is no triangular array of nodes for which Lagrange interpolation would be *uniformly* convergent for every continuous function. In response to this, Erdős and Turán (1937) showed that if one considers convergence *in the mean*, then there indeed exist triangular arrays of nodes – for example the zeros of orthogonal polynomials – on which convergence holds for every continuous function. More precisely, given a positive weight function w on $(-1,1)$, we have

$$\lim_{n\to\infty}\|\,f - L_n f\,\|_w = 0, \qquad \text{for all } f \in C[-1,1], \tag{1.1}$$

where

$$\|\,u\,\|_w^2 = \int_{-1}^{1} u^2(t)w(t)\,\mathrm{d}t, \tag{1.2}$$

and $L_n f$ is the Lagrange interpolation polynomial of degree $< n$ interpolating f at the n zeros $\tau_i = \tau_i^{(n)}$, $i = 1, 2, \ldots, n$, of $\pi_n(\,\cdot\,;w)$, the nth-degree polynomial orthogonal on $[-1,1]$ relative to the weight function w. Convergence of the related quadrature process, that is,

$$\lim_{n\to\infty}\int_{-1}^{1}[f(t) - (L_n f)(t)]w(t)\,\mathrm{d}t = 0 \qquad \text{for all } f \in C[-1,1], \tag{1.3}$$

also holds, since the quadrature rule implied by (1.3) is simply the Gaussian rule (see Section 0.1), which is known to converge for any continuous function.

With this as a backdrop, suppose we wish to improve on $L_n f$ by considering an extended set of $2n + 1$ nodes,

$$\tau_i^{(n)}, \qquad i = 1, 2, \ldots, n; \qquad \hat{\tau}_j, \qquad j = 1, 2, \ldots, n+1, \qquad (1.4)$$

the first n being as before the zeros of $\pi_n(\,\cdot\,; w)$, and forming the corresponding Lagrange interpolant $\hat{L}_{2n+1} f$ of degree $< 2n + 1$. Is it true that (1.1) and/or (1.3) still hold if $L_n f$ is replaced by $\hat{L}_{2n+1} f$?

The answer cannot be expected to be an unqualified 'yes', as the choice of the added nodes $\{\hat{\tau}_j\}$ has a marked influence on the convergence behaviour. A natural choice for these nodes is the set of zeros of $\pi_{n+1}(\,\cdot\,; w)$, for which it has recently been shown (see Criscuolo, Mastroianni and Nevai (1993), Theorem 3.2; and Mastroianni and Vértesi (1993), Theorem 2.3) that the analogue of (1.1), when w is a 'generalized Jacobi weight' (see Section 6.1, Example 6.2), holds if and only if the Jacobi parameters α, β are both strictly between -1 and 0. The analogue of (1.3) holds for any weight function w since the underlying quadrature rule turns out to be simply the $(n+1)$-point Gaussian rule for w (all nodes τ_i receive the weight zero).

Another interesting choice for the nodes $\hat{\tau}_j$, first proposed by Bellen (1981, 1988), is the set of zeros of $\hat{\pi}_{n+1}(\,\cdot\,) = \pi_{n+1}(\,\cdot\,; \pi_n^2 w)$,

$$\pi_{n+1}(\hat{\tau}_j; \pi_n^2 w) = 0, \qquad j = 1, 2, \ldots, n+1 \qquad (\pi_n(\,\cdot\,) = \pi_n(\,\cdot\,; w)). \qquad (1.5)$$

Here the polynomial $\hat{\pi}_{n+1}$ is the $(n+1)$st-degree polynomial of an infinite sequence of polynomials $\pi_m(\,\cdot\,; \pi_n^2 w)$, $m = 0, 1, 2, \ldots$, studied in Gautschi and Li (1993) and termed there orthogonal polynomials *induced* by π_n. Both questions (1.1) and (1.3), for $\hat{L}_{2n+1} f$, then become considerably more difficult, and no precise results are known except for the four Chebyshev weight functions $w^{(\alpha,\beta)}(t) = (1 - t)^\alpha (1 + t)^\beta$, $\alpha, \beta = \pm\frac{1}{2}$. For these it has been shown in Gautschi (1992) that (1.1) is false unless $\alpha = \beta = -\frac{1}{2}$, in which case $\pi_n \hat{\pi}_{n+1}$ is a constant multiple of the 2nd-kind Chebyshev polynomial of degree $2n + 1$, and hence (1.1) (for $\hat{L}_{2n+1} f$) is a consequence of the Erdős–Turán result. More recently (Gautschi and Li 1996), the analogue of (1.3) was established for all four Chebyshev weight functions by showing that the respective quadrature rules are positive and therefore convergent, by a classical result of Pólya (1933). In the case $\alpha = \beta = \frac{1}{2}$, for example, the weights of the quadrature rule are given by Gautschi and Li (1996, Theorem 1).

$$\lambda_i = \frac{\pi}{3n}, \qquad i = 1, 2, \ldots, n,$$
$$\mu_j = \frac{2\pi/3}{n + \frac{3}{9 - 8\hat{\tau}_j^2}}, \qquad j = 1, 2, \ldots, n+1.$$

For Jacobi weight functions $w = w^{(\alpha,\beta)}$, there are only conjectural results, obtained by extensive computation based on the methods of Section 7.2. From these it appears that the analogue of (1.1) for $\hat{L}_{2n+1}f$ holds in the Gegenbauer case $\underline{\alpha} \leq \alpha = \beta \leq \overline{\alpha}$, where $\underline{\alpha} = -.31$ and $\overline{\alpha} = 1.6$ (perhaps even in a slightly larger interval), and in the Jacobi case when $0 \leq \alpha, \beta \leq \overline{\alpha}$ (again possibly in some slightly larger domain; see Gautschi (1992, Conjectures 5.1–5.3). The case $\alpha < 0$ remains open. The analogue of (1.3) is conjectured to hold for Jacobi weight functions with $|\alpha| \leq \frac{1}{2}$, $|\beta| \leq \frac{1}{2}$ (Gautschi and Li 1996, Conjecture 3.1).

1.2. Rational interpolation

Given $N + 1$ distinct points $\{t_i\}_{i=0}^N$ on $\mathbb{R}$ and corresponding function values $f_i = f(t_i)$, $i = 0, 1, \ldots, N$, the problem now is to find a rational function

$$r_{m,n}(t) = \frac{p(t)}{q(t)}, \qquad m + n = N, \tag{1.6}$$

with q assumed monic of degree n and p of degree $\leq m$, such that

$$r_{m,n}(t_i) = f_i, \qquad i = 0, 1, \ldots, N. \tag{1.7}$$

To derive an algorithm, one starts from the interpolation conditions (1.7), written in the form

$$p(t_i) = f_i q(t_i), \qquad i = 0, 1, \ldots, N. \tag{1.8}$$

Now recall that the Nth divided difference of a function g can be represented in the form

$$[t_0, t_1, \ldots, t_N]g = \sum_{i=0}^N \frac{g(t_i)}{w_i}, \qquad w_i = \prod_{\substack{j=0 \\ j \neq i}}^N (t_i - t_j). \tag{1.9}$$

Letting $\psi_j(t) = t^j$, $j = 0, 1, \ldots, n - 1$, multiplying (1.8) by $\psi_j(t_i)/w_i$ and summing, yields

$$\sum_{i=0}^N \frac{\psi_j(t_i)p(t_i)}{w_i} = \sum_{i=0}^N \frac{\psi_j(t_i)f_i q(t_i)}{w_i};$$

hence, by (1.9),

$$[t_0, t_1, \ldots, t_N](\psi_j p) = [t_0, t_1, \ldots, t_N](\psi_j f q), \qquad j = 0, 1, \ldots, n - 1.$$

But $\psi_j p$ is a polynomial of degree $m + n - 1 < N$, hence the divided difference on the left vanishes. The same is therefore true of the divided difference on the right, that is,

$$\sum_{i=0}^N \frac{f_i}{w_i} q(t_i)\psi_j(t_i) = 0, \qquad j = 0, 1, \ldots, n - 1. \tag{1.10}$$

Defining the discrete measure $d\lambda_N$ to have support points $\{t_0, \ldots, t_N\}$, and jumps $\omega_i = f_i/w_i$ at t_i, we can write (1.10) as

$$\int_{\mathbb{R}} q(t)\psi(t)\,d\lambda_N(t) = 0, \qquad \text{all } \psi \in \mathbb{P}_{n-1}. \qquad (1.11)$$

Thus, $q(\,\cdot\,) = \pi_n(\,\cdot\,; d\lambda_N)$ is the nth-degree monic polynomial orthogonal with respect to the (indefinite) measure $d\lambda_N$.

The denominator $q(\,\cdot\,) = \pi_n(\,\cdot\,; d\lambda_N)$, when generated by methods to be discussed in Section 6, can be checked to see whether it vanishes at any of the points t_i and, thus, whether the existence of the rational interpolant (1.6) is in doubt.

If all function values are different from zero, then the numerator polynomial p or, more precisely, its monic companion, $p_{\text{mon}} \in \mathbb{P}_m$, can also be characterized as a discrete orthogonal polynomial. Indeed, it is orthogonal relative to the measure $d\lambda_N^{(-1)}$ having the same support points as $d\lambda_N$, but jumps $\omega_i^{(-1)} = f_i^{-1}/w_i$ instead of f_i/w_i. This follows immediately from (1.8) if we write it in the form

$$q(t_i) = f_i^{-1} p(t_i), \qquad i = 0, 1, \ldots, N, \qquad (1.12)$$

and apply the same reasoning as above to find

$$\int_{\mathbb{R}} p_{\text{mon}}(t)\varphi(t)\,d\lambda_N^{(-1)}(t) = 0, \qquad \text{all } \varphi \in \mathbb{P}_{m-1}. \qquad (1.13)$$

To obtain p itself, it suffices to multiply $p_{\text{mon}}(\,\cdot\,) = \pi_m(\,\cdot\,; d\lambda_N^{(-1)})$ by a suitable normalization factor c, for example, $c = f_0 q(t_0)/p_{\text{mon}}(t_0)$ (assuming, of course, that $q(t_0) \neq 0$, $p_{\text{mon}}(t_0) \neq 0$).

The procedure described is particularly attractive if *all* rational interpolants $r_{m,n}$ with $m+n = N$ are to be obtained, since the numerator and denominator of $r_{m,n}$, being orthogonal polynomials, can be generated efficiently by the three-term recurrence relation (*cf.* 0.2). Some caution, nevertheless, is advised because of possible build-up of computational errors. These are caused by the indefiniteness of the inner product $(\cdot, \cdot)_{d\lambda_N}$, in particular by the fact that the weights ω_i and $\omega_i^{(-1)}$ typically alternate in sign. One expects these errors to be more prevalent the larger the moduli of these weights, hence the smaller the interval $[t_0, t_N]$.

Notes to Section 1

1.1. The potential failure of $\hat{L}_{2n+1}f$ to converge in the mean to f for the special choices of nodes studied here must not so much be regarded as a critique of these choices, but rather as a reflection of the very large class – $C[-1, 1]$ – of functions f. Adding only a slight amount of regularity, for example Lipschitz continuity with a parameter larger than one half, would restore (mean) convergence. For smoother functions, numerical evidence presented in Gautschi (1992, Table 6.1) suggests very fast convergence.

An analogue of the Erdős–Turán result for a class of rational interpolants has been established in Van Assche and Vanherwegen (1993, Theorem 7).

Mean convergence of extended Lagrange interpolation with $\hat{\tau}_j$ the Gauss–Kronrod points is studied in Li (1994). Other types of extended Lagrange interpolation by polynomials are studied in Bellen (1981) for Lipschitz-continuous functions $f \in \operatorname{Lip} \gamma$, $\gamma > \frac{1}{2}$, and in Criscuolo, Mastroianni and Occorsio (1990, 1991) and Criscuolo, Mastroianni and Vértesi (1992) with a view toward uniform convergence; see also Criscuolo et al. (1993) and Mastroianni and Vértesi (1993). For yet other extended interpolation processes and their L_p-convergence for arbitrary continuous functions, see Mastroianni (1994).

1.2. There are well-established algorithms for constructing a rational interpolant when one exists; see, for instance, Stoer and Bulirsch (1980, Section 2.2) and Graves-Morris and Hopkins (1981). The approach described in this subsection, based on discrete orthogonal polynomials (though relative to an indefinite measure) can be traced back to Jacobi (1846) and has recently been advocated in Eğecioğlu and Koç (1989). A numerical example illustrating its weaknesses and strengths is given in Gautschi (1989).

2. Approximation

2.1. Constrained least squares approximation

The problem of least squares ties in with the early history of orthogonal polynomials. We thus begin by looking at the classical version of the problem.

Given a positive measure $d\lambda$ on the real line $\mathbb{R}$ and a function f defined on the support of $d\lambda$, we want to find a polynomial p of degree at most n minimizing the $L_{d\lambda}^2$-error,

$$\text{minimize } \int_{\mathbb{R}} [p(t) - f(t)]^2 \, d\lambda(t) : \qquad p \in \mathbb{P}_n. \tag{2.1}$$

Often, the measure $d\lambda$ is a discrete measure $d\lambda_N$ concentrated on N distinct points of $\mathbb{R}$, with $N > n$ (*cf.* footnote (^{1}) of Section 0.1). If not, we must assume that f is in $L_{d\lambda}^2$, and we will also assume that all polynomials are in $L_{d\lambda}^2$. On the space $\mathbb{P}$ (of all real polynomials), respectively $\mathbb{P}_{N-1}$ (if $d\lambda = d\lambda_N$), we introduce the inner product (0.4),

$$(u, v)_{d\lambda} = \int_{\mathbb{R}} u(t)v(t) \, d\lambda(t), \qquad u, v \in \mathbb{P} \text{ (resp. } u, v \in \mathbb{P}_{N-1}), \tag{2.2}$$

which renders these spaces true inner product spaces. There exist, therefore, unique polynomials

$$\pi_k(t; d\lambda) = t^k + \text{lower-degree terms}, \qquad k = 0, 1, 2, \ldots, \tag{2.3}$$

satisfying

$$(\pi_k, \pi_\ell)_{d\lambda} \begin{cases} = 0 & \text{if } k \neq \ell, \\ > 0 & \text{if } k = \ell. \end{cases} \tag{2.4}$$

These are the (monic) *orthogonal polynomials* relative to the measure $d\lambda$ (*cf.* Section 0.2). There are infinitely many of them if the support of $d\lambda$ is infinite, and exactly N of them ($0 \leq k \leq N-1$ in (2.3)) if $d\lambda = d\lambda_N$. The solution of (2.1) is then given by

$$p(t) = \sum_{k=0}^{n} c_k \pi_k(t; d\lambda), \qquad c_k = \frac{(f, \pi_k)_{d\lambda}}{(\pi_k, \pi_k)_{d\lambda}}, \tag{2.5}$$

the $(n+1)$st partial sum of the Fourier series of f in the orthogonal system $\{\pi_k\}$.

Suppose now that we wish to minimize (2.1) among all polynomials $p \in \mathbb{P}_n$ satisfying the constraints

$$p(s_j) = f(s_j), \qquad j = 0, 1, \ldots, m; \qquad m < n, \tag{2.6}$$

where s_j are given distinct points on $\mathbb{R}$ where f is defined. It is then natural to seek p of the form

$$p(t) = p_m(t; f) + s_m(t)\delta(t), \tag{2.7}$$

where

$$s_m(t) = \prod_{j=0}^{m} (t - s_j), \tag{2.8}$$

$p_m(\cdot; f)$ being the unique polynomial in $\mathbb{P}_m$ interpolating f at the points $\{s_j\}_0^m$ and δ a polynomial of degree $n - m - 1$. Every polynomial of the form (2.7) is indeed in $\mathbb{P}_n$ and satisfies the constraints (2.6). Conversely, every such polynomial can be written in the form (2.7). It thus remains to determine δ.

We have

$$\int_{\mathbb{R}} [p(t) - f(t)]^2 \, d\lambda(t) = \int_{\mathbb{R}} [p_m(t; f) + s_m(t)\delta(t) - f(t)]^2 \, d\lambda(t)$$

$$= \int_{\mathbb{R}} \left[\frac{f(t) - p_m(t; f)}{s_m(t)} - \delta(t) \right]^2 s_m^2(t) \, d\lambda(t),$$

so that our minimization problem (2.1), (2.6) becomes

$$\text{minimize} \int_{\mathbb{R}} [\Delta(t) - \delta(t)]^2 s_m^2(t) \, d\lambda(t) : \qquad \delta \in \mathbb{P}_{n-m-1}, \tag{2.9}$$

where

$$\Delta(t) := \frac{f(t) - p_m(t; f)}{s_m(t)} = [s_0, s_1, \ldots, s_m, t]f. \tag{2.10}$$

Here, the expression on the far right is the divided difference of f of order

$m + 1$ with respect to the points $s_0, s_1, \ldots, s_m, t$, and its equality with Δ is a consequence of the well-known remainder term of interpolation. We see that the desired polynomial δ is the solution of an *unconstrained* least squares problem, but for a new function, Δ, and a different measure, $s_m^2 \, d\lambda$. Therefore, the solution of the constrained least squares problem is given by (2.7) with

$$\delta(t) = \sum_{k=0}^{n-m-1} d_k \hat{\pi}_k(t), \qquad d_k = \frac{(\Delta, \hat{\pi}_k)_{s_m^2 \, d\lambda}}{(\hat{\pi}_k, \hat{\pi}_k)_{s_m^2 \, d\lambda}}, \qquad (2.11)$$

where

$$\hat{\pi}_k(\,\cdot\,) = \pi_k(\,\cdot\,; s_m^2 \, d\lambda). \qquad (2.12)$$

It is required, therefore, to construct the orthogonal polynomials relative to the measure $s_m^2 \, d\lambda$, assuming those for $d\lambda$ are known. This is an instance of a *modification problem*; its solution by 'modification algorithms' will be discussed in Section 7.2.

The same idea can be applied to least squares approximation by a rational function

$$r(t) = \frac{p(t)}{q(t)}, \qquad (2.13)$$

where q is a prescribed polynomial satisfying

$$q(t) > 0 \quad \text{for } t \in \text{supp}\,(d\lambda); \qquad q(s_j) \neq 0, \qquad j = 0, 1, \ldots, m. \quad (2.14)$$

One finds that

$$\text{minimize} \quad \int_{\mathbb{R}} \left[\frac{p(t)}{q(t)} - f(t) \right]^2 d\lambda(t) : \qquad p \in \mathbb{P}_n, \qquad (2.15)$$

subject to the constraints

$$\frac{p(s_j)}{q(s_j)} = f(s_j), \qquad j = 0, 1, \ldots, m, \qquad (2.16)$$

is now equivalent to

$$\text{minimize} \quad \int_{\mathbb{R}} [\Delta(t) - \delta(t)]^2 \, \frac{s_m^2(t)}{q^2(t)} \, d\lambda(t) : \qquad \delta \in \mathbb{P}_{n-m-1}, \qquad (2.17)$$

where

$$\Delta(t) = \frac{q(t)f(t) - p_m(t; qf)}{s_m(t)} = [s_0, s_1, \ldots, s_m, t](qf). \qquad (2.18)$$

With δ so obtained, the desired p in (2.13) is then given by

$$p(t) = p_m(t; qf) + s_m(t)\delta(t). \qquad (2.19)$$

The modification of the measure now involves not only multiplication but also division by a polynomial. This requires additional algorithms for generating the respective orthogonal polynomials, which will be the subject of Section 7.3.

2.2. Least squares approximation in Sobolev spaces

In order to approximate (in the least squares sense) not only functions, but also, simultaneously, some of their derivatives, we may pose the problem

$$\text{minimize } \int_{\mathbb{R}} \sum_{\sigma=0}^{s} [p^{(\sigma)}(t) - f^{(\sigma)}(t)]^2 \, d\lambda_\sigma(t) : \qquad p \in \mathbb{P}_n, \qquad (2.20)$$

where $d\lambda_0, \ldots, d\lambda_s$ are positive measures on $\mathbb{R}$ and each derivative $f^{(\sigma)}$ is defined on the support of the corresponding measure $d\lambda_\sigma$. The natural scenario in which to consider this problem is the Sobolev space

$$H_s(\mathbb{R}) = \{ f : \sum_{\sigma=0}^{s} \int_{\mathbb{R}} [f^{(\sigma)}]^2 \, d\lambda_\sigma < \infty \} \qquad (2.21)$$

of functions f whose successive derivatives of order $\sigma \le s$ are square integrable against the respective measures $d\lambda_\sigma$. If we assume that the measures $d\lambda_\sigma$ are such that the space $\mathbb{P}$ of polynomials is a subspace of $H_s(\mathbb{R})$, the problem (2.20) can be written as

$$\text{minimize } \| p - f \|_{H_s}^2 : \qquad p \in \mathbb{P}_n, \qquad (2.22)$$

where the norm $\| u \|_{H_s} = \sqrt{(u,u)_{H_s}}$ is defined in terms of the inner product

$$(u, v)_{H_s} = \sum_{\sigma=0}^{s} \int_{\mathbb{R}} u^{(\sigma)}(t) v^{(\sigma)}(t) \, d\lambda_\sigma(t). \qquad (2.23)$$

If $d\lambda_0$ has infinitely many points of increase, then, regardless of whether or not some or all of the other measures $d\lambda_\sigma$, $\sigma \ge 1$, are discrete, the inner product (2.23) is positive definite on $H_s(\mathbb{R})$ and therefore defines a unique set of (monic) orthogonal polynomials $\pi_k(\cdot) = \pi_k(\cdot; H_s)$, $k = 0, 1, 2, \ldots$, satisfying

$$(\pi_k, \pi_\ell)_{H_s} \begin{cases} = 0 & \text{if } k \ne \ell, \\ > 0 & \text{if } k = \ell. \end{cases} \qquad (2.24)$$

These are called *Sobolev orthogonal polynomials*. In terms of these functions, the solution of (2.20), as in (2.5), is given by a finite Fourier series,

$$p(t) = \sum_{k=0}^{n} c_k \pi_k(t; H_s), \qquad c_k = \frac{(f, \pi_k)_{H_s}}{(\pi_k, \pi_k)_{H_s}}. \qquad (2.25)$$

It is important to note that the inner product in (2.23), if $s > 0$, no longer satisfies the basic property (0.9), that is,

$$(tu, v)_{H_s} \ne (u, tv)_{H_s} \qquad (s > 0), \qquad (2.26)$$

which means that we can no longer expect the orthogonal polynomials to satisfy a simple three-term recurrence relation. The numerical computation of Sobolev orthogonal polynomials (not to speak of their algebraic and analytic properties!) is therefore inherently more complicated; we will give a brief account of this in Section 8.

A widely used choice of measures is

$$\mathrm{d}\lambda_\sigma(t) = \gamma_\sigma \, \mathrm{d}\lambda(t), \qquad \sigma = 0, 1, 2, \ldots, s, \tag{2.27}$$

where $\mathrm{d}\lambda$ is a (positive) 'base measure' and the $\gamma_\sigma > 0$ are positive constants with $\gamma_0 = 1$. The latter allow us to assign different weights to different derivatives. The most studied case, by far, is (2.27) with $s = 1$.

2.3. Moment-preserving spline approximation

Given a function f on $[0, \infty)$, we wish to approximate it by a spline function of degree m with n positive knots. The approximation is not to be sought in any of the usual L_p-metrics, but is to share with f as many of the initial moments as possible. This is a type of approximation favoured by physicists, since moments have physical meaning, and the approximation thus preserves physical properties.

The most general spline in question can be written in the form

$$s_{n,m}(t) = \sum_{\nu=1}^{n} a_\nu (\tau_\nu - t)_+^m, \tag{2.28}$$

where $m \geq 0$ is an integer, $u_+ = \max(0, u)$, a_ν are real numbers, and

$$0 < \tau_1 < \tau_2 < \cdots < \tau_n < \infty \tag{2.29}$$

are the knots of the spline. The arbitrary polynomial of degree m that one could add to (2.28) must be identically zero if the moments of $s_{n,m}$ are to be finite. Since we have $2n$ parameters to choose from – the n coefficients a_ν and the n knots τ_ν – we expect to be able to match the first $2n$ moments,

$$\int_0^\infty s_{n,m}(t) t^j \, \mathrm{d}t = \int_0^\infty f(t) t^j \, \mathrm{d}t, \qquad j = 0, 1, \ldots, 2n - 1. \tag{2.30}$$

This problem, not surprisingly, leads to a problem of Gaussian quadrature. Assume, indeed, for fixed $n \in \mathbb{N}$ and $m \in \mathbb{N}_0$, that

(i) $f \in C^{m+1}[\mathbb{R}_+]$,
(ii) $\int_0^\infty f(t) t^j \, \mathrm{d}t$ exists for $j = 0, 1, \ldots, 2n - 1$,
(iii) $f^{(\mu)}(t) = o(t^{-2n-\mu})$ as $t \to \infty$, for $\mu = 0, 1, \ldots, m$,

and define the measure

$$\mathrm{d}\lambda_m(t) = \frac{(-1)^{m+1}}{m!} \, t^{m+1} f^{(m+1)}(t) \, \mathrm{d}t \qquad \text{on } \mathbb{R}_+. \tag{2.31}$$

Then we have the following result.

Theorem 2 Given a function f on $[0, \infty)$ satisfying assumptions (i)–(iii), there is a unique spline function $s_{n,m}$, (2.28), matching the first $2n$ moments of f, (2.30), if and only if the measure $d\lambda_m$ in (2.31) admits a Gaussian quadrature formula

$$\int_0^\infty g(t)\, d\lambda_m(t) = \sum_{\nu=1}^n \lambda_\nu^G g(t_\nu^G) + R_{n,m}^G(g), \qquad R_{n,m}^G(\mathbb{P}_{2n-1}) = 0, \quad (2.32)$$

having distinct positive nodes

$$0 < t_1^G < t_2^G < \cdots < t_n^G. \tag{2.33}$$

If that is the case, then the desired spline $s_{n,m}$ is given by

$$\tau_\nu = t_\nu^G, \qquad a_\nu = \frac{\lambda_\nu^G}{(t_\nu^G)^{m+1}}, \qquad \nu = 1, 2, \ldots, n. \tag{2.34}$$

Proof. Since τ_ν is positive, substituting (2.28) in (2.30) yields

$$\sum_{\nu=1}^n a_\nu \int_0^{\tau_\nu} t^j (\tau_\nu - t)^m\, dt = \int_0^\infty t^j f(t)\, dt, \qquad j = 0, 1, \ldots, 2n - 1. \tag{2.35}$$

We now apply m (respectively $m + 1$) integrations by parts to the integrals on the left (respectively right) of (2.35). On the left, we obtain

$$m![(j+1)(j+2)\cdots(j+m)]^{-1} \sum_{\nu=1}^n a_\nu \int_0^{\tau_\nu} t^{j+m}\, dt$$

$$= m![(j+1)(j+2)\cdots(j+m)(j+m+1)]^{-1} \sum_{\nu=1}^n a_\nu \tau_\nu^{j+m+1}. \tag{2.36}$$

On the right, we carry out the first integration by parts in detail to exhibit the reasonings involved. We have, for any $b > 0$,

$$\int_0^b t^j f(t)\, dt = \frac{1}{j+1} t^{j+1} f(t) \Big|_0^b - \frac{1}{j+1} \int_0^b t^{j+1} f'(t)\, dt. \tag{2.37}$$

The integrated term clearly vanishes at $t = 0$ and tends to zero as $t = b \to \infty$ by assumption (iii) with $\mu = 0$, since $j + 1 \leq 2n$. The integral on the left converges as $b \to \infty$ by assumption (ii); the same is true, therefore, for the integral on the right. We conclude that

$$\int_0^\infty t^j f(t)\, dt = -\frac{1}{j+1} \int_0^\infty t^{j+1} f'(t)\, dt.$$

Continuing in this manner, using assumption (iii) to show convergence to zero of the integrated term at the upper limit (its value at $t = 0$ always being zero) and the existence of $\int_0^\infty t^{j+\mu} f^{(\mu)}(t)\, dt$ already established to infer the existence of $\int_0^\infty t^{j+\mu+1} f^{(\mu+1)}(t)\,dt$, $\mu = 1, 2, \ldots, m$, we arrive at

$$\int_0^\infty t^j f(t)\,dt = \frac{(-1)^{m+1}}{(j+1)(j+2)\cdots(j+m+1)} \int_0^\infty t^{j+m+1} f^{(m+1)}(t)\,dt.$$

In particular, this shows that the first $2n$ moments of $d\lambda_m$ all exist. Since the last expression obtained, by (2.35), must be equal to the one in (2.36), we see that (2.30) is equivalent to

$$\sum_{\nu=1}^n (a_\nu \tau_\nu^{m+1}) \tau_\nu^j = \int_0^\infty \frac{(-1)^{m+1}}{m!}\, t^{m+1} f^{(m+1)}(t) \cdot t^j\,dt,$$
$$j = 0, 1, \ldots, 2n-1.$$

These are precisely the conditions for τ_ν to be the nodes of the Gauss formula (2.32) and for $a_\nu \tau_\nu^{m+1}$ to be the respective weights. Both, if indeed they exist, are uniquely determined. $\square$

The measure $d\lambda_m$ in (2.31) is neither one of the classical measures nor is it necessarily positive, in general. Thus we need constructive methods that also work for sign-changing measures.

The simplest example is the exponential function, $f(t) = e^{-t}$, in which case

$$d\lambda_m(t) = \frac{1}{m!}\, t^{m+1} e^{-t}\,dt \qquad (f(t) = e^{-t}) \tag{2.38}$$

is a generalized Laguerre measure with parameter $\alpha = m+1$, hence indeed one of the classical measures. Examples of positive measures $d\lambda_m$ are furnished by completely monotone functions, that is, functions f satisfying

$$(-1)^k f^{(k)}(t) > 0 \qquad \text{on } \mathbb{R}_+, \qquad k = 0, 1, 2, \ldots . \tag{2.39}$$

The physically important example of the Maxwell velocity distribution, $f(t) = e^{-t^2}$, is an example leading to a sign-variable measure,

$$d\lambda_m(t) = \frac{1}{m!}\, t^{m+1} H_{m+1}(t) e^{-t^2}\,dt \qquad (f(t) = e^{-t^2}), \tag{2.40}$$

where H_{m+1} is the Hermite polynomial of degree $m+1$. If $m > 0$, then H_{m+1} has $\lfloor (m+1)/2 \rfloor$ positive zeros, hence the measure (2.40) changes sign that many times.

Although the spline $s_{n,m}$ was constructed to match the moments of f, it also provides a reasonably good pointwise approximation. Its error indeed can be shown to be related to the remainder $R_{n,m}^G$ of the Gauss formula (2.32) in the sense that for any $t > 0$ one has

$$f(t) - s_{n,m}(t) = R_{n,m}^G(h_{t,m}), \tag{2.41}$$

where

$$h_{t,m}(u) = u^{-(m+1)}(u-t)_+^m, \qquad 0 \le u < \infty. \tag{2.42}$$

From a known convergence theorem for Gauss quadrature on $[0, \infty)$ (*cf.* Freud (1971, Chapter 3, Theorem 1.1)) it follows, in particular, that for fixed m,

$$\lim_{n \to \infty} s_{n,m}(t) = f(t), \qquad t > 0,$$

if f satisfies the assumptions of Theorem 2 for all $n = 1, 2, 3, \ldots$ and if $d\lambda_m$ is a positive measure for which the moment problem is determined.

Similar approximation problems can be posed on a finite interval, which then give rise to generalized Gauss–Lobatto and Gauss–Radau quadrature for a measure $d\lambda_m$ which again depends on $f^{(m+1)}$.

Notes to Section 2

2.1. Least squares approximation by polynomials was considered as early as 1859 by Chebyshev (1859) in the case of discrete measures $d\lambda = d\lambda_N$. Although Chebyshev expressed the solution in the form (2.5), he did not refer to the polynomials $\pi_k(\,\cdot\,; d\lambda_N)$ as 'orthogonal polynomials' – a concept unknown at the time – but characterized them, as did other writers of the period, as denominators of certain continued fractions. A more recent treatment of discrete least squares approximation by polynomials, including computational and statistical aspects, is Forsythe (1957). The idea of reducing the constrained least squares problem for polynomials to an unconstrained one involving a new objective function and a new measure can be found in Walsh (1969, p. 320). For the extension to rational functions, see Lin (1988).

2.2. In the case of measures (2.27) with $s = 1$, the Sobolev-type least squares approximation problem (2.20) was first considered by Lewis (1947), largely, however, with a view toward analysing the error of approximation (via the Peano kernel, as it were). The respective Sobolev orthogonal polynomials were studied later by Althammer (1962) and Gröbner (1967) in the case of the Legendre measure, $d\lambda(t) = dt$ in (2.27). Other choices of measures $d\lambda_\sigma$ in (2.23), especially discrete ones for $\sigma \geq 1$, have been studied extensively in recent years. For surveys, see Marcellán, Alfaro and Rezola (1993), Marcellán, Pérez and Piñar (1995), and for a bibliography, Marcellán and Ronveaux (1995). Special pairs of measures $\{ d\lambda_0, d\lambda_1 \}$ in the case $s = 1$, termed 'coherent', are studied in Iserles, Koch, Nørsett and Sanz-Serna (1990; 1991) and shown to allow efficient evaluation not only of the Sobolev–Fourier coefficients c_k in (2.25), but also of the Sobolev polynomials $\pi_k(\cdot; II_1)$ themselves. For zeros of such polynomials, see Meijer (1993), and de Bruin and Meijer (1995).

An application of Sobolev-type least squares approximation to the solution of systems of linear algebraic equations is proposed in Moszyński (1992). Here, $s + 1$ is the dimension of the largest Jordan block in the matrix of the system.

2.3. Piecewise constant approximations on $\mathbb{R}_+$ to the Maxwell velocity distribution that preserve the maximum number of moments were used in computational plasma physics by Calder, Laframboise and Stauffer (1983), and Calder and Laframboise (1986), under the colourful name 'multiple-water-bag distributions'. The connection with Gaussian quadrature was pointed out in Gautschi (1984b). Since piecewise constant functions are a special case of polynomial spline functions, it is natural to extend the idea of moment-preserving approximation to spline functions of arbitrary degree. This was done in Gautschi and Milovanović (1986), where one can find Theorem 2 and the error formulae (2.41), (2.42), along with their proofs. In the same paper, the sign-variable measure (2.40) was examined numerically and shown to lead, on occasion, to Gauss formulae with negative, or even conjugate complex, nodes. The analogous approximation on a finite interval, mentioned at the end of Section 2.3, was studied in Frontini, Gautschi and Milovanović (1987). Further extensions can be found in Milovanović and Kovačević (1988, 1992), Micchelli (1988), Frontini and Milovanović (1989), Gori and Santi (1989, 1992) and Kovačević and Milovanović (1996), with regard to both the type of spline function and the type of approximation.

3. Quadrature

3.1. Gauss quadrature for rational functions

Traditionally, Gauss quadrature rules (*cf.* Section 0.1) are designed to integrate exactly (against some measure) polynomials up to a maximum degree. This makes sense if one integrates functions that are 'polynomial-like'. Here we are interested in integrating functions that have poles, perhaps infinitely many. In this case, the use of rational functions, in combination with polynomials, seems more appropriate. The rational functions should be chosen so as to match the most important poles of the given function. This gives rise to the following problem.

Let $d\lambda$ be a (usually positive) measure on $\mathbb{R}$, and let there be given M nonzero complex numbers $\zeta_1, \ldots, \zeta_M$ such that

$$\zeta_\mu \neq 0, \qquad 1 + \zeta_\mu t \neq 0 \qquad \text{on supp} (d\lambda), \qquad \mu = 1, 2, \ldots, M. \qquad (3.1)$$

For given integers m, n with $1 \leq m \leq 2n$, find an n-point quadrature rule that integrates exactly (against the measure $d\lambda$) the m rational functions

$$(1 + \zeta_\mu t)^{-s}, \qquad \mu = 1, 2, \ldots M, \qquad s = 1, 2, \ldots, s_\mu, \qquad (3.2)$$

where $s_\mu \geq 1$ and

$$\sum_{\mu=1}^{M} s_\mu = m, \qquad (3.3)$$

as well as polynomials of degree $\leq 2n - m - 1$. If $m = 2n$, a polynomial of degree -1 is understood to be identically zero. We then have the extreme case of $2n$ rational functions (with poles of multiplicities s_μ at $-1/\zeta_\mu$) being integrated exactly, but no nontrivial polynomials. The quadrature rule is then optimal for rational functions, just as the classical Gaussian rule is optimal for polynomials; *cf.* Section 0.1. The latter corresponds to the limit case $M = m = 0$.

In principle, it is straightforward to construct the desired quadrature rule according to the following theorem.

Theorem 3 Define

$$\omega_m(t) = \prod_{\mu=1}^{M} (1 + \zeta_\mu t)^{s_\mu}, \tag{3.4}$$

by (3.3) a polynomial of degree m. Assume that the measure $\mathrm{d}\lambda/\omega_m$ admits a (polynomial) n-point Gauss quadrature formula, that is,

$$\int_{\mathbb{R}} f(t) \frac{\mathrm{d}\lambda(t)}{\omega_m(t)} = \sum_{\nu=1}^{n} w_\nu^G f(t_\nu^G) + R_n^G(f), \qquad R_n^G(\mathbb{P}_{2n-1}) = 0, \tag{3.5}$$

and define

$$t_\nu = t_\nu^G, \qquad \lambda_\nu = w_\nu^G \omega_m(t_\nu^G), \qquad \nu = 1, 2, \ldots, n. \tag{3.6}$$

Then

$$\int_{\mathbb{R}} g(t) \, \mathrm{d}\lambda(t) = \sum_{\nu=1}^{n} \lambda_\nu g(t_\nu) + R_n(g), \tag{3.7}$$

where

$$R_n(g) = 0 \quad \text{if } g \in \mathbb{P}_{2n-m-1}, \text{ or } g(t) = (1 + \zeta_\mu t)^{-s}, \ 1 \le \mu \le M, \ 1 \le s \le s_\mu. \tag{3.8}$$

Once again, we are led to a modification problem that involves division by a polynomial, so that the algorithms of Section 7.3 become relevant.

Proof of Theorem 3. For $\mu = 1, 2, \ldots, M$; $s = 1, 2, \ldots, s_\mu$, define

$$q_{\mu,s}(t) = \frac{\omega_m(t)}{(1 + \zeta_\mu t)^s}.$$

Since $m \le 2n$ and $s > 1$, we have $q_{\mu,s} \in \mathbb{P}_{m-s} \subset \mathbb{P}_{2n-1}$, and therefore, by (3.5),

$$\int_{\mathbb{R}} \frac{\mathrm{d}\lambda(t)}{(1 + \zeta_\mu t)^s} = \int_{\mathbb{R}} q_{\mu,s}(t) \frac{\mathrm{d}\lambda(t)}{\omega_m(t)} - \sum_{\nu=1}^{n} w_\nu^G q_{\mu,s}(t_\nu^G)$$

$$= \sum_{\nu=1}^{n} w_\nu^G \frac{\omega_m(t_\nu^G)}{(1 + \zeta_\mu t_\nu^G)^s} = \sum_{\nu=1}^{n} \frac{\lambda_\nu}{(1 + \zeta_\mu t_\nu)^s},$$

where (3.6) has been used in the last step. This proves the assertion in the top line of (3.8).

To prove the bottom part of (3.8), let p be an arbitrary polynomial in

$\mathbb{P}_{2n-m-1}$. Then, since $p\omega_m \in \mathbb{P}_{2n-1}$, again by (3.5) and (3.6),

$$\int_{\mathbb{R}} p(t)\, d\lambda(t) = \int_{\mathbb{R}} p(t)\omega_m(t)\frac{d\lambda(t)}{\omega_m(t)}$$

$$= \sum_{\nu=1}^{n} w_\nu^G p(t_\nu^G)\omega_m(t_\nu^G) = \sum_{\nu=1}^{n} \lambda_\nu p(t_\nu).$$

$\square$

The existence of the Gaussian quadrature formula in Theorem 3 is assured if it exists for the measure $d\lambda$ and the polynomial ω_m has constant sign on supp $(d\lambda)$. This is typically the case if the complex poles $-1/\zeta_\mu$ (if any) occur in conjugate complex pairs and the real ones are all outside the support interval of $d\lambda$.

Quantum statistical distributions provide important examples of integrals amenable to rational Gauss-type quadrature. Thus, the Fermi–Dirac distribution gives rise to the generalized Fermi–Dirac integral

$$F_k(\eta, \theta) = \int_0^\infty \frac{t^k\sqrt{1 + \frac{1}{2}\theta t}}{e^{-\eta+t} + 1}\, dt, \qquad \theta \geq 0, \qquad \eta \in \mathbb{R}, \qquad (3.9)$$

where the k-values of physical interest are the half-integers $\frac{1}{2}$, $\frac{3}{2}$ and $\frac{5}{2}$. Similarly, Bose–Einstein distributions lead to the generalized Bose–Einstein integral

$$G_k(\eta, \theta) = \int_0^\infty \frac{t^k\sqrt{1 + \frac{1}{2}\theta t}}{e^{-\eta+t} - 1}\, dt, \qquad \theta \geq 0, \qquad \eta \leq 0, \qquad (3.10)$$

with the same values of k as before. For the integral in (3.9), the poles are located at

$$t = \eta \pm (2\mu - 1)\, i\pi, \qquad \mu = 1, 2, 3, \ldots, \qquad (3.11)$$

whereas for the one in (3.10) they are at

$$t = \eta \pm 2\mu\, i\pi, \qquad \mu = 0, 1, 2, \ldots . \qquad (3.12)$$

This suggests taking for the ζ_μ in (3.1) the negative reciprocals of (3.11) and (3.12), respectively. If in the integral (3.9) we match the first n pairs of complex poles, we are led to apply Theorem 3 with $m = 2n$ and

$$\omega_{2n}(t) = \prod_{\mu=1}^{n} [(1 + \xi_\mu t)^2 + \eta_\mu^2 t^2],$$

where ξ_μ and η_μ are the real and imaginary parts, respectively, of $\zeta_\mu = -(\eta + (2\mu - 1)\, i\pi)^{-1}$. Similarly for the integral (3.10), where we need to match the real pole (at η) and the first $n - 1$ pairs of complex poles. This calls for Theorem 3 with $m = 2n - 1$ and

$$\omega_{2n-1}(t) = (1 + \xi_0 t) \prod_{\mu=1}^{n-1} [(1 + \xi_\mu t)^2 + \eta_\mu^2 t^2],$$

where ξ_μ and η_μ are the real and imaginary parts of $\zeta_\mu = -(\eta + 2\mu i\pi)^{-1}$.

3.2. Slowly convergent series

It may seem strange, at first, to see infinite series dealt with in a section on quadrature. But infinite series are integrals relative to a discrete measure supported on the positive integers! It is not unnatural, therefore, to try to approximate such integrals by finite sums. We do this for a special class of series in which the general term can be expressed as the Laplace transform of some function evaluated at an integer. Such series exhibit notoriously slow convergence. We will show that they can be transformed into an integral containing a positive, but nonclassical, weight function and then apply Gauss quadrature to obtain an effective summation procedure.

Thus, suppose that

$$S = \sum_{k-1}^{\infty} a_k, \qquad a_k = (\mathcal{L}f)(k), \tag{3.13}$$

where $\mathcal{L}f$ is the Laplace transform of some (known!) function f, that is,

$$(\mathcal{L}f)(s) = \int_0^\infty e^{-st} f(t) \, dt. \tag{3.14}$$

Then by Watson's lemma (see, for example, Wong 1989, p. 20), if f is regular near the origin, except possibly for a branch point at $t = 0$, where $f(t) \sim t^\lambda$, $\lambda > 0$, as $t \to 0$, and if f grows at most exponentially at infinity, one has $a_k \sim k^{-\lambda-1}$ as $k \to \infty$, showing that convergence of the series (3.13) is slow unless λ is large. However, we can write

$$\begin{aligned} S &= \sum_{k=1}^{\infty} (\mathcal{L}f)(k) - \sum_{k=1}^{\infty} \int_0^\infty e^{-kt} f(t) \, dt \\ &= \int_0^\infty \sum_{k=1}^{\infty} e^{-(k-1)t} \cdot e^{-t} f(t) \, dt \\ &= \int_0^\infty \frac{1}{1 - e^{-t}} \cdot e^{-t} f(t) \, dt, \end{aligned}$$

assuming the interchange of summation and integration is legitimate. This yields the following integral representation:

$$S = \int_0^\infty \epsilon(t) \frac{f(t)}{t} \, dt \tag{3.15}$$

involving the weight function

$$\epsilon(t) = \frac{t}{e^t - 1} \qquad \text{on } [0, \infty). \qquad (3.16)$$

Such integrals occur frequently in solid state physics, where ϵ is known as Einstein's function. (Of course, ϵ is also the generating function of the Bernoulli numbers.)

There are two approaches that suggest themselves naturally for evaluating the integral (3.15). One is Gaussian quadrature relative to the weight function ϵ, if $f(t)/t$ is sufficiently regular, or, if not, with respect to some modified weight function. The other is rational Gauss quadrature of the type discussed in Section 3.1, writing

$$S = \int_0^\infty \frac{t}{1 - e^{-t}} \frac{f(t)}{t} \cdot e^{-t} \, dt, \qquad (3.17)$$

letting $e^{-t} \, dt = d\lambda(t)$, and matching as many of the poles at $\pm 2\mu \, i\pi$, $\mu = 1, 2, 3, \ldots$, as possible. Both approaches call for nonclassical orthogonal polynomials.

To give an example, consider the series

$$S = \sum_{k=1}^\infty \frac{k^{\nu - 1}}{(k + a)^m}, \qquad 0 < \nu < 1, \qquad m \geq 1, \qquad (3.18)$$

where a is a complex number with $\operatorname{Re} a > 0$, $\operatorname{Im} a \geq 0$. Writing the general term of the series as

$$k^{\nu - 1} \cdot (k + a)^{-m} = (\mathcal{L}f)(k),$$

we note that

$$k^{\nu - 1} = \left(\mathcal{L} \frac{t^{-\nu}}{\Gamma(1 - \nu)} \right)(k), \qquad (k + a)^{-m} = \left(\mathcal{L} \frac{t^{m-1}}{(m - 1)!} e^{-at} \right)(k),$$

so that the convolution theorem for Laplace transforms (see, for example, Widder 1941, Theorem 12.1a)

$$\mathcal{L}g \cdot \mathcal{L}h = \mathcal{L}g * h,$$

where

$$(g * h)(t) = \int_0^t g(\tau) h(t - \tau) \, d\tau,$$

yields

$$f(t) = \frac{1}{(m - 1)! \Gamma(1 - \nu)} \int_0^t e^{-a(t - \tau)} (t - \tau)^{m-1} \tau^{-\nu} \, d\tau.$$

After the change of variable $\tau = tu$, this becomes

$$f(t) = \frac{t^{m-\nu}e^{-at}}{(m-1)!\Gamma(1-\nu)} \int_0^1 e^{atu}(1-u)^{m-1}u^{-\nu}\,du.$$

The integral on the right, up to a constant factor, can be recognized as Kummer's function $M(\alpha, \beta, z)$ with parameters $\alpha = 1-\nu$, $\beta = m+1-\nu$ and variable $z = at$ (see Abramowitz and Stegun, eds, 1964, Equation 13.2.1). Thus,

$$f(t) = t^{1-\nu}g_{m-1}(t; a, \nu), \qquad (3.19)$$

where

$$g_n(t; a, \nu) = \frac{t^n e^{-at}}{\Gamma(n+2-\nu)}\, M(1-\nu, n+2-\nu, at), \qquad n = 0, 1, 2, \ldots . \quad (3.20)$$

It is known that Kummer's function satisfies a recurrence relation relative to its second parameter (Abramowitz and Stegun, eds, 1964, Equation 13.4.2), from which one gets for $g_n(\cdot) = g_n(\cdot\,; a, \nu)$ the three-term recurrence relation

$$g_{n+1}(t) - \frac{1}{n+1}\left\{\left(t + \frac{n+1-\nu}{a}\right)g_n(t) - \frac{t}{a}\,g_{n-1}(t)\right\}, \qquad n \geq 0,$$

$$g_{-1}(t) = \frac{t^{-1}}{\Gamma(1-\nu)}\ .$$

$$(3.21)$$

To compute g_{m-1} in (3.19), it is enough, therefore, to compute $g_0(t) = e^{-at}M(1-\nu, 2-\nu, at)/\Gamma(2-\nu)$ and then to apply (3.21). On the other hand, g_0 is expressible (Abramowitz and Stegun, eds, 1964, Equation 13.6.10) in terms of Tricomi's form of the incomplete gamma function (Abramowitz and Stegun, eds, 1964, Equation 6.5.4),

$$g_0(t; a, \nu) = e^{-at}\gamma^*(1-\nu, -at), \qquad (3.22)$$

where

$$\gamma^*(\lambda, z) = \frac{z^{-\lambda}}{\Gamma(\lambda)} \int_0^z e^{-t}t^{\lambda-1}\,dt. \qquad (3.23)$$

Since g_0 is known to be an entire function of all its variables (see Tricomi 1954, Chapter IV), it follows from (3.21) that each function $g_n(t)$ is an entire function of t. Putting (3.19) into (3.15), we thus finally arrive at

$$\sum_{k=1}^{\infty} \frac{k^{\nu-1}}{(k+a)^m} = \int_0^{\infty} t^{-\nu}\epsilon(t)\cdot g_{m-1}(t; a, \nu)\,dt,$$

$$\mathrm{Re}\,a > 0, \qquad 0 < \nu < 1, \qquad m \geq 1,$$

$$(3.24)$$

with ϵ given by (3.16) and g_{m-1} an entire function of t. We can now proceed evaluating the integral on the right as discussed above, either treating $t^{-\nu}\epsilon(t)$ as a weight function in ordinary Gaussian quadrature, or writing $t^{-\nu}\epsilon(t) = (t/(1-e^{-t}))\cdot t^{-\nu}e^{-t}$ and using $t^{-\nu}e^{-t}\,dt = d\lambda(t)$ in rational Gauss quadrature.

It is worth noting that in this way we can sum series of the more general form

$$S = \sum_{k=1}^{\infty} k^{\nu-1} r(k), \qquad 0 < \nu < 1, \tag{3.25}$$

where $r(k)$ is any rational function

$$r(s) = \frac{p(s)}{q(s)}, \qquad \deg p < \deg q. \tag{3.26}$$

It suffices to decompose r into partial fractions and to apply (3.24) to each of them. The parameter $-a$ in (3.24) then represents one of the zeros of q, and m its multiplicity. If the condition $\operatorname{Re} a > 0$ is not satisfied, we can sum a few of the initial terms directly until the condition holds for all remaining terms.

We remark that for series with alternating sign factors, that is,

$$S' = \sum_{k=1}^{\infty} (-1)^{k-1} a_k, \qquad a_k = (\mathcal{L}f)(k), \tag{3.27}$$

analogous techniques can be applied, with the result that

$$S' = \int_0^\infty f(t)\varphi(t)\,\mathrm{d}t, \tag{3.28}$$

where now

$$\varphi(t) = \frac{1}{e^t + 1} \tag{3.29}$$

is what is known in solid state physics as Fermi's function.

Notes to Section 3

3.1. Convergence of the quadrature rule (3.5), when $m = 2n$, $\operatorname{supp}(\mathrm{d}\lambda) = [-1, 1]$ and $\zeta_\mu \in (-1, 1)$ with $s_\mu = 1$, for functions f analytic in a domain containing the interval $[-1, 1]$ in its interior has been studied by López and Illán (1984). Theorem 3, in this case, is due to Van Assche and Vanherwegen (1993, Theorem 1). These authors also consider a quadrature rule of the type (0.1) with $\operatorname{supp}(\mathrm{d}\lambda) = [-1, 1]$ whose nodes are the zeros of the rational function $(1 + \zeta_n t)^{-1} + \sum_{\mu=1}^{n-1} c_\mu (1 + \zeta_\mu t)^{-1}$ orthogonal (in the measure $\mathrm{d}\lambda$) to 1 and to $(1 + \zeta_\mu t)^{-1}$, $\mu = 1, 2, \ldots, n - 1$, where $\zeta_\mu \in (-1, 1)$ are given parameters. This is no longer a 'Gaussian' formula, as would be the case for polynomials, but leads to polynomials orthogonal with respect to the measure $\mathrm{d}\lambda/(\omega_{n-1}\omega_n)$, where $\omega_m(t) = \prod_{\mu=1}^{m}(1+\zeta_\mu t)$. The use of conjugate complex parameters ζ_μ in the context of rational quadrature rules is considered in López and Illán (1987). Theorem 3 in the general form stated is from Gautschi (1993b), where one can also find numerical examples. The application of rational Gauss formulae to generalized Fermi–Dirac integrals (3.9) and Bose–Einstein integrals (3.10) is further discussed in Gautschi (1993c) and has proven to be very effective.

3.2. The use of Gaussian quadrature for the purpose of summing infinite series has already been proposed by Newbery (unpublished). Summation of series (3.13) and (3.27) involving the Laplace transform by means of Gaussian quadrature relative to Einstein and Fermi weight functions, respectively, was first proposed in Gautschi and Milovanović (1985). The technique has since been applied to series of the type (3.25), and to analogous series with alternating sign factors, in Gautschi (1991a), and was also used in Gautschi (1991b) to sum slowly convergent power series of interest in plate contact problems. For the latter, an alternative complementary treatment has been given in Boersma and Dempsey (1992). Series of the type (3.18) were encountered by Davis (1993) in his study of spirals, in particular in his attempt to smooth certain discrete spirals ascribed by him to the 4th-century BC mathematician Theodorus. The treatment given here is taken from Davis (1993, Appendix A), where one also finds numerical examples. Alternative approaches using special function theory can be found in Boersma and Dempsey (1992), and using Euler–Maclaurin summation in Lewanowicz (1994); see also Davis (1993, pp. 40–41). Series (3.13) and (3.27) in which the terms a_k are values $f(k)$ of certain analytic functions f are summed in Milovanović (1994) by Gaussian quadrature involving weight functions $\cosh^{-2}(t)$ and $\sinh(t)\cosh^{-2}(t)$ on $\mathbb{R}_+$. Applications to series of the type (3.18), also with alternating signs, and to the Riemann Zeta function, are given in Milovanović (1995).

PART II: COMPUTATION

4. Computation of Gauss-type quadrature rules

In many applications, as we have seen in Part I, the need for orthogonal polynomials arises via Gauss-type quadrature with respect to some measure $d\lambda$. We therefore begin by discussing the computational aspects of Gaussian quadrature rules.

4.1. Gaussian rules

We assume that $d\lambda$ is a positive measure whose support contains infinitely many points, and all moments of which exist. There then exists, for each integer $n \geq 1$, an n-point Gauss formula

$$\int_{\mathbb{R}} f(t)\,d\lambda(t) = \sum_{\nu=1}^{n} \lambda_\nu^G f(t_\nu^G) + R_n^G(f), \qquad R_n^G(\mathbb{P}_{2n-1}) = 0. \tag{4.1}$$

The connection with orthogonal polynomials is well known (*cf.* Section 0.1). The nodes t_ν^G are the zeros of $\pi_n(\,\cdot\,; d\lambda)$, while the weights λ_ν^G – also called the *Christoffel numbers* – can be expressed in various ways in terms of the same orthogonal polynomials. For purposes of computation, however, it is better to characterize both quantities in terms of an eigenvalue problem.

To describe this characterization, we recall (*cf.* Section 0.2) that every system of (monic) orthogonal polynomials $\pi_k(\,\cdot\,) = \pi_k(\,\cdot\,; \mathrm{d}\lambda)$ satisfies a three-term recurrence relation

$$\pi_{k+1}(t) = (t - \alpha_k)\pi_k(t) - \beta_k\pi_{k-1}(t), \qquad k = 0, 1, 2, \ldots,$$
$$\pi_{-1}(t) = 0, \qquad \pi_0(t) = 1, \tag{4.2}$$

where the coefficients $\alpha_k = \alpha_k(\mathrm{d}\lambda)$, $\beta_k = \beta_k(\mathrm{d}\lambda)$ are real numbers uniquely determined by the measure $\mathrm{d}\lambda$, and each β_k is positive. With the recursion coefficients α_k, β_k we associate an infinite, symmetric, tridiagonal matrix

$$J_\infty = J_\infty(\mathrm{d}\lambda) = \begin{bmatrix} \alpha_0 & \sqrt{\beta_1} & & & 0 \\ \sqrt{\beta_1} & \alpha_1 & \sqrt{\beta_2} & & \\ & \sqrt{\beta_2} & \alpha_2 & \sqrt{\beta_3} & \\ & & \ddots & \ddots & \ddots \\ 0 & & & & \end{bmatrix}, \tag{4.3}$$

the *Jacobi matrix* belonging to the measure $\mathrm{d}\lambda$. Its $n \times n$ leading principal minor matrix will be denoted by

$$J_n = J_n(\mathrm{d}\lambda) = [J_\infty(\mathrm{d}\lambda)]_{n \times n}. \tag{4.4}$$

The Gaussian nodes and weights can then be expressed in terms of the eigenvalues and eigenvectors of $J_n(\mathrm{d}\lambda)$ according to the following theorem.

Theorem 4 Let x_ν be the eigenvalues of $J_n(\mathrm{d}\lambda)$, and u_ν the corresponding normalized eigenvectors, so that

$$J_n(\mathrm{d}\lambda)u_\nu = x_\nu u_\nu, \qquad u_\nu^T u_\nu = 1, \qquad \nu = 1, 2, \ldots, n. \tag{4.5}$$

Then the Gaussian nodes t_ν^G and weights λ_ν^G in (4.1) are given by

$$t_\nu^G = x_\nu, \qquad \lambda_\nu^G = \beta_0 u_{\nu,1}^2, \qquad \nu = 1, 2, \ldots, n, \tag{4.6}$$

where $u_{\nu,1}$ is the first component of u_ν and $\beta_0 = \int_\mathbb{R} \mathrm{d}\lambda(t)$.

Thus, the Gauss formula can be generated by computing the eigenvalues and (first components of) eigenvectors of a symmetric tridiagonal matrix. This is a routine problem in numerical linear algebra and can be solved by powerful algorithms such as the QR algorithm with carefully selected shifts (see, for example, Parlett 1980, Sections 8.9–8.11). The approach via eigenvalues is generally more efficient than traditional methods based on polynomial rootfinding.

Note also that the positivity of the Gauss weights λ_ν^G is an immediate consequence of (4.6).

Proof of Theorem 4. Let $\tilde{\pi}_k(\,\cdot\,) = \tilde{\pi}_k(\,\cdot\,; \mathrm{d}\lambda)$ denote the normalized orthogonal polynomials, so that $\pi_k = \sqrt{(\pi_k, \pi_k)_{\mathrm{d}\lambda}}\, \tilde{\pi}_k$. Inserting this into (0.11), dividing by $\sqrt{(\pi_{k+1}, \pi_{k+1})_{d\lambda}}$, and using (0.12), we obtain

$$\tilde{\pi}_{k+1}(t) = (t - \alpha_k) \frac{\tilde{\pi}_k}{\sqrt{\beta_{k+1}}} - \beta_k \frac{\tilde{\pi}_{k-1}}{\sqrt{\beta_{k+1}\beta_k}} ,$$

or, multiplying through by $\sqrt{\beta_{k+1}}$ and rearranging,

$$t\tilde{\pi}_k(t) = \alpha_k \tilde{\pi}_k(t) + \sqrt{\beta_k}\tilde{\pi}_{k-1}(t) + \sqrt{\beta_{k+1}}\tilde{\pi}_{k+1}(t),$$
$$k = 0, 1, 2, \ldots, n - 1. \tag{4.7}$$

In terms of the Jacobi matrix $J_n = J_n(d\lambda)$ we can write these relations in vector form as

$$t\tilde{\pi}(t) = J_n\tilde{\pi}(t) + \sqrt{\beta_n}\tilde{\pi}_n(t)e_n, \tag{4.8}$$

where $\tilde{\pi}(t) = [\tilde{\pi}_0(t), \tilde{\pi}_1(t), \ldots, \tilde{\pi}_{n-1}(t)]^T$ and $e_n = [0, 0, \ldots, 0, 1]^T$ are vectors in $\mathbb{R}^n$. Since t_ν^G is a zero of $\tilde{\pi}_n$, it follows from (4.8) that

$$t_\nu^G \tilde{\pi}(t_\nu^G) = J_n\tilde{\pi}(t_\nu^G), \qquad \nu = 1, 2, \ldots, n. \tag{4.9}$$

This proves the first relation in (4.6), since $\tilde{\pi}$ is a nonzero vector, its first component being

$$\pi_0 = \beta_0^{-1/2} \tag{4.10}$$

To prove the second relation in (4.6), note from (4.9) that the normalized eigenvector u_ν is

$$u_\nu = \frac{1}{[\tilde{\pi}(t_\nu^G)^T \tilde{\pi}(t_\nu^G)]^{1/2}} \tilde{\pi}(t_\nu^G) = \left(\sum_{\mu-1}^{n} \tilde{\pi}_{\mu-1}^2(t_\nu^G) \right)^{-1/2} \tilde{\pi}(t_\nu^G).$$

Comparing the first component on the far left and right, and squaring, gives, by virtue of (4.10),

$$\frac{1}{\sum_{\mu=1}^{n} \tilde{\pi}_{\mu-1}^2(t_\nu^G)} = \beta_0 u_{\nu,1}^2, \qquad \nu = 1, 2, \ldots, n. \tag{4.11}$$

On the other hand, letting $f(t) = \tilde{\pi}_{\mu-1}(t)$ in (4.1), one gets, by orthogonality, using (4.10) again, that

$$\beta_0^{1/2}\delta_{\mu-1,0} = \sum_{\nu=1}^{n} \lambda_\nu^G \tilde{\pi}_{\mu-1}(t_\nu^G) \qquad (\delta_{\mu-1,0} = \text{Kronecker delta}),$$

or, in matrix form,

$$P\lambda^G = \beta_0^{1/2}e_1, \tag{4.12}$$

where $P \in \mathbb{R}^{n \times n}$ is the matrix of eigenvectors, $\lambda^G \in \mathbb{R}^n$ the vector of Gauss weights, and $e_1 = [1, 0, \ldots, 0]^T \in \mathbb{R}^n$. Since the columns of P are orthogonal,

or, reverting to monic polynomials and recalling that $\tilde{\pi}_{n-1}/\tilde{\pi}_n = \beta_n^{1/2}\pi_{n-1}/\pi_n$,

$$\alpha_n^* = a - \beta_n \frac{\pi_{n-1}(a)}{\pi_n(a)}. \tag{4.21}$$

(The denominator $\pi_n(a)$ does not vanish by the assumption on a.) Therefore,

$$\pi_{n+1}^*(t) = (t-a)\omega_n(t), \qquad \omega_n \in \mathbb{P}_n, \tag{4.22}$$

and, by (4.19), the zeros $t_0 = a$, $t_1, t_2, \ldots, t_n$ of π_{n+1}^* are the eigenvalues of J_{n+1}^*, with $\tilde{\pi}(a)$, $\tilde{\pi}(t_1), \ldots, \tilde{\pi}(t_n)$ the corresponding eigenvectors. We now show that $t_\nu = t_\nu^R$, $\nu = 1, 2, \ldots, n$, that is, except for a constant factor,

$$\omega_n(t) = \pi_n(t; \mathrm{d}\lambda_a). \tag{4.23}$$

By (4.18) we have indeed

$$
\begin{aligned}
\sqrt{\beta_{n+1}}\pi_{n+1}^*(t) &= (t-\alpha_n^*)\tilde{\pi}_n(t) - \sqrt{\beta_n}\tilde{\pi}_{n-1}(t) \\
&= (t-\alpha_n)\tilde{\pi}_n(t) - \sqrt{\beta_n}\tilde{\pi}_{n-1}(t) + (\alpha_n - \alpha_n^*)\tilde{\pi}_n(t) \\
&= \sqrt{\beta_{n+1}}\tilde{\pi}_{n+1}(t) + (\alpha_n - \alpha_n^*)\tilde{\pi}_n(t),
\end{aligned}
$$

where in the last step we have used (4.7) for $k = n$. There follows, for any $p \in \mathbb{P}_{n-1}$,

$$
\begin{aligned}
\sqrt{\beta_{n+1}}\int_{\mathbb{R}} \pi_{n+1}^*(t)p(t)\,\mathrm{d}\lambda(t) &= \sqrt{\beta_{n+1}}\int_{\mathbb{R}} \omega_n(t)p(t)\cdot(t-a)\mathrm{d}\lambda(t) \\
&= \int_{\mathbb{R}} [\sqrt{\beta_{n+1}}\tilde{\pi}_{n+1}(t) + (\alpha_n - \alpha_n^*)\tilde{\pi}_n(t)]p(t)\,\mathrm{d}\lambda(t) = 0,
\end{aligned}
$$

by the orthogonality of the $\tilde{\pi}_k$. This proves (4.23).

By reasonings virtually identical with those in the proof of Theorem 4, one finds that

$$\lambda_\nu^R = \beta_0 u_{\nu,1}, \qquad \nu = 0, 1, 2, \ldots, n, \tag{4.24}$$

where $u_{\nu,1}$ is the first component of the normalized eigenvector u_ν of J_{n+1}^* corresponding to the eigenvalue t_ν^R (where $t_0^R = a$). We thus have the following result.

Theorem 5 The Gauss–Radau nodes $t_0^R = a$ and $t_1^R, \ldots, t_n^R$ are the eigenvalues of the matrix $J_{n+1}^*(\mathrm{d}\lambda)$ in (4.20), where α_n^* is defined by (4.21). The Gauss–Radau weights λ_ν^R are given by (4.24), where $u_{\nu,1}$ is the first component of the normalized eigenvector u_ν of $J_{n+1}^*(\mathrm{d}\lambda)$ corresponding to the eigenvalue t_ν^R.

The same theorem also holds for Gauss–Radau formulae with the fixed node at the upper end of the support interval. That is, if $\mathrm{d}\lambda$ has a support bounded from above, the number a, both in the formulation of Theorem 5 and in (4.16) and (4.21), may be replaced by $b \geq \sup \operatorname{supp}(\mathrm{d}\lambda)$.

Computing α_n^* by (4.21) may raise some concern about the possibility of a large cancellation error. The example of the Jacobi measure $d\lambda^{(\alpha,\beta)}(t) = (1-t)^\alpha(1+t)^\beta\,dt$ on $[-1,1]$, however, suggests that this concern is unwarranted. In this case, say for $a = -1$, one indeed finds that

$$\beta_n\frac{\pi_{n-1}(-1)}{\pi_n(-1)} = -\frac{1+\frac{\alpha}{n}}{2\left(1+\frac{\alpha+\beta}{2n}\right)\left(1+\frac{\alpha+\beta+1}{2n}\right)} \qquad (d\lambda = d\lambda^{(\alpha,\beta)}),$$

which for $n \to \infty$ tends to $-\frac{1}{2}$, so that for large n at least, there is no danger of cancellation. It is also interesting to note that for the generalized Laguerre measure $d\lambda^{(\alpha)}(t) = t^\alpha e^{-t}\,dt$ on $[0,\infty)$, and $a = 0$, one has $\alpha_n^* = n$.

4.3. Gauss–Lobatto rules

Assuming that $d\lambda$ has bounded support, we write the Gauss–Lobatto formula in the form

$$\int_{\mathbb{R}} f(t)\,d\lambda(t) = \lambda_0^L f(a) + \sum_{\nu=1}^n \lambda_\nu^L f(t_\nu^L) + \lambda_{n+1}^L f(b) + R_n^L(f), \qquad (4.25)$$
$$R_n^L(\mathbb{P}_{2n+1}) = 0,$$

where $a \le \inf\,\mathrm{supp}(\,d\lambda)$ and $b \ge \sup\,\mathrm{supp}(\,d\lambda)$. We recall from Section 0.1 that the interior nodes t_ν^L are the zeros of $\pi_n(\,\cdot\,;\,d\lambda_{a,b})$, that is,

$$\pi_n(t_\nu^L;\,d\lambda_{a,b}) = 0, \qquad \nu = 1,2,\ldots,n, \qquad (4.26)$$

where $d\lambda_{a,b}(t) = (t-a)(b-t)\,d\lambda(t)$, and that with these nodes so determined, the formula (4.25) must be interpolatory, that is, have degree of exactness $n+1$. We proceed similarly as in Section 4.2, but adjoin to the n relations (4.7) not one, but two additional relations:

$$t\tilde{\pi}_n(t) = \alpha_n\tilde{\pi}_n(t) + \sqrt{\beta_n}\tilde{\pi}_{n-1}(t) + \sqrt{\beta_{n+1}^*}\pi_{n+1}^*(t),$$
$$t\pi_{n+1}^*(t) = \alpha_{n+1}^*\pi_{n+1}^*(t) + \sqrt{\beta_{n+1}^*}\tilde{\pi}_n(t) + \sqrt{\beta_{n+2}}\pi_{n+2}^*(t), \qquad (4.27)$$

where α_{n+1}^*, β_{n+1}^* are parameters to be determined and $\alpha_n = \alpha_n(d\lambda)$, $\beta_n = \beta_n(d\lambda)$, $\beta_{n+2} = \beta_{n+2}(d\lambda)$. We now define

$$J_{n+2}^* = J_{n+2}^*(\,d\lambda) = \begin{bmatrix} \alpha_0 & \sqrt{\beta_1} & & & & & 0 \\ \sqrt{\beta_1} & \alpha_1 & \sqrt{\beta_2} & & & & \\ & & \ddots & \ddots & \ddots & & \\ & & & \alpha_{n-1} & \sqrt{\beta_n} & \\ & & & \sqrt{\beta_n} & \alpha_n & \sqrt{\beta_{n+1}^*} \\ 0 & & & & \sqrt{\beta_{n+1}^*} & \alpha_{n+1}^* \end{bmatrix}, \qquad (4.28)$$

so that, with the usual notation

$$\tilde{\pi}(t) = [\tilde{\pi}_0(t),\ldots,\tilde{\pi}_n(t),\pi_{n+1}^*(t)]^T, \qquad e_{n+2} = [0,\ldots,0,1]^T \in \mathbb{R}^{n+2},$$

the relations (4.7) and (4.27) can be written in matrix form as

$$t\tilde{\pi}(t) = J^*_{n+2}\tilde{\pi}(t) + \sqrt{\beta_{n+2}}\pi^*_{n+2}(t)e_{n+2}. \tag{4.29}$$

We now choose α^*_{n+1}, β^*_{n+1} such that $\pi^*_{n+2}(a) = \pi^*_{n+2}(b) = 0$. By the second relation in (4.27) this requires

$$(t - \alpha^*_{n+1})\pi^*_{n+1}(t) - \sqrt{\beta^*_{n+1}}\tilde{\pi}_n(t) = 0 \qquad \text{for } t = a, b,$$

or, using the first relation in (4.27) to eliminate π^*_{n+1},

$$(t - \alpha^*_{n+1})[(t - \alpha_n)\tilde{\pi}_n(t) - \sqrt{\beta_n}\tilde{\pi}_{n-1}(t)] - \beta^*_{n+1}\tilde{\pi}_n(t) = 0 \qquad \text{for } t = a, b.$$

The expression in brackets, however, is $\sqrt{\beta_{n+1}}\tilde{\pi}_{n+1}(t)$; thus,

$$(t - \alpha^*_{n+1})\sqrt{\beta_{n+1}}\tilde{\pi}_{n+1}(t) - \beta^*_{n+1}\tilde{\pi}_n(t) = 0 \qquad \text{for } t = a, b.$$

Converting to monic polynomials, we obtain the 2×2 linear system

$$\begin{bmatrix} \pi_{n+1}(a) & \pi_n(a) \\ \pi_{n+1}(b) & \pi_n(b) \end{bmatrix} \begin{bmatrix} \alpha^*_{n+1} \\ \beta^*_{n+1} \end{bmatrix} = \begin{bmatrix} a\pi_{n+1}(a) \\ b\pi_{n+1}(b) \end{bmatrix}.$$

By assumption on a and b, we have $\operatorname{sgn}[\pi_{n+1}(a)\pi_n(b)] = (-1)^{n+1}$ and $\operatorname{sgn}[\pi_{n+1}(b)\pi_n(a)] = (-1)^n$, so that the determinant is nonzero and, in fact, has sign $(-1)^{n+1}$. The system, therefore, has a unique solution, namely

$$\begin{aligned} \alpha^*_{n+1} &= (a\pi_{n+1}(a)\pi_n(b) - b\pi_{n+1}(b)\pi_n(a))/\Delta_n, \\ \beta^*_{n+1} &= (b - a)\pi_{n+1}(a)\pi_{n+1}(b)/\Delta_n, \end{aligned} \tag{4.30}$$

where

$$\Delta_n = \pi_{n+1}(a)\pi_n(b) - \pi_{n+1}(b)\pi_n(a). \tag{4.31}$$

Since both Δ_n and $\pi_{n+1}(a)\pi_{n+1}(b)$ have the sign $(-1)^{n+1}$, we see that $\beta^*_{n+1} > 0$, so that π^*_{n+1} and π^*_{n+2} in (4.27) are uniquely determined real polynomials, and J^*_{n+2} in (4.28) a real symmetric tridiagonal matrix. Its eigenvalues, by (4.29), are the zeros of π^*_{n+2}, among them a and b. Writing

$$\pi^*_{n+2}(t) = (t - a)(b - t)\omega_n(t), \qquad \omega_n \in \mathbb{P}_n, \tag{4.32}$$

we now show that, up to a constant factor,

$$\omega_n(t) = \pi_n(t; \mathrm{d}\lambda_{a,b}), \tag{4.33}$$

so that the eigenvalues of J^*_{n+2} are precisely the nodes of the Gauss–Lobatto formula (4.25), including a and b (cf. (4.26)). Using in turn the second and first relation of (4.27), we have

$$\begin{aligned} \sqrt{\beta_{n+2}}\pi^*_{n+2}(t) &= (t - \alpha^*_{n+1})\pi^*_{n+1}(t) - \sqrt{\beta^*_{n+1}}\tilde{\pi}_n(t), \\ \sqrt{\beta^*_{n+1}\beta_{n+2}}\pi^*_{n+2}(t) &= (t - \alpha^*_{n+1})[(t - \alpha_n)\tilde{\pi}_n(t) - \sqrt{\beta_n}\tilde{\pi}_{n-1}(t)] - \beta^*_{n+1}\tilde{\pi}_n(t) \\ &= (t - \alpha^*_{n+1})\sqrt{\beta_{n+1}}\tilde{\pi}_{n+1}(t) - \beta^*_{n+1}\tilde{\pi}_n(t). \end{aligned}$$

It follows that π_{n+2}^* is orthogonal relative to the measure $d\lambda$ to polynomials of degree $< n$, which by (4.32) implies (4.33).

Since, again by (4.29), the eigenvectors of J_{n+2}^* are $\tilde{\pi}(t_\nu^L)$, $\nu = 0, 1, \ldots, n$, $n + 1$, where $t_0^L = a$, $t_{n+1}^L = b$, the now familiar argument (used previously in Sections 4.1 and 4.2) yields the following theorem.

Theorem 6   The Gauss–Lobatto nodes $t_0^L = a$, $t_{n+1}^L = b$ and $t_1^L, \ldots, t_n^L$ are the eigenvalues of the matrix $J_{n+2}^*(d\lambda)$ in (4.28), where α_{n+1}^*, β_{n+1}^* are defined by (4.30), (4.31). The Gauss–Lobatto weights λ_ν^L are given by

$$\lambda_\nu^L = \beta_0 u_{\nu,1}^2, \qquad \nu = 0, 1, 2, \ldots, n, n + 1, \qquad (4.34)$$

where $u_{\nu,1}$ is the first component of the normalized eigenvector u_ν of $J_{n+2}^*(d\lambda)$ corresponding to the eigenvalue t_ν^L.

Since, as already noted, the two terms defining Δ_n in (4.31) are of opposite sign, there is no cancellation in the computation of Δ_n, nor is there any in computing β_{n+1}^*. For α_{n+1}^* this may no longer be true (indeed, $\alpha_{n+1}^* = 0$ for symmetric measures!), but here it is more the absolute error than the relative error that matters.

The construction of Gauss-type quadrature formulae is just one of several instances illustrating the importance of the recursion coefficients $\alpha_k(d\lambda)$, $\beta_k(d\lambda)$ for computational purposes. It is for this reason that all our constructive methods for orthogonal polynomials are directed toward computing these coefficients.

Notes to Section 4

4.1.   The fact that Gauss quadrature nodes can be viewed as eigenvalues of a symmetric tridiagonal matrix – the Jacobi matrix – has long been known. The characterization of the Gauss weights in terms of eigenvectors seems more recent; it was noted in Wilf (1962, Chapter 2, Exercise 9) and previously, around 1954, by Goertzel (Wilf 1980), and has also been used by Gordon (1968). The importance of these characterizations for computational purposes has been emphasized by Golub and Welsch (1969), who give a detailed computational procedure based on Francis's QR algorithm. Alternative procedures that compute the Gauss nodes as zeros of orthogonal polynomials by Newton's method or other rootfinding methods not only require considerable care in the selection of initial approximations, but also tend to be slower (Gautschi 1979). Also of importance is the inverse problem (Boley and Golub 1987) – given the Gauss nodes and weights, find the corresponding Jacobi matrix – and its solution by Lanczos-type algorithms.

4.2, 4.3.   The eigenvalue techniques described for generating Gauss–Radau and Gauss–Lobatto quadrature rules are due to Golub (1973); our derivation slightly differs from the one in Golub (1973).

5. Moment-based methods

The classical approach of generating orthogonal polynomials is based on the moments of the given measure $d\lambda$:

$$\mu_k = \mu_k(d\lambda) = \int_{\mathbb{R}} t^k \, d\lambda(t), \qquad k = 0, 1, 2, \ldots . \tag{5.1}$$

The desired recursion coefficients can be expressed in terms of Hankel determinants in these moments,

$$\left.\begin{aligned}
\alpha_k(d\lambda) &= \frac{D'_{k+1}}{D_{k+1}} - \frac{D'_k}{D_k} \\
\beta_k(d\lambda) &= \frac{D_{k+1}D_{k-1}}{D_k^2}
\end{aligned}\right\} \qquad k = 0, 1, 2, \ldots, \tag{5.2}$$

where $D_0 = D_{-1} = 1$, $D_1 = \mu_0$, $D'_0 = 0$, $D'_1 = \mu_1$ and D_m, D'_m, $m \geq 2$, are determinants whose first row consists of $\mu_0, \mu_1, \ldots, \mu_{m-1}$ and $\mu_0, \mu_1, \ldots, \mu_{m-2}, \mu_m$, respectively (the others having the subscripts successively increased by 1). Likewise, the orthogonal polynomials themselves admit the determinantal representation

$$\pi_n(t; d\lambda) = \frac{1}{D_n} \begin{vmatrix} \mu_0 & \mu_1 & \cdots & \mu_{n-1} & \mu_n \\ \mu_1 & \mu_2 & \cdots & \mu_n & \mu_{n+1} \\ \cdots & \cdots & \cdots & \cdots & \cdots \\ \mu_{n-1} & \mu_n & \cdots & \mu_{2n-2} & \mu_{2n-1} \\ 1 & t & \cdots & t^{n-1} & t^n \end{vmatrix}. \tag{5.3}$$

The trouble with these formulae is that the coefficients α_k, β_k, and with them π_n, become extremely sensitive to small changes (such as rounding errors) in the moments as k increases. In other words, the (nonlinear) map

$$K_n: \qquad \mathbb{R}^{2n} \to \mathbb{R}^{2n} \qquad \mu \mapsto \rho, \tag{5.4}$$

which maps the moment vector $\mu = [\mu_0, \mu_1, \ldots, \mu_{2n-1}]^T$ to the vector $\rho = [\alpha_0, \ldots, \alpha_{n-1}, \beta_0, \ldots, \beta_{n-1}]^T$ of recursion coefficients becomes extremely ill conditioned. Therefore it is important to study the condition of such moment-related maps.

A natural idea to overcome this difficulty is to use *modified moments* instead. That is, given a system of polynomials $\{p_k\}$, one uses

$$m_k = m_k(d\lambda) = \int_{\mathbb{R}} p_k(t) \, d\lambda(t), \qquad k = 0, 1, 2, \ldots, \tag{5.5}$$

in place of μ_k. One then has a new map K_n,

$$K_n: \qquad \mathbb{R}^{2n} \to \mathbb{R}^{2n} \qquad m \mapsto \rho, \tag{5.6}$$

where $m = [m_0, m_1, \ldots, m_{2n-1}]^T$, which one hopes is better conditioned than the old map (5.4). We discuss the conditioning of these maps in Section 5.1. In Section 5.2 we develop an algorithm that implements the maps K_n in (5.4) and (5.6) when the polynomials p_k defining the modified moments (5.5) satisfy a three-term recurrence relation. An example will be given in Section 5.3.

5.1. The conditioning of moment maps

The analysis of the map K_n in (5.4) or (5.6) is facilitated if the map is thought of as a composition of two maps,

$$K_n = H_n \circ G_n, \tag{5.7}$$

where $G_n : \mathbb{R}^{2n} \to \mathbb{R}^{2n}$ maps μ (respectively m) into the Gaussian quadrature rule,

$$G_n : \mu \text{ (resp. } m) \mapsto \gamma, \qquad \gamma = [\lambda_1, \ldots, \lambda_n, t_1, \ldots, t_n]^T, \tag{5.8}$$

where $\lambda_\nu = \lambda_\nu^G$, $t_\nu = t_\nu^G$ (cf. (4.1)), and $H_n : \mathbb{R}^{2n} \to \mathbb{R}^{2n}$ maps the Gaussian quadrature rule into the recursion coefficients,

$$H_n : \qquad \gamma \mapsto \rho. \tag{5.9}$$

The reason for this is that the map H_n, as was seen at the end of Section 4.1, is well conditioned, and G_n is easier to analyse. For a direct study of the map K_n see, however, Fischer (1996).

Just as the sensitivity of a function $f \colon \mathbb{R} \to \mathbb{R}$ at a point x can be measured by the magnitude of the derivative f' at x, in the sense that a small change dx of x produces the change $df(x) - f'(x) dx$, we can measure the sensitivity of the map $G_n \colon \mathbb{R}^{2n} \to \mathbb{R}^{2n}$ at a given vector μ (respectively m) by the magnitude of the Fréchet derivative at μ (respectively m). For finite-dimensional maps, this derivative is nothing but the linear map defined by the Jacobian matrix. We thus define

$$\text{cond } G_n = \| \partial G_n \|, \tag{5.10}$$

where by ∂G_n we denote the Jacobian matrix of the map G_n, and where for $\| \cdot \|$ we can take any convenient matrix norm. Note that this concept of condition is based on absolute errors; one could refine it to deal with relative errors as well, but we shall not do so here.

5.1.1. We begin with the map G_n for ordinary moments. Since the Gauss formula (4.1) is exact for the first $2n$ monomials t^j, $j = 0, 1, \ldots, 2n - 1$, we have

$$\sum_{\nu=1}^{n} \lambda_\nu t_\nu^j = \int_{\mathbb{R}} t^j \, d\lambda(t) = \mu_j, \qquad j = 0, 1, \ldots, 2n - 1,$$

which can be written as

$$\Phi(\gamma) = \mu, \qquad \Phi_j(\gamma) = \sum_{\nu=1}^{n} \lambda_\nu t_\nu^j, \qquad j = 0, 1, \ldots, 2n-1. \qquad (5.11)$$

The map G_n consists in solving this (nonlinear) system for the unknown vector γ, given the vector μ. The Jacobian ∂G_n, therefore, is the inverse of the Jacobian $\partial\Phi$ of Φ. This latter is readily computed to be

$$\partial\Phi = \begin{bmatrix} 1 & \cdots & 1 & 0 & \cdots & 0 \\ t_1 & \cdots & t_n & \lambda_1 & \cdots & \lambda_n \\ t_1^2 & \cdots & t_n^2 & 2\lambda_1 t_1 & \cdots & 2\lambda_n t_n \\ \cdots & \cdots & \cdots & \cdots & \cdots & \cdots \\ t_1^{2n-1} & \cdots & t_n^{2n-1} & (2n-1)\lambda_1 t_1^{2n-2} & \cdots & (2n-1)\lambda_n t_n^{2n-2} \end{bmatrix} = TD_\lambda,$$

where T is the confluent Vandermonde matrix

$$T = \begin{bmatrix} 1 & \cdots & 1 & 0 & \cdots & 0 \\ t_1 & \cdots & t_n & 1 & \cdots & 1 \\ t_1^2 & \cdots & t_n^2 & 2t_1 & \cdots & 2t_n \\ \cdots & \cdots & \cdots & \cdots & \cdots & \cdots \\ t_1^{2n-1} & \cdots & t_n^{2n-1} & (2n-1)t_1^{2n-2} & \cdots & (2n-1)t_n^{2n-2} \end{bmatrix} \qquad (5.12)$$

and D_λ the diagonal matrix

$$D_\lambda = \operatorname{diag}(1, \ldots, 1, \lambda_1, \ldots, \lambda_n). \qquad (5.13)$$

Therefore,

$$\partial G_n = D_\lambda^{-1} T^{-1}. \qquad (5.14)$$

It is now convenient to work with the uniform vector and matrix norm $\|\cdot\| = \|\cdot\|_\infty$. Since $\sum_{\nu=1}^{n} \lambda_\nu = \mu_0$ implies $\lambda_\nu < \mu_0$, and $\lambda_\nu^{-1} > \mu_0^{-1}$, it follows readily from (5.14) that

$$\|\partial G_n\| > \min(1, \mu_0^{-1}) \|T^{-1}\|.$$

Since the factor on the right involving μ_0 is unimportant, we shall henceforth assume that $\mu_0 = 1$ (which amounts to a normalization of the measure $d\lambda$). To obtain a particularly simple result, we further assume that $d\lambda$ is supported on the positive real line,

$$\operatorname{supp}(d\lambda) \subset \mathbb{R}_+.$$

It then follows from norm estimates for the inverse confluent Vandermonde matrix (see Gautschi 1963) that

$$\|\partial G_n\| > \frac{\prod_{\nu=1}^{n}(1+t_\nu)^2}{\min_{1\le\nu\le n}\left\{(1+t_\nu)\prod_{\substack{\mu=1\\\mu\neq\nu}}^{n}(t_\nu - t_\mu)^2\right\}}.$$

By definition (5.10) of the condition of G_n, and because the $\{t_\nu\}$ are the zeros of $\pi_n(\,\cdot\,) = \pi_n(\,\cdot\,;\mathrm{d}\lambda)$, we can write this inequality more elegantly as

$$\operatorname{cond} G_n > \frac{\pi_n^2(-1)}{\min_{1\le\nu\le n}\{(1+t_\nu)[\pi_n'(t_\nu)]^2\}}. \qquad (5.15)$$

Elegant as this result may be, it is also quite disconcerting, since orthogonal polynomials are known to grow rapidly with the degree when the argument is outside the support interval. In (5.15), the argument is -1, a good distance away from $\mathbb{R}_+$, and squaring the polynomial does not help either! Since the denominator in (5.15) grows only moderately with n, we must conclude that G_n becomes rapidly ill conditioned as n increases.

To illustrate (5.15), consider the (normalized) Chebyshev measure $\mathrm{d}\lambda(t) = \frac{1}{\pi}[t(1-t)]^{-1/2}$ on $[0,1]$, for which $\pi_n = T_n^*$, the 'shifted' Chebyshev polynomial, except for normalization. It then follows from (5.15) by elementary calculations that

$$\operatorname{cond} G_n > \frac{(3+\sqrt{8})^n}{64n^2} \qquad (\pi_n = T_n^*).$$

The lower bound happens to grow at the same exponential rate as the (Turing) condition number of the $n \times n$ Hilbert matrix!

5.1.2. We consider now the map $G_n : m \to \gamma$, where $m \in \mathbb{R}^{2n}$ is the vector of modified moments (5.5). We assume that the polynomials p_k defining these modified moments are themselves orthogonal, but relative to a measure, $\mathrm{d}s$, over which we can exercise control,

$$p_k(\,\cdot\,) = \pi_k(\,\cdot\,;\mathrm{d}s), \qquad k = 0,1,2,\ldots. \qquad (5.16)$$

The hope is that by choosing $\mathrm{d}s$ 'close' to the target measure $\mathrm{d}\lambda$, there is little chance for things to go wrong during the 'short' transition from the p_k to the π_k.

In analysing the condition of G_n, one arrives at a more satisfying result if, instead of the modified moments m_k, one departs from the *normalized modified moments*

$$\tilde{m}_k = \frac{m_k}{\|p_k\|_{\mathrm{d}s}}, \qquad k = 0,1,2,\ldots; \qquad \|p_k\|_{\mathrm{d}s} = \sqrt{(p_k,p_k)_{\mathrm{d}s}}. \qquad (5.17)$$

We thus consider the map

$$\tilde{G}_n : \qquad \mathbb{R}^{2n} \to \mathbb{R}^{2n} \qquad \tilde{m} \mapsto \gamma, \qquad \tilde{m} = [\tilde{m}_0,\tilde{m}_1,\ldots,\tilde{m}_{2n-1}]^T. \qquad (5.18)$$

The preliminary map $m \mapsto \tilde{m}$ is a perfectly well-conditioned diagonal map, and therefore does not distort the condition of G_n.

Similarly, as in (5.11), the map $\tilde{G}_n$ amounts to solving the nonlinear system

$$F(\gamma) = \tilde{m}, \qquad F_j(\gamma) = s_j^{-1}\sum_{\nu=1}^n \lambda_\nu p_j(t_\nu), \qquad j = 0,1,\ldots,2n-1,$$

where $s_j = \| p_j \|_{ds}$, and

$$\partial \tilde{G}_n = (\partial F)^{-1}.$$

By an elementary computation,

$$\partial F = D_s^{-1} P D_\lambda,$$

where $D_s = \text{diag}(s_0, s_1, \ldots, s_{2n-1})$, $D_\lambda = \text{diag}(1, \ldots, 1, \lambda_1, \ldots, \lambda_n)$, and $P \in \mathbb{R}^{2n \times 2n}$ is a confluent Vandermonde matrix in the polynomials $\{p_k\}$, that is,

$$\text{row}_j P = [p_j(t_1), \ldots, p_j(t_n), p_j'(t_1), \ldots, p_j'(t_n)], \qquad j = 0, 1, \ldots, 2n-1. \tag{5.19}$$

Therefore,

$$\partial \tilde{G}_n = D_\lambda^{-1} P^{-1} D_s. \tag{5.20}$$

In order to invert the matrix P in (5.19), we let h_ν, k_ν be the fundamental Hermite interpolation polynomials of degree $2n-1$ associated with the Gaussian abscissae $t_1, t_2, \ldots, t_n$:

$$\begin{aligned} h_\nu(t_\mu) &= \delta_{\nu\mu}, & h_\nu'(t_\mu) &= 0; \\ k_\nu(t_\mu) &= 0, & k_\nu'(t_\mu) &= \delta_{\nu\mu}, \end{aligned} \tag{5.21}$$

and expand them in the polynomials $\{p_k\}$,

$$h_\nu(t) = \sum_{\mu=1}^{2n} a_{\nu\mu} p_{\mu-1}(t), \qquad k_\nu(t) = \sum_{\mu=1}^{2n} b_{\nu\mu} p_{\mu-1}(t), \qquad \nu = 1, 2, \ldots, n. \tag{5.22}$$

Letting

$$A = [a_{\nu\mu}], \qquad B = [b_{\nu\mu}],$$

we can write the interpolation conditions (5.21), in conjunction with (5.19), in the form

$$AP = [I, O], \qquad BP = [O, I],$$

that is,

$$\begin{bmatrix} A \\ B \end{bmatrix} P = \begin{bmatrix} I & O \\ O & I \end{bmatrix},$$

which shows that

$$P^{-1} = \begin{bmatrix} A \\ B \end{bmatrix}.$$

We are now ready to compute the norm of $\partial \tilde{G}_n$ in (5.20). This time it turns out to be convenient to use the Frobenius norm $\| \cdot \| = \| \cdot \|_F$. Since

$$\begin{aligned} (D_\lambda^{-1} P^{-1} D_s)_{\nu\mu} &= s_{\mu-1} a_{\nu\mu}, & (D_\lambda^{-1} P^{-1} D_s)_{\nu+n,\mu} &= \lambda_\nu^{-1} s_{\mu-1} b_{\nu\mu}, \\ \nu &= 1, 2, \ldots, n; & \mu &= 1, 2, \ldots, 2n, \end{aligned}$$

one indeed obtains

$$\| \partial \tilde{G}_n \|^2 = \sum_{\nu=1}^{n} \sum_{\mu=1}^{2n} s_{\mu-1}^2 \left(a_{\nu\mu}^2 + \frac{1}{\lambda_\nu^2} b_{\nu\mu}^2 \right) \tag{5.23}$$

from (5.20). On the other hand, by (5.22),

$$\int_{\mathbb{R}} h_\nu^2(t)\, ds(t) = \sum_{\mu,\kappa=1}^{2n} a_{\nu\mu} a_{\nu\kappa} \int_{\mathbb{R}} p_{\mu-1}(t) p_{\kappa-1}(t)\, ds(t) = \sum_{\mu=1}^{2n} s_{\mu-1}^2 a_{\nu\mu}^2,$$

where the last equation follows from the orthogonality of the p_k. Similarly,

$$\int_{\mathbb{R}} k_\nu^2(t)\, ds(t) = \sum_{\mu=1}^{2n} s_{\mu-1}^2 b_{\nu\mu}^2.$$

Hence, recalling (5.10), equation (5.23) finally yields

$$\text{cond } \tilde{G}_n = \left\{ \int_{\mathbb{R}} \sum_{\nu=1}^{n} \left[h_\nu^2(t) + \frac{1}{\lambda_\nu^2} k_\nu^2(t) \right] ds(t) \right\}^{1/2}. \tag{5.24}$$

This result clearly identifies the factors influencing the condition of $\tilde{G}_n$. On the one hand, we have the polynomial of degree $4n - 2$,

$$g_n(t; d\lambda) = \sum_{\nu=1}^{n} \left[h_\nu^2(t) + \frac{1}{\lambda_\nu^2} k_\nu^2(t) \right], \tag{5.25}$$

appearing in the integrand of (5.24), which depends only on the measure $d\lambda$ (through the Gaussian nodes $t_\nu = t_\nu^G$). On the other hand, there is integration with respect to the measure ds. It is a combination of both, namely the *magnitude of g_n on the support of ds*, which determines the magnitude of cond $\tilde{G}_n$.

We note from (5.21) and (5.25) that $g_n(\cdot) = g_n(\cdot; d\lambda)$ is strictly positive on $\mathbb{R}$ and satisfies

$$g_n(t_\nu) = 1, \qquad g_n'(t_\nu) = 0, \qquad \nu = 1, 2, \ldots, n. \tag{5.26}$$

(By themselves, these conditions of course do not yet determine g_n.) Ideally, one would like g_n to remain ≤ 1 throughout the support of ds, in which case cond $\tilde{G}_n$ would be bounded by $s_0 = (\int_{\mathbb{R}} ds(t))^{1/2}$, uniformly in n. Unfortunately, this is only rarely the case. One example in which this property is likely to hold, based on computation, is $d\lambda_k(t) = [(1 - k^2 t^2)(1 - t^2)]^{-1/2} dt$ on $[-1, 1]$, where $0 < k < 1$. For $k = 0$, it was shown in Fischer (1996) that $g_n \leq 1 + 2/\pi^2$ on $[-1, 1]$. In other cases, such as $d\lambda_\sigma(t) = t^\sigma \ln(1/t)$ on $[0, 1]$, where $\sigma > -1$, the property $g_n(t) \leq 1$ holds over part of the interval, whereas in the remaining part, g_n assumes relatively large peaks between consecutive nodes t_ν, but such that the integral in (5.24) (when $ds(t) = 1$) is still of acceptable magnitude.

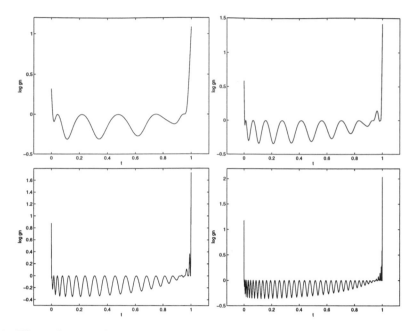

Fig. 1. The polynomial g_n, $n = 5, 10, 20, 40$, for the Maxwell measure with $c = 1$

An example of interest in quantum physics is the Maxwell velocity distribution

$$d\lambda(t) = e^{-t^2}\,dt \qquad \text{on } [0, c], \qquad 0 < c \leq \infty. \tag{5.27}$$

One finds by computation that g_n 'almost' satisfies $g_n \leq 1$ on $[0, c]$ when c is only moderately large, but develops larger and larger peaks, encroaching on an ever increasing portion of the interval, as c increases. This is illustrated in Fig. 1, which depicts $\log g_n$ for $n = 5, 10, 20, 40$ in the case $c = 1$, and in Fig. 2, where the analogous information is shown for $c = 5$. The respective condition numbers (when $ds(t) = dt$) are all less than 1 in the case $c = 1$, and range from 3.52×10^{12} to 8.57×10^{19} when $c = 5$. Fig. 2 is also representative for the case $c = \infty$. Arguably, Legendre moments ($ds(t) = dt$) are a poor choice in this case, but it has been observed in Gautschi (1996c) that even the best choice, $ds(t) = d\lambda(t)$, gives rise to very large condition numbers if c is large.

It has generally been our experience that cond $\tilde{G}_n$ becomes unacceptably large, even for moderately large n, when the support of $d\lambda$ is unbounded, as in the case $c = \infty$ of (5.27).

A final example of some interest in theoretical chemistry involves a measure $d\lambda$ of Chebyshev type supported on two separate intervals, say $[-1, -\xi]$ and $[\xi, 1]$, where $0 < \xi < 1$. Here, all nodes t_ν congregate on the two support intervals, at most one being located on the 'hole' $[-\xi, \xi]$ (see Szegő 1975, Theorem 3.41.2). As a consequence, g_n is likely to remain relatively small

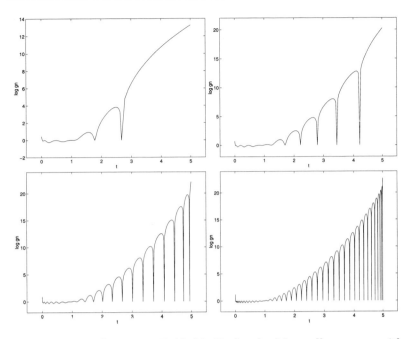

Fig. 2. The polynomial g_n, $n = 5, 10, 20, 40$, for the Maxwell measure with $c = 5$

(perhaps even ≤ 1) on the two support intervals, but may well become extremely large on the hole. To avoid a large condition number cond $\tilde{G}_n$, it is then imperative not to choose a measure ds for the modified moments that is supported on the whole interval $[-1, 1]$, but one that preferably has the same support as $d\lambda$.

5.2. The modified Chebyshev algorithm

We assumed in Section 5.1.2 that the polynomials p_k defining the modified moments (5.5) are themselves orthogonal. We now assume only that they satisfy a three-term recurrence relation

$$
\begin{aligned}
&p_{-1}(t) = 0, \qquad p_0(t) = 1, \\
&p_{k+1}(t) = (t - a_k)p_k(t) - b_k p_{k-1}(t), \qquad k = 0, 1, 2, \dots,
\end{aligned}
\tag{5.28}
$$

with known coefficients a_k, b_k, where the b_k need not necessarily be positive. This, in particular, encompasses the case $a_k = b_k = 0$, leading to $p_k(t) = t^k$, hence to ordinary moments (5.1).

To formulate an algorithm that implements the map $K_n : m \mapsto \rho$, we introduce 'mixed moments'

$$
\sigma_{k,\ell} = \int_{\mathbb{R}} \pi_k(t)p_\ell(t) \, d\lambda(t), \qquad k, \ell \geq -1,
\tag{5.29}
$$

and immediately observe that, by orthogonality, $\sigma_{k,\ell} = 0$ for $k > \ell$, and

$$\int_{\mathbb{R}} \pi_k^2(t)\,\mathrm{d}\lambda(t) = \int_{\mathbb{R}} \pi_k(t)tp_{k-1}(t)\,\mathrm{d}\lambda(t) = \sigma_{k,k}, \qquad k \geq 1. \tag{5.30}$$

The relation $\sigma_{k+1,k-1} = 0$, therefore, in combination with the recurrence relation (0.11) for the π_k, yields

$$0 = \int_{\mathbb{R}} [(t - \alpha_k)\pi_k(t) - \beta_k\pi_{k-1}(t)]p_{k-1}(t)\,\mathrm{d}\lambda(t) = \sigma_{k,k} - \beta_k\sigma_{k-1,k-1},$$

hence

$$\beta_k = \frac{\sigma_{k,k}}{\sigma_{k-1,k-1}}, \qquad k = 1, 2, 3, \ldots. \tag{5.31}$$

(Recall that $\beta_0 = m_0$ by convention.) Similarly, $\sigma_{k+1,k} = 0$ gives

$$0 = \int_{\mathbb{R}} \pi_k(t)tp_k(t)\,\mathrm{d}\lambda(t) - \alpha_k\sigma_{k,k} - \beta_k\sigma_{k-1,k}.$$

Using (5.28) in the form $tp_k(t) = p_{k+1}(t) + a_kp_k(t) + b_kp_{k-1}(t)$, we can write this as

$$0 = \sigma_{k,k+1} + (a_k - \alpha_k)\sigma_{k,k} - \beta_k\sigma_{k-1,k},$$

which, together with (5.31) and $\sigma_{-1,k} = 0$, yields

$$\begin{cases} \alpha_0 = a_0 + \dfrac{\sigma_{0,1}}{\sigma_{0,0}}, \\ \alpha_k = a_k + \dfrac{\sigma_{k,k+1}}{\sigma_{k,k}} - \dfrac{\sigma_{k-1,k}}{\sigma_{k-1,k-1}}, \qquad k = 1, 2, 3, \ldots. \end{cases} \tag{5.32}$$

With the αs and βs expressed by (5.32) and (5.31) in terms of the σs, it remains to compute $\sigma_{k,\ell}$. This can be done recursively, using the recurrence (0.11) for the π_k and (5.28) (with k replaced by ℓ) for the p_ℓ:

$$\begin{aligned} \sigma_{k,\ell} &= \int_{\mathbb{R}} [(t - \alpha_{k-1})\pi_{k-1}(t) - \beta_{k-1}\pi_{k-2}(t)]p_\ell(t)\,\mathrm{d}\lambda(t) \\ &= \int_{\mathbb{R}} \pi_{k-1}(t)[p_{\ell+1}(t) + a_\ell p_\ell(t) + b_\ell p_{\ell-1}(t)]\,\mathrm{d}\lambda(t) \\ &\quad - \alpha_{k-1}\sigma_{k-1,\ell} - \beta_{k-1}\sigma_{k-2,\ell} \\ &= \sigma_{k-1,\ell+1} - (\alpha_{k-1} - a_\ell)\sigma_{k-1,\ell} - \beta_{k-1}\sigma_{k-2,\ell} + b_\ell\sigma_{k-1,\ell-1}. \end{aligned}$$

The algorithm is now complete: to compute α_k, β_k for $k = 0, 1, \ldots, n - 1$, one first initializes

$$\begin{aligned} \sigma_{-1,\ell} &= 0, & \ell &= 1, 2, \ldots, 2n - 2, \\ \sigma_{0,\ell} &= m_\ell, & \ell &= 0, 1, \ldots, 2n - 1, \\ \alpha_0(\,\mathrm{d}\lambda) &= a_0 + \frac{m_1}{m_0}, & \beta_0(\,\mathrm{d}\lambda) &= m_0, \end{aligned} \tag{5.33}$$

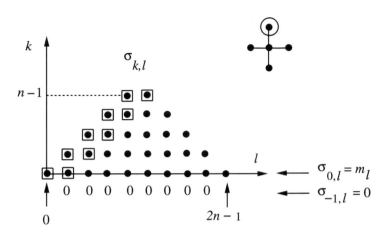

Fig. 3. The modified Chebyshev algorithm, schematically

and then continues, for $k = 1, 2, \ldots, n-1$, with

$$
\sigma_{k,\ell} = \sigma_{k-1,\ell+1} - (\alpha_{k-1} - a_\ell)\sigma_{k-1,\ell} - \beta_{k-1}\sigma_{k-2,\ell} + b_\ell\sigma_{k-1,\ell-1},
$$
$$
\ell = k, k+1, \ldots, 2n-k-1,
$$
$$
\alpha_k(\mathrm{d}\lambda) = a_k + \frac{\sigma_{k,k+1}}{\sigma_{k,k}} - \frac{\sigma_{k-1,k}}{\sigma_{k-1,k-1}}, \qquad \beta_k(\mathrm{d}\lambda) = \frac{\sigma_{k,k}}{\sigma_{k-1,k-1}}. \tag{5.34}
$$

Given the first $2n$ modified moments $m_0, m_1, \ldots, m_{2n-1}$ and the first $2n-1$ coefficients $a_0, a_1, \ldots, a_{2n-2}$ and $b_0, b_1, \ldots, b_{2n-2}$, this generates the first n coefficients $\alpha_0, \alpha_1, \ldots, \alpha_{n-1}$ and $\beta_0, \beta_1, \ldots, \beta_{n-1}$ via a trapezoidal array of auxiliary quantities $\sigma_{k,\ell}$ depicted schematically (for $n = 5$) in Fig. 3. The computing stencil in Fig. 3 indicates the location of the five entries in the array that are involved in the relation (5.34). The circled entry in the stencil is the one the algorithm computes in terms of the other four. The entries in boxes are used to compute the α_k and β_k. The complexity of the algorithm is clearly $\mathcal{O}(n^2)$.

It is interesting to observe that in the special case of a discrete measure $\mathrm{d}\lambda_N$ and ordinary moments (that is, $a_k = b_k = 0$), algorithm (5.34) was already known to Chebyshev (1859). We therefore call (5.34) the *modified Chebyshev algorithm*. The modified moments required can sometimes be computed in closed form or by a judicious application of recurrence formulae, or else can be approximated by a suitable discretization, similarly as in Section 6.1 in another context.

We remark that by virtue of (5.30), the algorithm (5.34) also provides the normalization constants $\sigma_{k,k} = (\pi_k, \pi_k)_{\mathrm{d}\lambda}$.

Table 1. *Errors in the $\alpha_k s$ and $\beta_k s$*

k	err α_k	err β_k
2	4.2×10^{-13}	7.6×10^{-13}
5	4.2×10^{-9}	1.2×10^{-10}
8	4.3×10^{-6}	3.8×10^{-6}
11	1.3×10^{0}	3.2×10^{-1}

5.3. An example

We illustrate the advantage of modified over classical moments in the case of the measure

$$\mathrm{d}\lambda_\sigma(t) = t^\sigma \ln(1/t)\,\mathrm{d}t \qquad \text{on } [0,1], \qquad \sigma > -1. \qquad (5.35)$$

We expect this advantage to be rather noticeable here, since, as was already observed in Section 5.1.2, the map $\tilde{G}_n\colon \tilde{m} \mapsto \gamma$ based on (normalized) Legendre moments is quite well conditioned in this case, even for large n, in contrast to the map $G_n\colon \mu \mapsto \gamma$, which rapidly becomes ill conditioned as n increases (*cf.* Section 5.1.1).

The classical moments for $\mathrm{d}\lambda_\sigma$ are simple enough,

$$\mu_k(\mathrm{d}\lambda_\sigma) = \frac{1}{\sigma + 1 + k}, \qquad k = 0, 1, 2, \ldots, \qquad (5.36)$$

whereas the modified moments with respect to the Legendre polynomials on $[0,1]$ (that is, $a_k = \frac{1}{2}$ for $k \geq 0$ and $b_0 = 1$, $b_k = (4(4 - k^{-2}))^{-1}$ for $k \geq 1$ in (5.28)) are more complicated, but still easy to compute:

$$\frac{(2k)!}{k!^2}\, m_k(\mathrm{d}\lambda_\sigma) = \begin{cases} (-1)^{k-\sigma}\, \frac{\sigma!^2(k-\sigma-1)!}{(k+\sigma+1)!}, & 0 \leq \sigma < k, \qquad \sigma \in \mathbb{N}, \\ \frac{1}{\sigma+1}\left\{ \frac{1}{\sigma+1} + \sum_{r=1}^{k}\left(\frac{1}{\sigma+1+r} - \frac{1}{\sigma+1-r} \right) \right\} \prod_{r=1}^{k} \frac{\sigma+1-r}{\sigma+1+r}, \\ \hspace{6cm} \text{otherwise.} \end{cases}$$

$$(5.37)$$

Applying the modified Chebyshev algorithm in single precision (machine precision $\approx 7 \times 10^{-15}$) for the case $\sigma = 0$, using the ordinary moments (5.36) (that is, $a_k = b_k = 0$), one obtains the recursion coefficients α_k, β_k with relative errors shown in Table 1. As can be seen, the accuracy deteriorates rapidly, there being no significance left by the time $k = 11$. In contrast, the use of modified moments (5.37) allows us to compute the first 100 (*sic*) recursion coefficients to an accuracy of at least 12 decimal digits.

Unfortunately, such a dramatic improvement in accuracy is not always realizable. In particular, for measures $\mathrm{d}\lambda$ with unbounded support, even the modified version of Chebyshev's algorithm, as already mentioned, must be expected to become quite susceptible to error growth. It all depends on the condition of the underlying (nonlinear) map $\tilde{G}_n$.

Notes to Section 5

The numerical condition of the classical moment map $G_n : \mu \mapsto \gamma$ was studied in Gautschi (1968); the lower bound (5.15) for the condition number rephrases one of the basic results of Gautschi (1968). For the growth of the condition number of the Hilbert matrix, mentioned at the end of Section 5.1.1, see Todd (1954). Although the explicit expressions (5.2) for the recursion coefficients are extremely sensitive to rounding errors, with the use of high-precision arithmetic they can be applied to validate the accuracy of Gaussian quadrature formulae; see Gautschi (1983) for an example.

The idea of using modified moments to generate orthogonal polynomials was first advanced by Sack and Donovan (1969, 1971/2), who developed an algorithm similar to the one in (5.34). The latter was derived by Wheeler (1974) independently of the work of Chebyshev (1859), where the same algorithm was obtained in the case of discrete measures and classical moments. Another algorithm, based on the Cholesky decomposition of a Gram matrix, is given in Gautschi (1970), but is not competitive with the modified Chebyshev algorithm, since it has complexity $\mathcal{O}(n^3)$. The reference Gautschi (1970), however, contains the first analysis of the condition of the underlying moment map, using the L_1-norm for vectors and matrices. The analysis given in Section 5.1.2, based on the more convenient Frobenius norm, is taken from Gautschi (1982a), where (in Section 3.1) one also finds the use of more refined condition numbers based on relative errors. The example of the Maxwell distribution (5.27) is taken from Gautschi (1991c); other illustrations of the basic formula (5.24) for the condition of the map $\tilde{G}_n$ can be found in Gautschi (1984c) and Gautschi (1985). The properties (5.26) of the function g_n in (5.25) suggest the distinction between 'strong' and 'weak' Gaussian nodes, the former being more likely than the latter to develop severe ill conditioning. For this, and an application to Jacobi polynomials, see Gautschi (1986a). The example at the end of Section 5.1.2 is taken from Wheeler (1984) and Gautschi (1984a); see also Gautschi (1985, Example 4.3) for further details. For the example in Section 5.3, *cf.* Gautschi (1994, Example 3.2).

6. Discretization methods

These methods, as the name implies, involve a preliminary discretization of the given measure $d\lambda$, that is, one approximates $d\lambda$ by a discrete N-point Dirac measure,

$$d\lambda(t) \approx d\lambda_N(t) := \sum_{k=1}^{N} w_k \delta(t - t_k)\, dt. \tag{6.1}$$

This is often done by a suitable quadrature formula (more on this in Section 6.1):

$$\int_{\mathbb{R}} p(t)\,\mathrm{d}\lambda(t) \approx \sum_{k=1}^{N} w_k p(t_k) =: \int_{\mathbb{R}} p(t)\,\mathrm{d}\lambda_N(t). \qquad (6.2)$$

The desired recursion coefficients are then approximated by

$$\left.\begin{array}{l} \alpha_k(\mathrm{d}\lambda) \approx \alpha_k(\mathrm{d}\lambda_N) \\ \beta_k(\mathrm{d}\lambda) \approx \beta_k(\mathrm{d}\lambda_N) \end{array}\right\} \qquad k = 0, 1, \ldots, n-1. \qquad (6.3)$$

Assuming $\mathrm{d}\lambda$ is a positive measure, and $w_k > 0$ in (6.1), one can show that for any fixed k,

$$\left.\begin{array}{l} \alpha_k(\mathrm{d}\lambda_N) \longrightarrow \alpha_k(\mathrm{d}\lambda) \\ \beta_k(\mathrm{d}\lambda_N) \longrightarrow \beta_k(\mathrm{d}\lambda) \end{array}\right\} \qquad \text{as } N \to \infty, \qquad (6.4)$$

provided the discretization process (6.2) has the property that

$$\int_{\mathbb{R}} p(t)\,\mathrm{d}\lambda_N(t) \to \int_{\mathbb{R}} p(t)\,\mathrm{d}\lambda(t) \qquad \text{as } N \to \infty \qquad (6.5)$$

for any polynomial p. Thus, by choosing a quadrature rule in (6.2) that is convergent for polynomials, we can obtain the coefficients α_k, β_k, $0 \leq k \leq n-1$, to any desired accuracy, by selecting N sufficiently large. More precisely, one selects a sequence $N_1 < N_2 < N_3 < \cdots$ of integers N (for a specific choice, see Gautschi 1994, Equation (4.16)) and iterates until

$$\max_{0 \leq k \leq n-1} \left| \frac{\beta_k(\mathrm{d}\lambda_{N_{i+1}}) - \beta_k(\mathrm{d}\lambda_{N_i})}{\beta_k(\mathrm{d}\lambda_{N_{i+1}})} \right| \leq \varepsilon,$$

where ε is a preassigned error tolerance. The convergence criterion is based on the relative errors in the β-coefficients, which is possible because the β_k are known to be positive. The α-coefficients are expected to converge at a similar speed (at least in the sense of absolute errors), as their definition is similar to that of the β_k (*cf.* (0.12)).

In Section 6.1 we indicate some possible ways of discretizing the measure $\mathrm{d}\lambda$. Once the discrete measure is at hand, it remains to compute its first n recursion coefficients, that is, the approximations on the right of (6.3). We will discuss two methods in Sections 6.2 and 6.3.

6.1. Discretization of the measure

Suppose the measure $\mathrm{d}\lambda$ has the form

$$\mathrm{d}\lambda(t) = w(t)\,\mathrm{d}t \qquad \text{on } [a, b], \qquad (6.6)$$

where $[a, b]$ is a finite or infinite interval and w an appropriate weight function. The first step, in general, is the decomposition of $[a, b]$ into a finite number of (possibly overlapping) subintervals,

$$[a, b] = \bigcup_{i=1}^{m} [a_i, b_i] \qquad (m \geq 1), \tag{6.7}$$

and to rewrite integrals such as those on the left of (6.2) as

$$\int_{\mathbb{R}} p(t) w(t) \, dt = \sum_{i=1}^{m} \int_{a_i}^{b_i} p(t) w_i(t) \, dt, \tag{6.8}$$

where w_i is an appropriate weight function on $[a_i, b_i]$. For example, the weight function w may be the sum $w = w_1 + w_2$ of two weight functions on $[a, b]$ that we wish to treat individually. In that case, one would take $[a_1, b_1] = [a_2, b_2] = [a, b]$ and associate w_1 with $[a_1, b_1]$ and w_2 with $[a_2, b_2]$. Alternatively, we may simply want to use a composite quadrature rule to approximate the integral, in which case (6.7) is a partition of $[a, b]$ and $w_i(t) = w(t)$ for each i. Still another example is a weight function w which is already supported on a union of disjoint intervals; in this case, (6.7) would be the same union, or possibly a refined union where some of the subintervals are further partitioned.

However (6.7) and (6.8) are constructed, the desired discretization (6.2) is now obtained by approximating each integral on the right of (6.8) by an appropriate quadrature rule,

$$\int_{a_i}^{b_i} p(t) w_i(t) \, dt \approx Q_i p, \qquad Q_i p = \sum_{r=1}^{N_i} w_{r,i} p(t_{r,i}), \tag{6.9}$$

for example a Gaussian rule for the weight function w_i. This yields

$$\int_{\mathbb{R}} p(t) w(t) \, dt \approx \sum_{i=1}^{m} \sum_{r=1}^{N_i} w_{r,i} p(t_{r,i}), \tag{6.10}$$

a formula of the type (6.2) with $N = \sum_{i=1}^{m} N_i$.

There is enough flexibility in this approach – choosing the subdivision (6.7), the local weight functions w_i in (6.8), and the quadrature rules in (6.9) – to come up with an effective scheme of discretization, that is, one that not only converges in the sense of (6.5), but converges reasonably fast. Further variations, of course, are possible. In particular, it is straightforward to adapt the approach to deal with measures containing, in addition to an absolutely continuous component (6.6), a discrete point spectrum, say

$$d\lambda(t) = w(t) \, dt + \sum_{j} \omega_j \delta(t - \tau_j) \, dt. \tag{6.11}$$

One only has to add $\sum_j \omega_j p(\tau_j)$ to (6.10).

Example 6.1. A good example of the kind of discretization indicated above is furnished by the measure

$$d\lambda(t) = t^\mu K_0(t)\,dt \qquad \text{on } [0,\infty), \qquad \mu > -1, \tag{6.12}$$

where K_0 is the modified Bessel function.

It is important, here, that one find a discretization that does justice to the special properties of the weight function $w(t) = t^\mu K_0(t)$, in particular its behaviour for small and large t. For the factor K_0, this behaviour can be described by

$$K_0(t) = \begin{cases} R(t) + I_0(t)\ln(1/t), & 0 < t \le 1, \\ t^{-1/2}e^{-t}S(t), & 1 \le t < \infty, \end{cases} \tag{6.13}$$

where R, S are well-behaved smooth functions, and I_0 is the 'regular' modified Bessel function. All three functions can be accurately evaluated on their respective intervals by rational approximations (Russon and Blair 1969). Therefore,

$$\int_0^\infty p(t)\,d\lambda(t) = \int_0^1 t^\mu [R(t)p(t)]\,dt + \int_0^1 t^\mu \ln(1/t)[I_0(t)p(t)]\,dt \\ + \int_1^\infty e^{-t}[t^{\mu-1/2}S(t)p(t)]\,dt. \tag{6.14}$$

This suggests a decomposition (6.7) with $m = 3$, namely $[0,\infty) = [0,1] \cup [0,1] \cup [1,\infty)$, weight functions $w_1(t) = t^\mu$, $w_2(t) = t^\mu \ln(1/t)$ and $w_3(t) = e^{-t}$, and for Q_i the corresponding Gaussian quadrature rules, after the last integral in (6.14) has been rewritten as

$$\int_1^\infty e^{-t}[t^{\mu-1/2}S(t)p(t)]\,dt = e^{-1}\int_0^\infty e^{-t}[(1+t)^{\mu-1/2}S(1+t)p(1+t)]\,dt.$$

The first and last Gauss formulae are classical – Gauss–Jacobi and Gauss–Laguerre – and are easily generated by the method of Section 4.1. The second is nonclassical, but can be generated by the same method, once the recursion coefficients for the respective orthogonal polynomials have been generated by the modified Chebyshev algorithm, as discussed in Sections 5.2 and 5.3.

Example 6.2. We call *generalized Jacobi measure* a measure of the form

$$d\lambda(t) = \varphi(t)(1-t)^\alpha(1+t)^\beta \prod_{i=2}^m |t - a_i|^{\gamma_i}, \qquad t \in (-1,1), \tag{6.15}$$

where φ is a smooth function, $m \ge 2$, $-1 < a_2 < \cdots < a_m < 1$, and

$$\gamma_1 = \beta > -1; \qquad \gamma_i > -1, \qquad i = 2,\dots,m; \qquad \gamma_{m+1} = \alpha > -1. \tag{6.16}$$

Here, the natural decomposition is

$$[-1,1] = \bigcup_{i=1}^m [a_i, b_i], \qquad a_1 = -1, \qquad b_i = a_{i+1}, \qquad a_{m+1} = 1,$$

and the appropriate weight function w_i on $[a_i, b_i]$ is the Jacobi weight with parameters γ_i, γ_{i+1}, transformed to the interval $[a_i, b_i]$. One then obtains a formula similar to (6.8), except that on the right, $p(t)$ has to be replaced by

$$p(t)\varphi(t) \prod_{j=1, j\neq i, j\neq i+1}^{m+1} |t - a_j|^{\gamma_j}, \qquad a_i \leq t \leq b_i.$$

This function is free of singularities in $[a_i, b_i]$, so that its Gauss–Jacobi quadrature with weight function w_i will converge – and reasonably fast at that, unless one of the a_j is very close to either a_i or b_i (and γ_j not an integer).

It may not always be possible to come up with natural discretizations as in these examples. In that case, one may try to apply a standard quadrature rule to each integral on the right of (6.8), paying no special attention to the weight function w_i and treat it as part of the integrand. Since w_i may have singularities at the endpoints of $[a_i, b_i]$, it is imperative that an open quadrature formula be used; stability considerations furthermore favour Chebyshev nodes, and convergence considerations an interpolatory formula. Taking the same number of nodes for each Q_i, we are thus led to choose, on the canonical interval $[-1, 1]$, the N^F-point *Fejér rule*, that is, the interpolatory quadrature rule

$$\int_{-1}^{1} f(t)\, dt \approx Q_{N^F} f, \qquad Q_{N^F} f = \sum_{r=1}^{N^F} w_r^F f(t_r^F), \qquad (6.17)$$

where $t_r^F = \cos((2r - 1)\pi/2N^F)$ are the Chebyshev points. The weights are expressible in trigonometric form as

$$w_r^F = \frac{2}{N^F}\left(1 - 2\sum_{s=1}^{\lfloor N^F/2 \rfloor} \frac{\cos(2s\theta_r^F)}{4s^2 - 1}\right), \qquad t_r^F = \cos\theta_r^F, \qquad (6.18)$$

and are known to be all positive (Fejér 1933). Furthermore, the rule converges as $N^F \to \infty$, even in the presence of singularities, provided they occur at the endpoints and are monotone and integrable (Gautschi 1967). The rule (6.17) is now applied to each integral on the right of (6.8) by transforming the interval $[-1, 1]$ to $[a_i, b_i]$ via some monotone function ϕ_i (a linear function if $[a_i, b_i]$ is finite) and letting $f(t) = p(t)w_i(t)$:

$$\int_{a_i}^{b_i} p(t)w_i(t)\, dt = \int_{1}^{1} p(\phi_i(\tau))w_i(\phi_i(\tau))\phi_i'(\tau)\, d\tau$$

$$\approx \sum_{r=1}^{N^F} w_r^F w_i(\phi_i(t_r^F))\phi_i'(t_r^F) \cdot p(\phi_i(t_r^F)).$$

Thus, in effect, we take in (6.9)

$$t_{r,i} = \phi_i(t_r^F), \qquad w_{r,i} = w_r^F w_i(\phi_i(t_r^F))\phi_i'(t_r^F), \qquad i = 1, 2, \ldots, m. \quad (6.19)$$

Suitable functions ϕ_i are $\phi_i(t) = (1+t)/(1-t)$ if the interval $[a_i, b_i]$ is half-infinite, say of the form $[0, \infty)$, and similarly for intervals $[a, \infty)$ and $(-\infty, b]$, and $\phi_i(t) = t/(1-t^2)$ if $[a_i, b_i] = (-\infty, \infty)$.

6.2. Orthogonal reduction method

Assuming now that a discrete measure (6.1) has been constructed, with (positive) weights w_k and abscissae t_k, we denote by $\sqrt{w}$ the column vector whose components are $\sqrt{w_k}$, and by D_t the diagonal matrix with the t_k on the diagonal. Since for any function p,

$$\int_{\mathbb{R}} p(t)\, d\lambda_N(t) = \sum_{k=1}^{N} w_k p(t_k) \tag{6.20}$$

(*cf.* (6.2)), we may interpret (6.20) as a 'Gauss formula' for the measure $d\lambda_N$. From (4.15) it then follows that there exists an orthogonal matrix $Q_1 \in \mathbb{R}^{N \times N}$ such that

$$\begin{bmatrix} 1 & 0^T \\ 0 & Q_1^T \end{bmatrix} \begin{bmatrix} 1 & \sqrt{w}^T \\ \sqrt{w} & D_t \end{bmatrix} \begin{bmatrix} 1 & 0^T \\ 0 & Q_1 \end{bmatrix} = \\ \begin{bmatrix} 1 & \sqrt{\beta_0(d\lambda_N)}\, e_1^T \\ \sqrt{\beta_0(d\lambda_N)}\, e_1 & J_N(d\lambda_N) \end{bmatrix}, \tag{6.21}$$

where $e_1 = [1, 0, \ldots, 0]^T \in \mathbb{R}^N$ is the first coordinate vector and $J_N(d\lambda_N)$ the Jacobi matrix of order N for the measure $d\lambda_N$ (*cf.* (4.4)). It is the latter that we wish to obtain.

Observe that (6.21) has the form

$$Q^T A Q = T, \tag{6.22}$$

where all matrices are $(N+1) \times (N+1)$, Q is orthogonal and T symmetric tridiagonal with positive elements on the side diagonals. It is then well known (see, for instance, Parlett 1980, p. 113) that Q and T in (6.22) are uniquely determined by A and the first column of Q. Since the latter in (6.21) is e_1, and the former $\begin{bmatrix} 1 & \sqrt{w}^T \\ \sqrt{w} & D_t \end{bmatrix}$, we see that knowledge of w and D_t, that is, of $d\lambda_N$, uniquely determines the desired $J_N(d\lambda_N)$ and $\beta_0(d\lambda_N)$ by the orthogonal similarity transformation (6.21). A method that accomplishes this transformation is *Lanczos's algorithm*. There are various versions of this algorithm, a particularly elegant one consisting of a sequence of elementary orthogonal similarity transformations of Givens type designed to successively push the elements bordering the diagonal matrix D_t in (6.21) towards the diagonal. It is not necessary to carry the transformation to completion; it can be terminated once the submatrix $J_n(d\lambda_N)$ has been produced, which is all that is needed. Also, in spite of the square roots of the weights appearing on the left of (6.21), it is not required in the resulting algorithm that all weights be of the same (positive) sign, since only their squares enter into the algorithm.

6.3. The Stieltjes procedure

This is based on the explicit formulae (see (0.12))

$$\alpha_k(\,\mathrm{d}\lambda) = \frac{(t\pi_k, \pi_k)\,\mathrm{d}\lambda}{(\pi_k, \pi_k)\,\mathrm{d}\lambda}\,, \qquad k = 0, 1, 2, \ldots,$$

$$\beta_0(\,\mathrm{d}\lambda) = (\pi_0, \pi_0)\,\mathrm{d}\lambda, \qquad \beta_k(\,\mathrm{d}\lambda) = \frac{(\pi_k, \pi_k)\,\mathrm{d}\lambda}{(\pi_{k-1}, \pi_{k-1})\,\mathrm{d}\lambda}\,, \qquad k = 1, 2, 3, \ldots,$$

$$\tag{6.23}$$

where $\pi_k(\,\cdot\,) = \pi_k(\,\cdot\,; \mathrm{d}\lambda)$. One applies (6.23) for $\mathrm{d}\lambda = \mathrm{d}\lambda_N$ in tandem with the basic recurrence relation (see (0.11))

$$\pi_{k+1}(t) = (t - \alpha_k)\pi_k(t) - \beta_k\pi_{k-1}(t), \qquad k = 0, 1, 2, \ldots,$$
$$\pi_{-1}(t) = 0, \qquad \pi_0(t) = 1. \tag{6.24}$$

Note that all inner products in (6.23) are finite sums when $\mathrm{d}\lambda = \mathrm{d}\lambda_N$, so that they are easily computed once the π_k are known. Since $\pi_0 = 1$, we can thus compute α_0, β_0 from (6.23). Having obtained α_0, β_0, we then use (6.24) with $k = 0$ to compute π_1 for all $\{t_1, \ldots, t_N\}$ to obtain the values of π_1 needed to reapply (6.23) with $k = 1$. This yields α_1, β_1, which in turn can be used in (6.24) to obtain the values of π_2 needed to return to (6.23) for computing α_2, β_2. In this way, alternating between (6.23) and (6.24), we can 'bootstrap' ourselves up to any desired order of the recursion coefficients. The procedure is now commonly referred to as the *Stieltjes procedure*.

Although the recurrence relation (6.24) may develop the phenomenon of pseudostability mentioned at the end of Section 0.2, as k approaches N, this normally causes no problem for the Stieltjes procedure since the maximum order $n-1$ desired for the recursion coefficients α_k, β_k is usually much smaller than the integer N eventually needed for convergence in (6.4). The onset of pseudostability is thus avoided. On the other hand, suitable scaling of the weights w_k may be required to stay clear of overflow or underflow. No such problems occur with the Lanczos method, which, moreover, has been observed to be typically about twice as fast as the Stieltjes procedure. For these reasons, one normally prefers orthogonal reduction methods over the Stieltjes procedure.

Notes to Section 6

6.1. The idea of discretizing the measure to approximate the recursion coefficients, and the use of Fejér's quadrature rule (6.17) in this context, goes back to Gautschi (1968). The convergence property (6.4), (6.5) is proved in Gautschi (1968, Theorem 4). The idea has been further developed along the lines of Section 6.1 in Gautschi

(1982a) and is implemented in the computer routine `mcdis` of Gautschi (1994). Example 6.1 is taken from Gautschi (1982a, Example 4.10) and is of interest in the asymptotic approximation of oscillatory integral transforms (Wong 1982).

6.2, 6.3. A Lanczos-type algorithm of the type mentioned at the end of Section 6.2 can be found in Gragg and Harrod (1984) and is used in the routine `lancz` of Gautschi (1994). The bootstrap procedure of Section 6.3 was briefly mentioned by Stieltjes (1884) and also forms the basis of the procedures in Forsythe (1957). For the phenomenon of pseudostability mentioned at the end of Section 6.3, see Gautschi (1993a) and Gautschi (1996b).

7. Modification algorithms

The idea of (and need for) looking at orthogonal polynomials relative to modified measures goes back to Christoffel (1858), who multiplied the measure $d\lambda$ by a polynomial $u(t) = \prod_{\lambda=1}^{\ell}(t - u_\lambda)$, where all u_λ are outside the support interval (the smallest interval containing supp($d\lambda$)); he represented the polynomial $u(t)\pi_n(t; u \, d\lambda)$ in determinantal form as a linear combination of $\pi_n(t; d\lambda), \ldots, \pi_{n+\ell}(t; d\lambda)$. This is now known as *Christoffel's theorem*. More recently, Uvarov (1959, 1969) extended Christoffel's result to measures multiplied by a rational function $u(t)/v(t)$, where $v(t) = \prod_{\mu=1}^{m}(t - v_\mu)$, expressing $u(t)\pi_n(t; (u/v) \, d\lambda)$ again in determinantal form as a linear combination of $\pi_{n-m}(t; d\lambda), \ldots, \pi_{n+\ell}(t; d\lambda)$ if $m \leq n$, and of $\pi_0(t; d\lambda), \ldots, \pi_{n+\ell}(t; d\lambda)$ if $m > n$. We have called this (Gautschi 1982b) the *generalized Christoffel theorem*.

While these theorems are mathematically elegant, they do not lend themselves easily to computational purposes. What is more useful is trying to compute the recursion coefficients $\alpha_k(d\hat{\lambda})$, $\beta_k(d\hat{\lambda})$ for the modified measure $d\hat{\lambda} = (u/v) \, d\lambda$ in terms of those for $d\lambda$, which we assume are known. This need not be accomplished all at once, but can be carried out in elementary steps: multiply or divide by one linear complex factor $t - z$ at a time, or else, if we prefer to compute in the real domain, multiply or divide by either a linear real factor $t - x$, or a quadratic real factor $(t - x)^2 + y^2$. Thus, the problem we wish to consider is the following. Given the recursion coefficients $\alpha_k(d\lambda)$, $\beta_k(d\lambda)$ for the measure $d\lambda$, compute the recursion coefficients $\alpha_k(d\hat{\lambda})$, $\beta_k \, d(\hat{\lambda})$ for the measures $d\hat{\lambda} = u \, d\lambda$ and $d\hat{\lambda} = d\lambda/v$, where $u(t)$ and $v(t)$ are elementary real factors of the type $t - x$ or $(t - x)^2 + y^2$, $x \in \mathbb{R}$, $y \in \mathbb{R}$.

We begin in Section 7.1 with the theory of quasi-definite measures and kernel polynomials, which lies at the heart of modification algorithms for linear and quadratic factors. The latter are discussed in Section 7.2. In Section 7.3 we develop algorithms for linear and quadratic divisors. The division algorithms, finally, are applied in Section 7.4 to construct the rational Gauss quadrature formulae that were discussed in Section 3.1.

7.1. Quasi-definite measures and kernel polynomials

It is convenient, in this subsection, to allow $d\lambda$ to be any real or complex-valued measure on R having finite moments of all orders,

$$\mu_r = \mu_r(d\lambda) = \int_{\mathbb{R}} t^r \, d\lambda(t), \qquad r = 0, 1, 2, \ldots . \tag{7.1}$$

The measure $d\lambda$ is called *quasi-definite* if all Hankel determinants D_n in the moments are nonzero, that is,

$$D_n = \det \begin{bmatrix} \mu_0 & \mu_1 & \cdots & \mu_{n-1} \\ \mu_1 & \mu_2 & \cdots & \mu_n \\ \cdots & \cdots & \cdots & \cdots \\ \mu_n & \mu_{n+1} & \cdots & \mu_{2n-1} \end{bmatrix} \neq 0, \qquad n = 1, 2, 3, \ldots . \tag{7.2}$$

If $d\lambda$ is quasi-definite, there exists a unique system $\{\pi_k\}_{k=0}^{\infty}$ of (monic) orthogonal polynomials $\pi_k(\cdot) = \pi_k(\cdot; d\lambda)$ relative to the measure $d\lambda$, which satisfy the three-term recurrence relation (0.11) with coefficients $\alpha_k = \alpha_k(d\lambda)$, $\beta_k = \beta_k(d\lambda)$ that are now complex-valued in general, but with $\beta_k \neq 0$. The measure $d\lambda$ is called *positive definite* if $\int_{\mathbb{R}} p(t) \, d\lambda(t) > 0$ for every polynomial $p(t) \not\equiv 0$ that is nonnegative on $\text{supp}(d\lambda)$. Equivalently, $d\lambda$ is positive definite if all moments (7.1) are real and $D_n > 0$ for all $n \geq 1$.

For arbitrary $z \in \mathbb{C}$, and for $\alpha_k = \alpha_k(d\lambda)$, $\beta_k = \beta_k(d\lambda)$, $\beta_0 = 0$, let

$$\left. \begin{array}{l} \alpha_k = z + q_k + e_{k-1} \\ \beta_k = e_{k-1} q_{k-1} \end{array} \right\} \qquad k = 0, 1, 2, \ldots ; e_{-1} = q_{-1} = 0. \tag{7.3}$$

Lemma 1 Let $d\lambda$ be quasi-definite and $\pi_k(\cdot) = \pi_k(\cdot; d\lambda)$.

(a) If $\pi_n(z) \neq 0$ for all $n = 1, 2, 3, \ldots$, then the relations (7.3) uniquely determine $q_0, e_0, q_1, e_1, \ldots$ in this order, and

$$q_k = - \frac{\pi_{k+1}(z)}{\pi_k(z)}, \qquad k = 0, 1, 2, \ldots . \tag{7.4}$$

(b) If $\pi_{\ell+1}(z) = 0$ for some $\ell \geq 0$, and $\pi_k(z) \neq 0$ for all $k \leq \ell$, then q_k, e_k are uniquely determined by (7.3) for $k < \ell$, while $q_\ell = 0$ and e_ℓ is undefined.

Proof. (a) The quantities $q_0, e_0, q_1, e_1, \ldots$ are uniquely defined if and only if $q_k \neq 0$ for all $k \geq 0$. It suffices, therefore, to prove (7.4). For $k = 0$, this follows from the first relation in (7.3) with $k = 0$:

$$q_0 = \alpha_0 - z = -(z - \alpha_0) = - \frac{\pi_1(z)}{\pi_0(z)} .$$

Proceeding by induction, assume (7.4) true for $k - 1$. Then, by (7.3),

$$q_k = \alpha_k - z - e_{k-1} = \alpha_k - z - \frac{\beta_k}{q_{k-1}} = \alpha_k - z + \beta_k \frac{\pi_{k-1}(z)}{\pi_k(z)} ,$$

hence

$$q_k = -\frac{1}{\pi_k(z)}\{(z - \alpha_k)\pi_k(z) - \beta_k\pi_{k-1}(z)\} = -\frac{\pi_{k+1}(z)}{\pi_k(z)},$$

where the recurrence relation (0.11) has been used in the last step.

(b) The argument in the proof of (a) establishes (7.4) for all $k \leq \ell$, from which the assertion follows immediately. $\square$

Consider now

$$d\hat{\lambda}(t) = (t - z)\,d\lambda(t), \qquad z \in \mathbb{C}.$$

If $d\lambda$ is quasi-definite, and z satisfies the assumption of Lemma 1(a), then $d\hat{\lambda}$ is also quasi-definite (Chihara 1978, Chapter I, Theorem 7.1), and hence gives rise to a sequence of (monic) orthogonal polynomials $\hat{\pi}_k(\,\cdot\,; z) = \pi_k(\,\cdot\,; d\hat{\lambda})$, $k = 0, 1, 2, \ldots$. These are called the *kernel polynomials*. They are given explicitly in terms of the polynomials $\pi_k(\,\cdot\,) = \pi_k(\,\cdot\,; d\lambda)$ by

$$\hat{\pi}_n(t; z) = \frac{1}{t - z}\left[\pi_{n+1}(t) - \frac{\pi_{n+1}(z)}{\pi_n(z)}\,\pi_n(t)\right], \qquad k = 0, 1, 2, \ldots, \qquad (7.5)$$

as is readily verified.

Let $\hat{\alpha}_k = \alpha_k(d\hat{\lambda})$, $\hat{\beta}_k = \beta_k(d\hat{\lambda})$ be the recursion coefficients for the kernel polynomials $\hat{\pi}_k(\,\cdot\,) = \hat{\pi}_k(\,\cdot\,; z)$,

$$\begin{aligned}\hat{\pi}_{k+1}(t) = (t - \hat{\alpha}_k)\hat{\pi}_k(t) - \hat{\beta}_k\hat{\pi}_{k-1}(t), \qquad k = 0, 1, 2, \ldots, \\ \hat{\pi}_{-1}(t) = 0, \qquad \hat{\pi}_0(t) = 1,\end{aligned} \qquad (7.6)$$

where the dependence on z has been suppressed. The following theorem shows how the coefficients $\hat{\alpha}_k$, $\hat{\beta}_k$ can be generated in terms of the quantities q_k, e_k of Lemma 1.

Theorem 7 Let $d\lambda$ be quasi-definite and $z \in \mathbb{C}$ be such that $\pi_n(z; d\lambda) \neq 0$ for all n. Let q_k, e_k, be the quantities uniquely determined by (7.3). Then

$$\left.\begin{aligned}\hat{\alpha}_k &= z + q_k + e_k \\ \hat{\beta}_k &= q_k e_{k-1}\end{aligned}\right\} \qquad k = 0, 1, 2, \ldots . \qquad (7.7)$$

In (7.7), $\hat{\beta}_0$ receives the value zero; it could be assigned any other convenient value such as the customary $\hat{\beta}_0 = \int_{\mathbb{R}} d\hat{\lambda}(t)$. In that case, $\hat{\beta}_0 = \int_{\mathbb{R}}(t - z)\,d\lambda(t) = \int_{\mathbb{R}}(t - \alpha_0 + \alpha_0 - z)\,d\lambda(t) = (\alpha_0 - z)\beta_0$, since $t - \alpha_0 = \pi_1(t; d\lambda)$ and $\int_{\mathbb{R}}\pi_1(t)\,d\lambda(t) = 0$.

Proof of Theorem 7. By (7.5) and (7.4) we can write

$$\hat{\pi}_k(t) = \frac{1}{t - z}[\pi_{k+1}(t) + q_k\pi_k(t)], \qquad (7.8)$$

or, solved for π_{k+1},

$$\pi_{k+1}(t) = (t - z)\hat{\pi}_k(t) - q_k\pi_k(t), \qquad k = 0, 1, 2, \ldots . \tag{7.9}$$

The three-term recurrence relation for the $\{\pi_k\}$, with the coefficients α_k, β_k written in the form (7.3), yields

$$\pi_{k+1}(t) = (t - z)\pi_k(t) - (q_k + e_{k-1})\pi_k(t) - e_{k-1}q_{k-1}\pi_{k-1}(t),$$

from which

$$\frac{\pi_{k+1}(t) + q_k\pi_k(t)}{t - z} = \pi_k(t) - e_{k-1}\frac{\pi_k(t) + q_{k-1}\pi_{k-1}(t)}{t - z},$$

or, by (7.8),

$$\hat{\pi}_k(t) = \pi_k(t) - e_{k-1}\hat{\pi}_{k-1}(t), \qquad k = 0, 1, 2, \ldots . \tag{7.10}$$

Replacing k by $k+1$ in (7.10) and applying first (7.9), and then again (7.10), we get

$$\begin{aligned}
\hat{\pi}_{k+1}(t) &= \pi_{k+1}(t) - e_k\hat{\pi}_k(t) \\
&= (t - z)\hat{\pi}_k(t) - q_k\pi_k(t) - e_k\hat{\pi}_k(t) \\
&= (t - z)\hat{\pi}_k(t) - q_k[\hat{\pi}_k(t) + e_{k-1}\hat{\pi}_{k-1}(t)] \quad e_k\hat{\pi}_k(t),
\end{aligned}$$

that is,

$$\hat{\pi}_{k+1}(t) = (t - z - q_k - e_k)\hat{\pi}_k(t) - q_k e_{k-1}\hat{\pi}_{k-1}(t), \atop k = 0, 1, 2, \ldots . \tag{7.11}$$

The assertion (7.7) now follows by comparing (7.11) with (7.6). $\square$

7.2. Linear and quadratic factors

We assume from now on that $d\lambda$ is a positive measure. The support of $d\lambda$ may extend to infinity at one end, when dealing with linear factors, but will be arbitrary otherwise.

Consider first modification by a linear factor,

$$d\hat{\lambda}(t) = (t - x)\,d\lambda(t), \qquad x \in \mathbb{R}\backslash I_{\text{supp}}(d\lambda),$$

where, as indicated, x is any real number outside the 'support interval' $I_{\text{supp}}(d\lambda)$ of $d\lambda$, that is, outside the smallest interval containing the support of $d\lambda$. Then $d\hat{\lambda}$ is positive definite if x is to the left of this interval, and negative definite otherwise. In either case, $\pi_n(x; d\lambda) \neq 0$ for all n, since the zeros of π_n are known to lie in the support interval. Theorem 7, therefore, applies with $z = x$ and, together with the remark immediately after Theorem 7, and (7.3), produces the following algorithm for calculating the first

n recursion coefficients of $\{\hat{\pi}_k\}$ from those of $\{\pi_k\}$:

$$
\left.
\begin{aligned}
e_{-1} &= 0 \\
q_k &= \alpha_k - e_{k-1} - x \\
\hat{\beta}_k &= q_k \cdot \begin{cases} \beta_0 & \text{if } k = 0 \\ e_{k-1} & \text{if } k > 0 \end{cases} \\
e_k &= \beta_{k+1}/q_k \\
\hat{\alpha}_k &= x + q_k + e_k
\end{aligned}
\right\} \qquad k = 0, 1, \ldots, n-1.
$$
(7.12)

Note that we need β_n in addition to α_k, β_k, $k = 0, 1, \ldots, n-1$, to obtain the first n recursion coefficients $\hat{\alpha}_k$, $\hat{\beta}_k$, $k = 0, 1, \ldots, n-1$. Numerical experience seems to indicate that the nonlinear recursion (7.12) is quite stable. In cases where the coefficients $\hat{\alpha}_k$ tend rapidly to zero, it is true that they can be obtained only to full absolute accuracy, not relative accuracy. This, however, should not impair the accuracy in the recursive computation of $\hat{\pi}_k$ by (7.6).

There is a similar, but more complicated, algorithm for modification by a quadratic factor,

$$
\mathrm{d}\hat{\lambda}(t) = ((t-x)^2 + y^2)\,\mathrm{d}\lambda(t), \qquad x \in \mathbb{R}, \qquad y > 0, \qquad (7.13)
$$

which can be obtained by two successive applications of linear (complex) factors $t - z$ and $t - \bar{z}$, where $z = x + iy$. A particularly elegant algorithm is known when $y = 0$ in (7.13). In terms of the Jacobi matrices of $\mathrm{d}\lambda$ and $\mathrm{d}\hat{\lambda}$, it consists in applying one QR step with the shift x: if

$$
\begin{aligned}
J_{n+1}(\mathrm{d}\lambda) - xI_{n+1} &= QR, \\
Q \text{ orthogonal, } R \text{ upper triangular, } \operatorname{diag} R &\geq 0,
\end{aligned}
\qquad (7.14)
$$

then

$$
J_n(\mathrm{d}\hat{\lambda}) = (RQ + xI_{n+1})_{n \times n}. \qquad (7.15)
$$

Thus, having completed the QR step applied to the Jacobi matrix of order $n+1$ for the measure $\mathrm{d}\lambda$, one discards the last row and last column to obtain the Jacobi matrix of order n for the modified measure $\mathrm{d}\hat{\lambda}$. This algorithm, too, appears to be quite stable.

7.3. Linear and quadratic divisors

Consider first division by a linear divisor,

$$
\mathrm{d}\hat{\lambda}(t) = \frac{\mathrm{d}\lambda(t)}{t - x}, \qquad x \in \mathbb{R} \backslash I_{\mathrm{supp}}(\mathrm{d}\lambda), \qquad (7.16)
$$

where x is assumed real, outside the support interval of $\mathrm{d}\lambda$. Here again, there exists a nonlinear algorithm of the type (7.12) (indeed, a reversal thereof), but it is quite unstable unless x is very close to the support interval of $\mathrm{d}\lambda$. Although such values of x are not without interest in applications, we shall not develop the algorithm here and refer instead to Gautschi (1982b).

For other values of x, and particularly for measures with bounded support (*cf.* the remark at the end of Section 5.3), we recommend applying the modified Chebyshev algorithm, using the orthogonal polynomials $p_k(\cdot) = \pi_k(\cdot\,; d\lambda)$ as the polynomial system defining the modified moments, that is, letting

$$m_k = \int_{\mathbb{R}} \pi_k(t; d\lambda)\,\frac{d\lambda(t)}{t - x}\,, \qquad k = 0, 1, 2, \ldots\,. \tag{7.17}$$

We shall assume again that the recursion coefficients $\alpha_k = \alpha_k(d\lambda)$, $\beta_k = \beta_k(d\lambda)$ are known. Under mild assumptions on the measure $d\lambda$ (for instance, if $I_{\mathrm{supp}}(d\lambda)$ is a finite interval), the sequence $\{m_k\}$ is a minimal solution of the basic recurrence relation

$$\begin{aligned} y_{k+1} &= (x - \alpha_k)y_k - \beta_k y_{k-1}, \qquad k = 0, 1, 2, \ldots, \\ y_{-1} &= -1, \end{aligned} \tag{7.18}$$

where $\alpha_k = \alpha_k(d\lambda)$, $\beta_k = \beta_k(d\lambda)$. Its first $N + 1$ members can then be computed by the following algorithm: select $\nu > N$ and recur backwards by means of

$$r_\nu^{(\nu)} = 0, \qquad r_{k-1}^{(\nu)} = \frac{\beta_k}{x - \alpha_k - r_k^{(\nu)}}\,, \qquad k = \nu, \nu - 1, \ldots, 0. \tag{7.19}$$

Then compute

$$m_{-1}^{(\nu)} = -1, \qquad m_k^{(\nu)} = r_{k-1}^{(\nu)} m_{k-1}^{(\nu)}, \qquad k = 0, 1, \ldots, N. \tag{7.20}$$

The algorithm converges in the sense that

$$m_k = \lim_{\nu \to \infty} m_k^{(\nu)}. \tag{7.21}$$

Thus, applying (7.19) and (7.20) for ν sufficiently large, we can compute m_k to any desired accuracy.

A similar algorithm works for division by a quadratic divisor, say

$$d\hat{\lambda}(t) = \frac{d\lambda(t)}{(t - x)^2 + y^2}\,, \qquad x \in \mathbb{R}, \qquad y > 0, \tag{7.22}$$

if one notes that

$$\frac{1}{(t - x)^2 + y^2} = \frac{1}{2\,iy}\left(\frac{1}{t - z} - \frac{1}{t - \bar{z}}\right), \qquad z = x + iy,$$

hence

$$m_k = \int_{\mathbb{R}} \pi_k(t; d\lambda)\,\frac{d\lambda(t)}{(t - x)^2 + y^2} = \frac{\operatorname{Im} f_k(z)}{\operatorname{Im} z}\,, \tag{7.23}$$

where

$$f_k(z) = \int_{\mathbb{R}} \pi_k(t; d\lambda)\,\frac{d\lambda(t)}{t - z}\,. \tag{7.24}$$

This again is a minimal solution of (7.18), where x is to be replaced by z, and therefore the same algorithm applies as in (7.19)–(7.20) with x replaced by z.

7.4. Application to rational Gauss quadrature

We have seen in Section 3.1 that the construction of rational Gauss-type quadrature rules requires the computation of (ordinary) Gaussian quadrature formulae relative to a measure that involves division by a polynomial. These can be generated by the eigenvalue techniques discussed in Section 4.1, once the recursion coefficients of the required orthogonal polynomials have been obtained. This in turn can be accomplished by methods discussed in Sections 7.2 and 7.3.

We will assume in the rational quadrature rule (3.5) that the divisor polynomial ω_m is positive on the support interval of $d\lambda$.

The problem, therefore, is to generate the first n recursion coefficients $\hat{\alpha}_k = \alpha_k(d\hat{\lambda})$, $\hat{\beta}_k = \beta_k(d\hat{\lambda})$, $k = 0, 1, \ldots, n-1$, for the modified measure

$$d\hat{\lambda}(t) = \frac{d\lambda(t)}{\omega_m(t)} , \tag{7.25}$$

assuming the coefficients known for $d\lambda$. Here, ω_m is a polynomial of degree m,

$$\omega_m(t) = \prod_{\mu=1}^{M} (1 + \zeta_\mu t)^{s_\mu}, \qquad \sum_{\mu=1}^{M} s_\mu = m, \tag{7.26}$$

with ζ_μ distinct real or complex numbers such that ω_m is positive on the support interval of $d\lambda$.

A possible solution of the problem is based on the following observation. Suppose $d\Lambda_N$ is a discrete N-point measure, say

$$\int_{\mathbb{R}} p(t)\, d\Lambda_N(t) = \sum_{k=1}^{N} W_k p(T_k), \tag{7.27}$$

with coefficients W_k not necessarily all positive, and suppose further that it provides a quadrature formula for the measure $d\hat{\lambda}$ having degree of exactness $2n-1$, that is,

$$\int_{\mathbb{R}} p(t)\, d\hat{\lambda}(t) = \sum_{k=1}^{N} W_k p(T_k), \qquad \text{all } p \in \mathbb{P}_{2n-1}, \qquad d\hat{\lambda} = \frac{d\lambda}{\omega_m} . \tag{7.28}$$

Then the first n recursion coefficients for $d\hat{\lambda}$ are identical with those for $d\Lambda_N$:

$$\begin{aligned} \alpha_k(d\hat{\lambda}) &= \alpha_k(d\Lambda_N), \\ \beta_k(d\hat{\lambda}) &= \beta_k(d\Lambda_N), \end{aligned} \qquad k = 0, 1, \ldots, n-1. \tag{7.29}$$

This follows immediately from the inner product representation (0.12) of the coefficients on the left of (7.29), since all inner products are integrals (with respect to $d\hat\lambda$) over polynomials of degree $\leq 2n - 1$ and are thus integrated exactly by the formula (7.28). To generate the coefficients on the right of (7.29), we can now apply either the Stieltjes procedure of Section 6.3 or the Lanczos method (of Section 6.2); for the latter, see the remark at the end of Section 6.2.

It remains to show how a formula of the type (7.28) can be constructed. We first look at the simplest case where the polynomial ω_m in (7.26) has all $s_\mu = 1$ (hence $M = m$) and $\zeta_\mu = \xi_\mu$ are all real. Expanding its reciprocal into partial fractions,

$$\frac{1}{\omega_m(t)} = \frac{1}{\prod_{\nu=1}^m (1 + \xi_\nu t)} = \sum_{\nu=1}^m \frac{c_\nu}{t + (1/\xi_\nu)} ,$$

where

$$c_\nu = \frac{\xi_\nu^{m-2}}{\prod_{\substack{\mu=1 \\ \mu \neq \nu}}^m (\xi_\nu - \xi_\mu)} , \qquad \nu = 1, 2, \ldots, m,$$

we then have

$$\int_{\mathbb{R}} p(t)\, d\hat\lambda(t) = \sum_{\nu=1}^m \int_{\mathbb{R}} p(t)\, \frac{c_\nu\, d\lambda(t)}{t + (1/\xi_\nu)} . \qquad (7.30)$$

Each integral on the right now involves modification of the measure $d\lambda$ by a linear divisor. The first n recursion coefficients of the modified measure can therefore be obtained by the procedure of Section 7.3 (using the modified Chebyshev algorithm), which then enables us to compute the respective n-point Gauss formula

$$\int_{\mathbb{R}} p(t)\, \frac{c_\nu\, d\lambda(t)}{t + (1/\xi_\nu)} = \sum_{r=1}^n w_r^{(\nu)} p(t_r^{(\nu)}), \qquad p \in \mathbb{P}_{2n-1}, \qquad (7.31)$$

by the techniques of Section 4.1. Inserting (7.31) in (7.30) then yields

$$\int_{\mathbb{R}} p(t)\, d\hat\lambda = \sum_{\nu=1}^m \sum_{r=1}^n w_r^{(\nu)} p(t_r^{(\nu)}), \qquad p \in \mathbb{P}_{2n-1},$$

the desired quadrature formula (7.28), with $N = mn$ and

$$\begin{aligned} T_{(\nu-1)n+r} &= t_r^{(\nu)}, \\ W_{(\nu-1)n+r} &= w_r^{(\nu)}, \end{aligned} \qquad \nu = 1, 2, \ldots, m; \qquad r = 1, 2, \ldots, n. \qquad (7.32)$$

Analogous procedures apply to other polynomials ω_m, for example to those for which the ζ_μ occur in $m/2$ pairs of conjugate complex numbers: $\zeta_\nu = \xi_\nu + i\eta_\nu, \zeta_{\nu+m/2} = \bar\zeta_\nu, \nu = 1, 2, \ldots, m/2$, where $\xi_\nu \in \mathbb{R}$, $\eta_\nu > 0$, and m is even.

An elementary computation then yields the partial fraction decomposition

$$\frac{1}{\omega_m(t)} = \sum_{\nu=1}^{m/2} \frac{c_\nu + d_\nu t}{\left(t + \frac{\xi_\nu}{\xi_\nu^2 + \eta_\nu^2}\right)^2 + \left(\frac{\eta_\nu}{\xi_\nu^2 + \eta_\nu^2}\right)^2}, \qquad t \in \mathbb{R}, \qquad (7.33)$$

where

$$c_\nu = \frac{1}{\eta_\nu}\left(\frac{\xi_\nu}{\xi_\nu^2 + \eta_\nu^2}\operatorname{Im} p_\nu + \frac{\eta_\nu}{\xi_\nu^2 + \eta_\nu^2}\operatorname{Re} p_\nu\right),$$
$$d_\nu = \frac{1}{\eta_\nu}\operatorname{Im} p_\nu$$

and

$$p_\nu = \prod_{\substack{\mu=1\\\mu\neq\nu}}^{m/2} \frac{(\xi_\nu + i\eta_\nu)^2}{(\xi_\nu - \xi_\mu)^2 - (\eta_\nu^2 - \eta_\mu^2) + 2i\eta_\nu(\xi_\nu - \xi_\mu)},$$

with $p_1 = 1$ if $m = 2$. One can proceed as before, except that the modification of the measure $d\lambda$ now involves a quadratic divisor (see (7.33)) and, if $d_\nu \neq 0$, in addition a linear factor. Thus, not only the methods of Section 7.3, but also those of Section 7.2 come into play.

The procedures described here, since they rely on the modified Chebyshev algorithm to execute the division algorithm of Section 7.3, work best if the support of $d\lambda$ is a finite interval. For measures with unbounded support, methods based on discretization (see Section 6.1) will be more effective, but possibly also more expensive.

Notes to Section 7

7.1. A good reference for the theory of quasi-definite measures and kernel polynomials is Chihara (1978, Chapter I). Lemma 1 and Theorem 7 are from Gautschi (1982b). Kernel polynomials also play an important role in numerical linear algebra in connection with iterative methods for solving linear algebraic systems and eigenvalue problems; for these applications, see Stiefel (1958). The proof of Theorem 7 indeed follows closely an argumentation used in Stiefel (1958), but does not require the assumption of a positive definite measure.

7.2. The algorithm (7.12) for modification by a linear factor is due to Galant (1971); an extension to quadratic factors (7.13) is given in Gautschi (1982b). The procedure (7.14), (7.15) based on QR methodology is due to Kautsky and Golub (1983). See also Buhmann and Iserles (1992) for an alternative proof.

7.3, 7.4. The treatment of linear and quadratic divisors follows Gautschi (1981b), where further details, in particular regarding the recursion algorithm (7.19), (7.20), can be found. For other, algebraic methods and a plausibility argument for the instability noted at the beginning of Section 7.3, see Galant (1992). The application to rational Gauss quadrature is taken from Gautschi (1993b).

8. Orthogonal polynomials of Sobolev type

As already mentioned in Section 2.2, the computation of orthogonal polynomials in the Sobolev space H_s of (2.21), involving the inner product

$$(u, v)_{H_s} = \sum_{\sigma=0}^{s} \int_{\mathbb{R}} u^{(\sigma)}(t) v^{(\sigma)}(t) \, \mathrm{d}\lambda_\sigma(t), \qquad (8.1)$$

is complicated by the lack of symmetry of this inner product with respect to
multiplication by t (see (2.26)). This means that we can no longer expect
a three-term recurrence relation to hold, or even a recurrence relation of
constant order. On the other hand, it is certainly true, as for any sequence
of monic polynomials whose degrees increase by 1 from one member to the
next, that

$$\pi_{k+1}(t) = t\pi_k(t) - \sum_{j=0}^{k} \beta_j^k \pi_{k-j}(t), \qquad k = 0, 1, 2, \dots, \qquad (8.2)$$

for suitable coefficients β_j^k. We may thus pose the problem of computing
$\{\beta_j^k\}_{0 \le j \le k}$ for $k = 0, 1, \dots, n - 1$, which will allow us to generate the first
$n + 1$ polynomials $\pi_0, \pi_1, \dots, \pi_n$ by (8.2). Moreover, the zeros of π_n are
computable as eigenvalues of the $n \times n$ Hessenberg matrix

$$B_n = \begin{bmatrix} \beta_0^0 & \beta_1^1 & \beta_2^2 & \cdots & \beta_{n-2}^{n-2} & \beta_{n-1}^{n-1} \\ 1 & \beta_0^1 & \beta_1^2 & \cdots & \beta_{n-3}^{n-2} & \beta_{n-2}^{n-1} \\ 0 & 1 & \beta_0^2 & \cdots & \beta_{n-4}^{n-2} & \beta_{n-3}^{n-1} \\ \cdots & \cdots & \cdots & \cdots & \cdots & \cdots \\ 0 & 0 & 0 & \cdots & \beta_0^{n-2} & \beta_1^{n-1} \\ 0 & 0 & 0 & \cdots & 1 & \beta_0^{n-1} \end{bmatrix}. \qquad (8.3)$$

In Section 8.1 we briefly describe how moment information can be used
to develop a 'modified Chebyshev algorithm' for Sobolev orthogonal polynomials, and in Section 8.2 show how Stieltjes's idea can be adapted for the
same purpose. Special inner products (8.1) of Sobolev type sometimes lead
to simpler recurrence relations. An instance of this is described in Section
8.3.

8.1. Algorithm based on moment information

In analogy to (5.5), we define modified moments for all $s + 1$ measures $\mathrm{d}\lambda_\sigma$,
but for simplicity use the same system of polynomials $\{p_k\}$ for each,

$$m_k^{(\sigma)} = \int_{\mathbb{R}} p_k(t) \, \mathrm{d}\lambda_\sigma(t), \qquad k = 0, 1, 2, \dots; \qquad \sigma = 0, 1, \dots, s. \qquad (8.4)$$

As in Section 5.2, we assume these polynomials to satisfy a three-term recurrence relation

$$p_{-1}(t) = 0, \qquad p_0(t) = 1,$$
$$p_{k+1}(t) = (t - a_k)p_k(t) - b_k p_{k-1}(t), \qquad k = 0, 1, 2, \ldots, \tag{8.5}$$

where the coefficients a_k, b_k are given real numbers. The objective is, for given $n \geq 1$, to compute the coefficients $\{\beta_j^k\}_{0 \leq j \leq k}$ in (8.2) for $k = 0, 1, \ldots, n-1$, using the recursion coefficients a_j, b_j, $0 \leq j \leq 2n-2$ in (8.5) and the modified moments $m_j^{(0)}$, $0 \leq j \leq 2n - 1$, and $m_j^{(\sigma)}$, $0 \leq j \leq 2n - 2$ (if $n \geq 2$), $\sigma = 1, 2, \ldots, s$.

It is possible to accomplish this task with the help of an algorithm that resembles the modified Chebyshev algorithm of Section 5.2. Like the latter, it uses 'mixed moments' $\sigma_{k,\ell} = (\pi_k, \pi_\ell)_{H_s}$, but now relative to the Sobolev inner product in H_s. These, in turn, require for their computation 'mixed derivative moments' $\mu_{k,\ell,\sigma}^{(i,j)} = (\pi_k^{(i)}, p_\ell^{(j)})_{\mathrm{d}\lambda_\sigma}$, $\sigma = 1, \ldots, s; i, j \leq \sigma$, relative to the individual inner products $(u, v)_{\mathrm{d}\lambda_\sigma} = \int_{\mathbb{R}} u(t)v(t) \, \mathrm{d}\lambda_\sigma(t)$, $\sigma \geq 1$. Accordingly, there will be a tableau containing the mixed moments $\sigma_{k,\ell}$, very much like the tableau in Fig. 3, and for each i, j and σ another auxiliary tableau containing the mixed derivative moments, which has a similar trapezoidal shape, but with height $n-2$ instead of $n-1$. Each quantity in these tableaux is computed recursively in terms of the three nearest quantities on the next lower level, and in terms of all quantities vertically below. The initialization of these tableaux calls for the modified moments (8.4), since $\sigma_{0,\ell} = m_\ell^{(0)}$ and $\mu_{0,\ell,\sigma}^{(0,0)} = m_\ell^{(\sigma)}$, $\sigma \geq 1$, but the complete initialization of all the quantities $\mu_{0,\ell,\sigma}^{(i,j)}$ is a rather involved process. Once the tableau for the $\sigma_{k,\ell}$ has been completed, one obtains first

$$\beta_0^0 = \frac{\sigma_{0,1}}{\sigma_{0,0}} + a_0,$$

and then, successively, for $k = 1, 2, \ldots, n-1$,

$$\beta_0^k = \frac{\sigma_{k,k+1}}{\sigma_{k,k}} + a_k - \frac{\sigma_{k-1,k}}{\sigma_{k-1,k-1}},$$

$$\beta_{k-j}^k = \frac{\sigma_{j,k+1}}{\sigma_{j,j}} + a_k \frac{\sigma_{j,k}}{\sigma_{j,j}} + b_k \frac{\sigma_{j,k-1}}{\sigma_{j,j}} - \frac{\sigma_{j-1,k}}{\sigma_{j-1,j-1}} - \sum_{\ell=j}^{k-1} \beta_{\ell-j}^\ell \frac{\sigma_{\ell,k}}{\sigma_{\ell,\ell}},$$

$$j = k - 1, k - 2, \ldots, 1 \, (\text{ if } k \geq 2),$$

$$\beta_k^k = \frac{\sigma_{0,k+1}}{\sigma_{0,0}} + a_k \frac{\sigma_{0,k}}{\sigma_{0,0}} + b_k \frac{\sigma_{0,k-1}}{\sigma_{0,0}} - \sum_{\ell=0}^{k-1} \beta_\ell^\ell \frac{\sigma_{\ell,k}}{\sigma_{\ell,\ell}},$$

where a_k, b_k are the coefficients in (8.5).

The algorithm is considerably more complicated than the modified Chebyshev algorithm of Section 5.2 – its complexity, indeed, is $\mathcal{O}(n^3)$ rather than $\mathcal{O}(n^2)$ – but this seems to reflect an inherently higher level of difficulty.

8.2. Stieltjes-type algorithm

The procedure sketched in Section 8.1 employs only *rational* operations on the data, which is one of the reasons why the resulting algorithm is so complicated. Allowing also *algebraic* operations (that is, solving algebraic equations) permits a simpler and more transparent (though not necessarily more efficient) approach. Basically, one expresses $-\beta_j^k$ in (8.2) as the Fourier–Sobolev coefficients of $\pi_{k+1} - t\pi_k(t)$, that is,

$$\beta_j^k = \frac{(t\pi_k, \pi_{k-j})_{H_s}}{\| \pi_{k-j} \|_{H_s}^2}, \qquad j = 0, 1, \ldots, k, \tag{8.6}$$

and evaluates the inner products in both numerator and denominator by numerical integration. If $k \leq n-1$, then all inner products involve polynomials of degree less than $2n$, and hence can be computed exactly by n-point Gaussian quadrature rules relative to the measures $\mathrm{d}\lambda_\sigma$. It is in the generation of these Gaussian rules where algebraic processes are required. The polynomials intervening in (8.6), and their derivatives, are computed recursively by (8.2) and its differentiated version, employing the coefficients β_j^k already computed. Thus, initially, (see (0.12))

$$\beta_0^0 = \frac{(t, 1)_{\mathrm{d}\lambda_0}}{(1, 1)_{\mathrm{d}\lambda_0}} = \alpha_0(\mathrm{d}\lambda_0),$$

which allows us to obtain π_1 by (8.2). In turn, this allows us to compute $\{\beta_j^1\}_{0 \leq j \leq 1}$ by (8.6), and hence, via (8.2), to obtain π_2. Continuing in this manner yields the following 'bootstrapping' procedure:

$$\beta_0^0 \overset{(8.2)}{\longmapsto} \pi_1 \overset{(8.6)}{\longmapsto} \{\beta_j^1\}_{0 \leq j \leq 1} \overset{(8.2)}{\longmapsto} \pi_2 \overset{(8.6)}{\longmapsto} \cdots \overset{(8.6)}{\longmapsto} \{\beta_j^{n-1}\}_{0 \leq j \leq n-1} \overset{(8.2)}{\longmapsto} \pi_n.$$

8.3. Special inner products

While symmetry with respect to multiplication by t in general does not hold for the inner product (8.1), a more general symmetry property may hold, namely

$$(hu, v)_{H_s} = (u, hv)_{H_s}, \tag{8.7}$$

where h is a polynomial of degree ≥ 1. This, however, implies, as is shown in Evans, Littlejohn, Marcellán, Markett and Ronveaux (1995), that all measures $\mathrm{d}\lambda_\sigma$, $\sigma \geq 1$, must be of Dirac type. On the other hand, there then exists a $(2m+1)$-term recurrence relation of the form

$$h(t)\pi_k(t; H_s) = \sum_{j=k-m}^{k+m} \omega_{kj}\pi_j(t; H_s), \tag{8.8}$$

where m is the smallest degree among polynomials h satisfying (8.7) and h in (8.8) is a polynomial of that minimum degree.

If, for example,

$$(u, v)_{H_s} = \int_{\mathbb{R}} u(t)v(t)\,d\lambda(t) + u^{(s)}(c)v^{(s)}(c), \tag{8.9}$$

where $d\lambda$ is a positive measure, s an integer ≥ 1, and $c \in \mathbb{R}$, then clearly

$$h(t) = (t - c)^{s+1} \tag{8.10}$$

satisfies (8.7) and is a polynomial of minimum degree $m = s + 1$ in (8.8). In this case,

$$\pi_k(\cdot; H_s) = \pi_k(\cdot; d\lambda), \qquad k = 0, 1, \ldots, s, \tag{8.11}$$

as follows easily from (8.9). Moreover, there is an alternative expansion of the polynomial on the left of (8.8), namely

$$h(t)\pi_k(t; H_s) = \sum_{j=k-m}^{k+m} \theta_{kj}\pi_j(t; d\lambda), \tag{8.12}$$

where h is as in (8.10) and $m = s + 1$. The coefficients in (8.8), as well as those in (8.12), can be computed with some effort, but the resulting procedure appears to be quite robust.

The two expansions above, together with (8.11), suggest the following two methods for computing the Sobolev-type orthogonal polynomials belonging to the inner product (8.9). In Method I, one computes π_{k+s+1} by solving (8.8) for π_{k+s+1}, noting that $\omega_{k,k+s+1} = 1$ (since the π_k are monic). Thus,

$$\pi_{k+s+1}(t; H_s) = (t-c)^{s+1}\pi_k(t; H_s) - \sum_{j=k-s-1}^{k+s} \omega_{kj}\pi_j(t; H_s), \qquad k = 0, 1, 2, \ldots, \tag{8.13}$$

where (8.11) is used on the right, when appropriate, and where $\omega_{kj} = 0$ if $j < 0$. In Method II, one computes π_k directly from (8.12),

$$\pi_k(t; H_s) = \frac{1}{(t - c)^{s+1}} \sum_{j=k-s-1}^{k+s+1} \theta_{kj}\pi_j(t; d\lambda), \tag{8.14}$$

where again $\theta_{kj} = 0$ if $j < 0$, and this time the polynomials $\pi_j(\cdot; d\lambda)$ on the right are generated by the basic three-term recurrence relation. Method I, curiously enough, may develop huge errors at a certain distance from c, either on one, or both, sides of c. Apparently, there is consistent cancellation at work, but the inherent reasons for this are not known. Some caution in the use of Method I is therefore indicated. Method II is more reliable, except in the immediate neighbourhood of $t = c$ (where it is safe to use Method I).

9. Software

A software package, called ORTHPOL, has been written, that implements all the procedures discussed above and a few others; see Gautschi (1994). Here is a brief description of the principal components of the package.

`recur`	generates the recursion coefficients for the classical orthogonal polynomials (of Legendre, Chebyshev, Jacobi, Laguerre and Hermite)
`cheb`	implements the modified Chebyshev algorithm (see Section 5.2)
`sti`	implements the Stieltjes procedure for discrete measures (see Section 6.3)
`lancz`	implements Lanczos's algorithm for discrete measures (see Section 6.2)
`mcdis`	implements the discretization procedure sketched in Section 6.1
`mccheb`	implements a version of the modified Chebyshev algorithm (not described in this article) that uses approximate values of the modified moments obtained by a discretization process similar to the one used in Section 6.1
`chri`	implements the nonlinear modification algorithms of Section 7, as well as modification by a QR step (see Section 7.2)
`gchri`	implements the modified moment procedure for linear and quadratic divisors (see Section 7.3)
`gauss`	generates Gauss quadrature formulae via eigenvalues and eigenvectors of the Jacobi matrix (see Section 4.1)
`radau`	generates Gauss–Radau formulae (see Section 4.2)
`lob`	generates Gauss–Lobatto formulae (see Section 4.3)

Numerical experience reported in this article and elsewhere is based on the use of one or a combination of these routines. Routines for rational Gauss quadrature rules and Sobolev orthogonal polynomials have also been written, but are not yet ready for publication.

Notes to Section 9

Historically, the first major effort of computing Gauss quadrature rules on electronic computers was made in the mid- and late 1950s. Davis and Rabinowitz (1956) computed Gauss–Legendre rules with up to 48 points to an accuracy of 20–21 decimal digits, and went up to 96-point rules in Davis and Rabinowitz (1958). Gauss–Laguerre rules were computed by Rabinowitz and Weiss (1959), and Gauss–Lobatto rules by Rabinowitz (1960). For a summary, as of 1981, of the major tables of Gaussian rules and computer programs for generating them, see Gautschi (1981a, Section 5.4). More recent software that includes also Gauss–Kronrod rules and other quadrature methods can be found in Piessens, de Doncker-Kapenga, Überhuber and Kahaner (1983); see also NAG (1991).

The software package in Gautschi (1994) is the first that includes routines for generating Gauss-type formulae and orthogonal polynomials not only for classical but also for essentially arbitrary measures. The package is public domain, and can be received via e-mail by sending the following message to netlib@netlib.org:

send 726 from toms

Alternatively, one can access the package via a WWW browser, using the following URL:

http://www.netlib.org/toms/726

The routines `recur` and `gauss` were instrumental in computations assisting de Branges in his famous proof of the Bieberbach conjecture (Gautschi 1986*b*).

REFERENCES

M. Abramowitz and I. A. Stegun, eds (1964), *Handbook of Mathematical Functions*, NBS Appl. Math. Ser., **55**, U.S. Government Printing Office, Washington, D.C.

P. Althammer (1962), 'Eine Erweiterung des Orthogonalitätsbegriffes bei Polynomen und deren Anwendung auf die beste Approximation', *J. Reine Angew. Math.* **211**, 192–204.

R. Askey and M. Ismail (1984), 'Recurrence relations, continued fractions and orthogonal polynomials', *Memoirs AMS* **49**, no. 300, Amer. Math. Soc., Providence, RI.

A. Bellen (1981), 'A note on mean convergence of Lagrange interpolation', *J. Approx. Theory* **33**, 85–95.

A. Bellen (1988), 'Alcuni problemi aperti sulla convergenza in media dell'interpolazione Lagrangiana estesa', *Rend. Istit. Mat. Univ. Trieste* **20**, Fasc. suppl., 1–9.

J. Boersma and J. P. Dempsey (1992), 'On the numerical evaluation of Legendre's chi-function', *Math. Comp.* **59**, 157–163.

D. Boley and G. H. Golub (1987), 'A survey of matrix inverse eigenvalue problems', *Inverse Problems* **3**, 595–622.

C. F. Borges (1994), 'On a class of Gauss-like quadrature rules', *Numer. Math.* **67**, 271–288.

K. Bowers and J. Lund, eds (1989), *Computation and Control*, Progress in Systems and Control Theory **1**, Birkhäuser, Boston.

M. G. de Bruin and H. G. Meijer (1995), 'Zeros of orthogonal polynomials in a non-discrete Sobolev space', *Ann. Numer. Math.* **2**, 233–246.

M. D. Buhmann and A. Iserles (1992), 'On orthogonal polynomials transformed by the QR algorithm', *J. Comput. Appl. Math.* **43**, 117–134.

A. C. Calder and J. G. Laframboise (1986), 'Multiple-water-bag simulation of inhomogeneous plasma motion near an electrode', *J. Comput. Phys.* **65**, 18–45.

A. C. Calder, J. G. Laframboise and A. D. Stauffer (1983), 'Optimum step-function approximation of the Maxwell distribution', unpublished manuscript.

F. Caliò, W. Gautschi and E. Marchetti (1986), 'On computing Gauss–Kronrod quadrature formulae', *Math. Comp.* **47**, 639–650.

P. L. Chebyshev (1859), 'Sur l'interpolation par la méthode des moindres carrés', *Mém. Acad. Impér. Sci. St. Petersbourg* (7) **1**, No. 15, 1–24. [*Œuvres I*, 473–498.]

T. S. Chihara (1978), *An Introduction to Orthogonal Polynomials*, Gordon and Breach, New York.

E. B. Christoffel (1858), 'Über die Gaußische Quadratur und eine Verallgemeinerung derselben', *J. Reine Angew. Math.* **55**, 61–82. [*Ges. Math. Abhandlungen I*, 42–50.]

G. Criscuolo, G. Mastroianni and P. Nevai (1993), 'Mean convergence of derivatives of extended Lagrange interpolation with additional nodes', *Math. Nachr.* **163**, 73–92.

G. Criscuolo, G. Mastroianni and D. Occorsio (1990), 'Convergence of extended Lagrange interpolation', *Math. Comp.* **55**, 197–212.

G. Criscuolo, G. Mastroianni and D. Occorsio (1991), 'Uniform convergence of derivatives of extended Lagrange interpolation', *Numer. Math.* **60**, 195–218.

G. Criscuolo, G. Mastroianni and P. Vértesi (1992), 'Pointwise simultaneous convergence of extended Lagrange interpolation with additional knots', *Math. Comp.* **59**, 515–531.

P. J. Davis (1993), *Spirals: From Theodorus to Chaos*, A K Peters, Wellesley, MA.

P. J. Davis and P. Rabinowitz (1956), 'Abscissas and weights for Gaussian quadratures of high order', *J. Res. Nat. Bur. Standards* **56**, 35–37.

P. J. Davis and P. Rabinowitz (1958), 'Additional abscissas and weights for Gaussian quadratures of high order. Values for $n = 64$, 80, and 96', *J. Res. Nat. Bur. Standards* **60**, 613–614.

J. Dombrowski and P. Nevai (1986), 'Orthogonal polynomials, measures and recurrence relations', *SIAM J. Math. Anal.* **17**, 752–759.

O. Eğecioğlu and Ç. K. Koç (1989), 'A fast algorithm for rational interpolation via orthogonal polynomials', *Math. Comp.* **53**, 249–264.

P. Erdős and P. Turán (1937), 'On interpolation I: quadrature- and mean-convergence in the Lagrange-interpolation', *Ann. of Math.* **38**, 142–155.

W. D. Evans, L. L. Littlejohn, F. Marcellán, C. Markett and A. Ronveaux (1995), 'On recurrence relations for Sobolev orthogonal polynomials', *SIAM J. Math. Anal.* **26**, 446–467.

L. Fejér (1933), 'Mechanische Quadraturen mit positiven Cotesschen Zahlen', *Math. Z.* **37**, 287–309.

H.-J. Fischer (1996), 'On the condition of orthogonal polynomials via modified moments', *Z. Anal. Anwendungen*, to appear.

G. E. Forsythe (1957), 'Generation and use of orthogonal polynomials for data-fitting with a digital computer', *J. Soc. Indust. Appl. Math.* **5**, 74–88.

G. Freud (1971), *Orthogonal Polynomials*, Pergamon, New York.

R. W. Freund, G. H. Golub and N. M. Nachtigal (1991), 'Iterative solution of linear systems', *Acta Numerica*, Cambridge University Press, 57–100.

M. Frontini and G. V. Milovanović (1989), 'Moment-preserving spline approximation on finite intervals and Turán quadratures', *Facta Univ. Ser. Math. Inform.* **4**, 45–56.

M. Frontini, W. Gautschi and G. V. Milovanović (1987), 'Moment-preserving spline approximation on finite intervals', *Numer. Math.* **50**, 503–518.

D. Galant (1971), 'An implementation of Christoffel's theorem in the theory of orthogonal polynomials', *Math. Comp.* **25**, 111–113.

D. Galant (1992), 'Algebraic methods for modified orthogonal polynomials', *Math. Comp.* **59**, 541–546.

W. Gautschi (1963), 'On inverses of Vandermonde and confluent Vandermonde matrices. II', *Numer. Math.* **5**, 425–430.

W. Gautschi (1967), 'Numerical quadrature in the presence of a singularity', *SIAM J. Numer. Anal.* **4**, 357–362.

W. Gautschi (1968), 'Construction of Gauss–Christoffel quadrature formulas', *Math. Comp.* **22**, 251–270.

W. Gautschi (1970), 'On the construction of Gaussian quadrature rules from modified moments', *Math. Comp.* **24**, 245–260.

W. Gautschi (1979), 'On generating Gaussian quadrature rules', in *Numerische Integration* (G. Hämmerlin, ed.), ISNM **45**, Birkhäuser, Basel, 147–154.

W. Gautschi (1981*a*), 'A survey of Gauss–Christoffel quadrature formulae', in *E. B. Christoffel: The Influence of his Work in Mathematics and the Physical Sciences* (P. L. Butzer and F. Fehér, eds), Birkhäuser, Basel, 72–147.

W. Gautschi (1981*b*), 'Minimal solutions of three-term recurrence relations and orthogonal polynomials', *Math. Comp.* **36**, 547–554.

W. Gautschi (1982*a*), 'On generating orthogonal polynomials', *SIAM J. Sci. Statist. Comput.* **3**, 289–317.

W. Gautschi (1982*b*), 'An algorithmic implementation of the generalized Christoffel theorem', in *Numerical Integration* (G. Hämmerlin, ed.), ISNM **57**, Birkhäuser, Basel, 89–106.

W. Gautschi (1983), 'How and how not to check Gaussian quadrature formulae', *BIT* **23**, 209–216.

W. Gautschi (1984*a*), 'On some orthogonal polynomials of interest in theoretical chemistry', *BIT* **24**, 473–483.

W. Gautschi (1984*b*), 'Discrete approximations to spherically symmetric distributions', *Numer. Math.* **44**, 53–60.

W. Gautschi (1984*c*), 'Questions of numerical condition related to polynomials', in *Studies in Numerical Analysis* (G. H. Golub, ed.), Studies in Mathematics **24**, The Mathematical Association of America, 140–177.

W. Gautschi (1985), 'Orthogonal polynomials – constructive theory and applications', *J. Comput. Appl. Math.* **12/13**, 61–76.

W. Gautschi (1986*a*), 'On the sensitivity of orthogonal polynomials to perturbations in the moments', *Numer. Math.* **48**, 369–382.

W. Gautschi (1986*b*), 'Reminiscences of my involvement in de Branges's proof of the Bieberbach conjecture', in *The Bieberbach Conjecture* (A. Baernstein II, D. Drasin, P. Duren and A. Marden, eds), Math. Surveys Monographs, no. 21, Amer. Math. Soc., Providence, RI, 205–211.

W. Gautschi (1988), 'Gauss–Kronrod quadrature – a survey', in *Numerical Methods and Approximation Theory III* (G. V. Milovanović, ed.), Faculty of Electronic Engineering, Univ. Niš, Niš, 39–66.

W. Gautschi (1989), 'Orthogonality – conventional and unconventional – in numerical analysis', in *Computation and Control* (K. Bowers and J. Lund, eds), Progress in Systems and Control Theory, v. 1, Birkhäuser, Boston, 63–95.

W. Gautschi (1991*a*), 'A class of slowly convergent series and their summation by Gaussian quadrature', *Math. Comp.* **57**, 309–324.

W. Gautschi (1991*b*), 'On certain slowly convergent series occurring in plate contact problems', *Math. Comp.* **57**, 325–338.

W. Gautschi (1991*c*), 'Computational problems and applications of orthogonal polynomials', in *Orthogonal Polynomials and Their Applications* (C. Brezinski, L. Gori and A. Ronveaux, eds), IMACS Annals Comput. Appl. Math. **9**, Baltzer, Basel, 61–71.

W. Gautschi (1992), 'On mean convergence of extended Lagrange interpolation', *J. Appl. Comput. Math.* **43**, 19–35.

W. Gautschi (1993*a*), 'Is the recurrence relation for orthogonal polynomials always stable?', *BIT* **33**, 277–284.

W. Gautschi (1993*b*), 'Gauss-type quadrature rules for rational functions', in *Numerical Integration IV* (H. Brass and G. Hämmerlin, eds), ISNM **112**, Birkhäuser, Basel, 111–130.

W. Gautschi (1993*c*), 'On the computation of generalized Fermi–Dirac and Bose–Einstein integrals', *Comput. Phys. Comm.* **74**, 233–238.

W. Gautschi (1994), 'Algorithm 726: ORTHPOL – a package of routines for generating orthogonal polynomials and Gauss-type quadrature rules', *ACM Trans. Math. Software* **20**, 21–62.

W. Gautschi (1996*a*), 'On the computation of special Sobolev-type orthogonal polynomials', *Ann. Numer. Math.*, to appear.

W. Gautschi (1996*b*), 'The computation of special functions by linear difference equations', in *Proc. 2nd Internat. Conf. on Difference Equations and Applications* (S. Elaydi, G. Ladas and I. Györi, eds), Gordon and Breach, Newark, NJ, to appear.

W. Gautschi (1996*c*), 'Moments in quadrature problems', *Comput. Math. Appl.*, Ser. B, to appear.

W. Gautschi and S. Li (1993), 'A set of orthogonal polynomials induced by a given orthogonal polynomial', *Aequationes Math.* **46**, 174–198.

W. Gautschi and S. Li (1996), 'On quadrature convergence of extended Lagrange interpolation', *Math. Comp.*, to appear.

W. Gautschi and G. V. Milovanović (1985), 'Gaussian quadrature involving Einstein and Fermi functions with an application to summation of series', *Math. Comp.* **44**, 177–190.

W. Gautschi and G. V. Milovanović (1986), 'Spline approximations to spherically symmetric distributions', *Numer. Math.* **49**, 111–121.

W. Gautschi and M. Zhang (1995), 'Computing orthogonal polynomials in Sobolev spaces', *Numer. Math.* **71**, 159–183.

G. H. Golub (1973), 'Some modified matrix eigenvalue problems', *SIAM Rev.* **15**, 318–334.

G. H. Golub and C. F. Van Loan (1989), *Matrix Computations*, 2nd edn, The Johns Hopkins University Press, Baltimore.

G. H. Golub and J. H. Welsch (1969), 'Calculation of Gauss quadrature rules', *Math. Comp.* **23**, 221–230.

R. G. Gordon (1968), 'Error bounds in equilibrium statistical mechanics', *J. Mathematical Phys.* **9**, 655–663.

L. Gori Nicolò-Amati and E. Santi (1989), 'On a method of approximation by means of spline functions', Proc. Internat. Symp. Approx., Optim. and Computing, Dalian, China.

L. Gori and E. Santi (1992), 'Moment-preserving approximations: a monospline approach', *Rend. Mat.* (7) **12**, 1031–1044.

W. B. Gragg and W. J. Harrod (1984), 'The numerically stable reconstruction of Jacobi matrices from spectral data', *Numer. Math.* **44**, 317–335.

P. R. Graves-Morris and T. R. Hopkins (1981), 'Reliable rational interpolation', *Numer. Math.* **36**, 111–128.

W. Gröbner (1967), 'Orthogonale Polynomsysteme, die gleichzeitig mit $f(x)$ auch deren Ableitung $f'(x)$ approximieren', in *Funktionalanalysis, Approximationstheorie, Numerische Mathematik* (L. Collatz, G. Meinardus and H. Unger, eds), ISNM **7**, Birkhäuser, Basel, 24–32.

L. A. Hageman and D. M. Young (1981), *Applied Iterative Methods*, Academic Press, New York.

A. Iserles, P. E. Koch, S. P. Nørsett and J. M. Sanz-Serna (1990), 'Orthogonality and approximation in a Sobolev space', in *Algorithms for Approximation II* (J. C. Mason and M. G. Cox, eds), Chapman and Hall, London, 117–124.

A. Iserles, P. E. Koch, S. P. Nørsett and J. M. Sanz-Serna (1991), 'On polynomials orthogonal with respect to certain Sobolev inner products', *J. Approx. Theory* **65**, 151–175.

C. G. J. Jacobi (1826), 'Ueber Gauß neue Methode, die Werthe der Integrale näherungsweise zu finden', *J. Reine Angew. Math.* **1**, 301–308.

C. G. J. Jacobi (1846), 'Über die Darstellung einer Reihe gegebener Werthe durch eine gebrochene rationale Funktion', *J. Reine Angew. Math.* **30**, 127–156. [*Math. Werke*, vol. 1, 287–316.]

J. Kautsky and S. Elhay (1984), 'Gauss quadratures and Jacobi matrices for weight functions not of one sign', *Math. Comp.* **43**, 543–550.

J. Kautsky and G. H. Golub (1983), 'On the calculation of Jacobi matrices', *Linear Algebra Appl.* **52/53**, 439–455.

M. A. Kovačević and G. V. Milovanović (1996), 'Spline approximation and generalized Turán quadratures', *Portugal. Math.*, to appear.

C. Lanczos (1950), 'An iteration method for the solution of the eigenvalue problem of linear differential and integral operators', *J. Res. Nat. Bur. Standards* **45B**, 225–280.

D. P. Laurie (1996), 'Calculation of Gauss–Kronrod quadrature rules', *Math. Comp.*, to appear.

S. Lewanowicz (1994), 'A simple approach to the summation of certain slowly convergent series', *Math. Comp.* **63**, 741–745.

D. C. Lewis (1947), 'Polynomial least square approximations', *Amer. J. Math.* **69**, 273–278.

S. Li (1994), 'On mean convergence of Lagrange–Kronrod interpolation', in *Approximation and Computation* (R. V. M. Zahar, ed.), ISNM **119**, Birkhäuser, Boston, 383–396.

J.-C. Lin (1988), 'Rational L^2-approximation with interpolation', Ph.D. thesis, Purdue University.

G. López Lagomasino and J. Illán (1984), 'A note on generalized quadrature formulas of Gauss–Jacobi type', in *Constructive Theory of Functions*, Publ. House Bulgarian Acad. Sci., Sofia, 513–518.

G. López Lagomasino and J. Illán Gonzalez (1987), 'Sobre los métodos interpolatorios de integración numérica y su conexión con la aproximación racional', *Rev. Ciencias Matém.* **8**, no. 2, 31–44.

F. Marcellán, T. E. Pérez and M. A. Piñar (1995), 'Orthogonal polynomials on weighted Sobolev spaces: the semiclassical case', *Ann. Numer. Math.* **2**, 93–122.

F. Marcellán and A. Ronveaux (1990), 'On a class of polynomials orthogonal with respect to a discrete Sobolev inner product', *Indag. Math. (N.S.)* **1**, 451–464.

F. Marcellán and A. Ronveaux (1995), 'Orthogonal polynomials and Sobolev inner products: a bibliography', Laboratoire de Physique Mathématique, Facultés Universitaires N.D. de la Paix, Namur, Belgium.

F. Marcellán, M. Alfaro and M. L. Rezola (1993), 'Orthogonal polynomials on Sobolev spaces: old and new directions', *J. Comput. Appl. Math.* **48**, 113–131.

G. Mastroianni (1994), 'Approximation of functions by extended Lagrange interpolation', in *Approximation and Computation* (R. V. M. Zahar, ed.), ISNM **119**, Birkhäuser, Boston, 409–420.

G. Mastroianni and P. Vértesi (1993), 'Mean convergence of Lagrange interpolation on arbitrary systems of nodes', *Acta Sci. Math. (Szeged)* **57**, 429–441.

H. G. Meijer (1993), 'Coherent pairs and zeros of Sobolev-type orthogonal polynomials', *Indag. Math. (N.S.)* **4**, 163–176.

C. A. Micchelli (1988), 'Monosplines and moment preserving spline approximation', in *Numerical Integration III* (H. Brass and G. Hämmerlin, eds), ISNM **85**, Birkhäuser, Basel, 130–139.

G. V. Milovanović (1994), 'Summation of series and Gaussian quadratures', in *Approximation and Computation* (R. V. M. Zahar, ed.), Birkhäuser, Boston, 459–475.

G. V Milovanović (1995), 'Summation of series and Gaussian quadratures, II', *Numer. Algorithms* **10**, 127–136.

G. V. Milovanović and M. A. Kovačević (1988), 'Moment-preserving spline approximations and Turán quadratures', in *Numerical Mathematics Singapore 1988* (R. P. Agarwal, Y. M. Chow and S. J. Wilson, eds), ISNM **86**, Birkhäuser, Basel, 357–365.

G. V. Milovanović and M. A. Kovačević (1992), 'Moment-preserving spline approximation and quadrature', *Facta Univ. Ser. Math. Inform.* **7**, 85–98.

G. Monegato (1982), 'Stieltjes polynomials and related quadrature rules', *SIAM Rev.* **24**, 137–158.

K. Moszyński (1992), 'Remarks on polynomial methods for solving systems of linear algebraic equations', *Appl. Math.* **37**, 419–436.

NAG Fortran Library Manual, Mark 15, Vol. 10, NAG Inc., 1400 Opus Place, Suite 200, Downers Grove, IL 60515–5702.

I. P. Natanson (1964/65), *Constructive Function Theory*, Vols 1–III, Ungar Publ. Co., New York.

A. C. R. Newbery (unpublished), 'Gaussian principles applied to summation,' unpublished manuscript.

S. E. Notaris (1994), 'An overview of results on the existence or nonexistence and the error term of Gauss–Kronrod quadrature formulae', in *Approximation and Computation* (R. V. M. Zahar, ed.), ISNM **119**, Birkhäuser, Boston, 485–496.

B. N. Parlett (1980), *The Symmetric Eigenvalue Problem*, Prentice-Hall, Englewood Cliffs, NJ.

R. Piessens, E. de Doncker-Kapenga, C. W. Überhuber and D. K. Kahaner (1983), *QUADPACK: A Subroutine Package for Automatic Integration*, Springer Ser. Comput Math., vol. 1, Springer, Berlin.

G. Pólya (1933), 'Über die Konvergenz von Quadraturverfahren', *Math. Z.* **37**, 264–286.

P. Rabinowitz (1960), 'Abscissas and weights for Lobatto quadrature of high order', *Math. Comp.* **14**, 47–52.

P. Rabinowitz and G. Weiss (1959), 'Tables of abscissas and weights for numerical evaluation of integrals of the form $\int_0^\infty e^{-x} x^n f(x)\,\mathrm{d}x$', *Math. Tables Aids Comp.* **13**, 285–294.

A. E. Russon and J. M. Blair (1969), 'Rational function minimax approximations for the Bessel functions $K_0(x)$ and $K_1(x)$', Rep. AECL–3461, Atomic Energy of Canada Limited, Chalk River, Ontario.

R. A. Sack and A. F. Donovan (1969), 'An algorithm for Gaussian quadrature given generalized moments', Dept. Math., Univ. of Salford, Salford, UK.

R. A. Sack and A. F. Donovan (1971/72), 'An algorithm for Gaussian quadrature given modified moments', *Numer. Math.* **18**, 465–478.

E. L. Stiefel (1958), 'Kernel polynomials in linear algebra and their numerical applications', in *Further Contributions to the Solution of Simultaneous Linear Equations and the Determination of Eigenvalues*, NBS Appl. Math. Ser., **49**, U.S. Government Printing Office, Washington, D.C., 1–22.

T. J. Stieltjes (1884), 'Quelques recherches sur la théorie des quadratures dites mécaniques', *Ann. Sci. École Norm. Paris*, Sér. 3, **1**, 409–426. [*Œuvres I*, 377–396.]

J. Stoer and R. Bulirsch (1980), *Introduction to Numerical Analysis*, 2nd edn, Springer, New York.

G. Szegő (1975), *Orthogonal Polynomials*, Colloq. Publ. **23**, 4th edn, Amer. Math. Soc., Providence, RI.

J. Todd (1954), 'The condition of the finite segments of the Hilbert matrix', NBS Appl. Math. Ser., v. 39, U.S. Government Printing Office, Washington, D.C., 109–116.

F. G. Tricomi (1954), *Funzioni Ipergeometriche Confluenti*, Edizioni Cremonese, Rome.

V. B. Uvarov (1959), 'Relation between polynomials orthogonal with different weights' (Russian), *Dokl. Akad. Nauk SSSR* **126**, 33–36.

V. B. Uvarov (1969), 'The connection between systems of polynomials that are orthogonal with respect to different distribution functions' (Russian), *Ž. Vyčisl. Mat. i Mat. Fiz.* **9**, 1253–1262. [English translation in *U.S.S.R. Comput. Math. and Math. Phys.* **9** (1969), No. 6, 25–36.]

W. Van Assche and I. Vanherwegen (1993), 'Quadrature formulas based on rational interpolation', *Math. Comp.* **61**, 765–783.

J. L. Walsh (1969), *Interpolation and Approximation by Rational Functions in the Complex Domain*, Colloq. Publ. **20**, 5th edn, Amer. Math. Soc., Providence, RI.

J. C. Wheeler (1974), 'Modified moments and Gaussian quadrature', *Rocky Mountain J. Math.* **4**, 287–296.

J. C. Wheeler (1984), 'Modified moments and continued fraction coefficients for the diatomic linear chain', *J. Chem. Phys.* **80**, 472–476.

D. V. Widder (1941), *The Laplace Transform*, Princeton University Press.

H. S. Wilf (1962), *Mathematics for the Physical Sciences*, Wiley, New York.

H. S. Wilf (1980), personal communication.

R. Wong (1982), 'Quadrature formulas for oscillatory integral transforms', *Numer. Math.* **39**, 351–360.

R. Wong (1989), *Asymptotic Approximations of Integrals*, Academic Press, Boston.

M. Zhang (1994), 'Sensitivity analysis for computing orthogonal polynomials of Sobolev type', in *Approximation and Computation* (R. V. M. Zahar, ed.), ISNM **119**, Birkhäuser, Boston, 563–576.

Acta Numerica (1996), *pp.* 121–148

Automatic grid generation

William D. Henshaw
Scientific Computing Group
Computing, Information and Communications Division
Los Alamos National Laboratory
Los Alamos, NM 87545, USA
E-mail: henshaw@lanl.gov

Current methods for the automatic generation of grids are reviewed. The approaches to grid generation that are discussed include Cartesian, multi-block-structured, overlapping and unstructured. Emphasis is placed on those methods that can create high-quality grids appropriate for the solution of equations of a hyperbolic nature, such as those that arise in fluid dynamics. Numerous figures illustrate the different grid generation techniques.

CONTENTS

1   Introduction   121
2   Basic steps in grid generation   127
3   Cartesian grid generation   128
4   Multi-block-structured grid generation   130
5   Overlapping grid generation   134
6   Unstructured grid generation   139
7   Conclusions   145
References   145

1. Introduction

The intent of this paper is to give a brief review of current methods for the automatic generation of grids for the solution of problems from computational fluid dynamics (CFD), computational electromagnetics and other fields where the solutions are hyperbolic in nature. These applications require the generation of high quality grids with a large number of grid points. It is often the case that the geometry may change with time or it may be necessary to refine the mesh adaptively. It is thus essential that the grid generation algorithms be fast, since the grid may have to be regenerated at every step of a time-dependent simulation. Various popular methods for structured and unstructured grid generation will be described. Figures will illustrate the current state of the technology. Grid generation capabilities have improved greatly in recent years. However, it is perhaps not an exaggeration to say that the construction of a grid is currently the most difficult and time-consuming aspect of determining an accurate solution to a problem on a complicated domain. Indeed, starting from scratch, with some description of the geometry, the time to generate a grid is measured in *weeks* rather than hours.

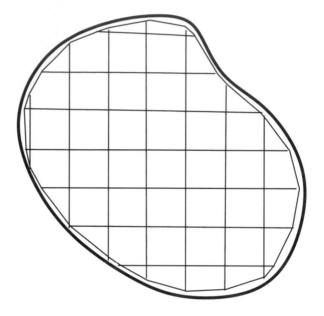

Fig. 1. A Cartesian grid.

Early computational grids were often *Cartesian grids*, or cut-out grids (Fig. 1), whereby the region was covered by a single rectangular grid and the portions of the grid lying outside the region were *cut out*, leaving some irregular cells. This approach was replaced by boundary-conforming grids, whereby a rectangular grid was mapped onto the region, with boundaries corresponding to a coordinate line. Such curvilinear grids that are transformations of a rectangular grid will be called *logically rectangular grids*. Grids that conformed to boundaries improved solution accuracy and made it easier to apply boundary conditions. As computers became faster and more complicated problems were attempted, it became apparent that this single-block approach was not flexible enough to handle complicated geometries. This led to the introduction of the multi-block approach, where the domain was partitioned into blocks and within each block a logically rectangular grid was constructed (Fig. 2). In time, however, it became apparent that this approach was still not sufficiently flexible, and was difficult to automate. Therefore some other approach was needed. One way to add additional flexibility, while still

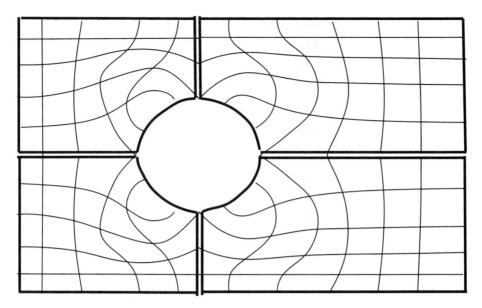

Fig. 2. A multi-block-structured grid divides the region into logically rectangular blocks.

retaining the logically rectangular structure, was the use of overlapping grids in which the component grids are allowed to overlap. There has also been renewed interest in the Cartesian grid approach, using adaptive mesh refinement to improve boundary resolution. Recently, the main interest and focus of research has been in unstructured meshes, which allow complete freedom in grid point placement, although at the expense of speed and memory usage. With little doubt, unstructured meshes offer the best hope for a completely automatic mesh generation program. Completely unstructured grids are not without their difficulties for CFD, and perhaps a hybrid method, combining the unstructured approach with locally structured grids (to resolve boundary layers, for example), will turn out to be the most effective.

The purpose of grid generation is to create a discrete representation for a domain. This entails distributing points throughout the domain. There are two main classes of grids, structured and unstructured. In a structured grid the points covering the domain result from the transformation of a logically rectangular square (or cube in three dimensions[*]). The grid points can be stored as an array $\mathbf{x}(i_1, i_2)$ and the neighbours of a given grid point are simply found as the neighbours in index space, $\mathbf{x}(i_1 \pm 1, i_2 \pm 1)$. In an unstructured grid, on the other hand, the points are connected to one another

---

[*] Throughout this paper, the terminology will be for two-dimensional grids (quadrilaterals, triangles), but the remarks will usually apply equally well to three-dimensional grids (hexahedra, tetrahedra).

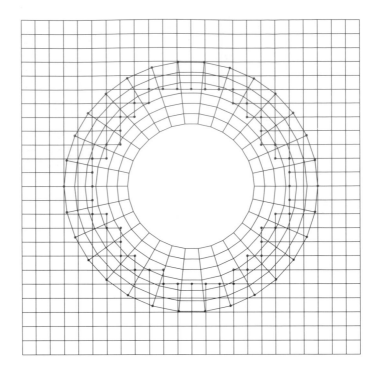

Fig. 3. An overlapping grid consists of logically rectangular blocks that overlap;
some blocks have cut-out regions.

in a general manner; the connectivity information must be explicitly saved. The grid points might be saved as a list $\mathbf{x}_i$, and there would be other lists giving information about neighbours. Of course, the partition of grid types into structured and unstructured is not entirely appropriate, since some grids consist of a *set* of structured grids and other *hybrid* grids have both unstructured and structured parts.

Grids are used to solve equations, typically partial differential equations (PDEs) and integral equations. The computer programs that solve these equations, hereafter referred to as *solvers*, typically discretize a continuous equation with finite-difference, finite-element, finite-volume, spectral-element or boundary-integral methods. At the grid generation level, it is usually not important which particular solver will use the grid: rather the *style* of the grid is most relevant. That is, it is important whether the grid is structured or unstructured, whether the grid elements are triangles or quadrilaterals, or whether the grid is multi-block-structured or overlapping (see Fig. 3). The unstructured triangular grid (see Fig. 4) can be used either with a finite-element solver or a finite-volume solver, just as a multi-block-structured grid can be used by a finite-difference solver or by finite-element solver for quad-

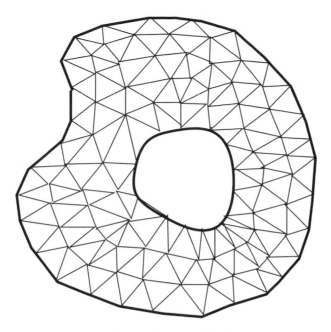

Fig. 4. An unstructured triangular grid is very flexible for representing geometry.

rilaterals. Despite these remarks, it is not uncommon to see references to a *finite-element mesh* (which usually just means an unstructured grid).

The errors in solving a PDE on a grid depend strongly on the quality of the grid. The *quality* of a grid is a relative concept and depends on the actual equations that will be solved, as well as the numerical method that will be used. In principle, a given problem could be repeatedly solved with different grids (with the same number of grid points) and the error in the numerical solution could be measured as a function of the grid. The smaller the error, the better the quality of the grid. Some adaptive methods do indeed redistribute points to try and minimize the error. When creating a grid initially, however, the grid is usually generated with some general principles in mind, such as keeping the cell size varying smoothly, and resolving boundary layers, if appropriate. Generally speaking, the solution of equations with wave-like behaviour (hyperbolic) require *smoother* grids than the solution of elliptic equations. The smoothness of a grid is hard to define in general but relates to the local variation of the cells. An elliptic problem can be accurately solved on a relatively poor-quality grid since the effects of any non-smoothness in the grid will be smoothed out by the elliptic operator. In contrast, hyperbolic operators provide no smoothing effects. To understand this further, note that the properties of the grid, such as the variations in the grid point positions, appear, implicitly or explicitly, in the discrete equations used in the solver. Consider the solution of the one-dimensional wave equation

$$\frac{\partial u}{\partial t} + \frac{\partial u}{\partial x} = 0.$$

If the grid points are allowed to vary according to the parameterization $x = X(r)$ (that is the grid points are equally spaced in r), then the equation for $v(r, t) = u(X(r), t)$ becomes

$$\frac{\partial v}{\partial t} + \frac{1}{X_r} \frac{\partial v}{\partial r} = 0.$$

It is now clear that, if the parameterization is not smooth, then X_r will not be smooth, and this will be reflected in the discrete solution. A grid that is not smooth can distort waves and cause spurious reflections, rather similar to the effect of a wave passing through a non-uniform medium. Higher-order accurate methods are also popular for the solution of wave-like problems, for both efficiency and accuracy reasons. Higher-order methods will in general require higher-quality grids than lower-order methods.

It is important to realize that solvers written for one type of grid will typically not work on other types of grids. Although a structured grid can always be turned into an unstructured grid and used with an unstructured solver, the unstructured solver would not usually take advantage of the structured nature of the grid. There is, however, increasing interest in *hybrid* grids and hybrid grid solvers. Hybrid grids range from those that are primarily unstructured triangles, with some structured quadrilaterals to resolve a boundary layer, to those that are primarily structured blocks, with triangles used to merge the blocks.

The number of grid points required for many three-dimensional problems is extremely large. For typical big simulations there are on the order of 10^6 grid points, this number being limited only by computer memory and speed. Viscous fluid flow computations over an entire aircraft could easily use orders of magnitude more grid points. Points are required not only to represent complicated geometries (as illustrated by some of the figures in this paper) but also to resolve rapidly varying features of the solution (shocks, boundary layers, vortex shedding).

The following grid generation methods will be discussed in more detail in the rest of this paper:

- Cartesian
- multi-block-structured
- overlapping
- unstructured.

The area of adaptive mesh refinement, a large and active field in itself, will only be briefly mentioned here. Further information can be found in many of the references. There are a number of issues that must be considered when evaluating the appropriateness of a given type of grid or grid generator:

- the speed of generating a grid
- the robustness of grid generation
- the quality of the generated grid
- the ability to construct grids from standard computer-aided-design specifications
- the level to which the grid generation is automatic – how much user intervention is required and how many tuning parameters are there?
- the support for adaptivity and moving geometries – can grids be regenerated quickly?
- the speed of the solver on the resulting grid
- the effectiveness of the approach on both parallel and serial architectures.

Generally, unstructured grid generators tend to be more robust and automatic, while structured grid generators create higher quality grids for which faster and more efficient solvers can be written. Further remarks on these issues will be made when the different approaches are discussed.

The field of grid generation is expanding rapidly. Many excellent references have been unavoidably omitted from this review and apologies are due to the authors. For further information, the reader is referred to the books by Thompson, Warsi and Mastin (1985), George (1991), Knupp and Steinberg (1993), and Castillo (1991); the conference proceedings edited by Weatherhill et al. (1994), Arcilla, Häuser, Eiseman and Thompson (1991), and Babuška, Flaherty, Henshaw, Hopcroft, Oliger and Tezduyar (1995); and the review papers by Löhner (1987), and Eiseman (1985). Some other excellent sources of information are Robert Schneiders' *Finite Element Mesh Generation* site on the World Wide Web:

`http://www-users.informatik.rwth-aachen.de/~roberts/meshgeneration.html`

and Steven Owen's *Meshing Research Corner* site:

`http://www.ce.cmu.edu:8000/user/sowen/www/mesh.html`

These sites include information about both unstructured and structured mesh generation, and pointers to a variety of public-domain and commercial grid generation packages.

2. Basic steps in grid generation

There are some basic steps in constructing a grid that are common to many of the grid generation approaches.

- As a first step in the grid generation process, the geometry of the region to be discretized must be defined, that is, the surfaces that make up the boundary of the region must be described. The geometry can be represented in many ways, such as with analytic shapes (spheres, cylinders), splines, NURBS (non-uniform rational b-splines), and interpolation methods. The geometry may be constructed within a computer-aided-design (CAD) system or within the grid generation system itself. Many CAD systems emphasize solid modelling using analytic shapes and do not cater particularly well to the creation of grids for flow problems. As a result, many grid-generation packages provide some level of CAD support.

- Given the representation of the surface (as a NURB, for example), it is often necessary to reparameterize the surface. This step is referred to as *constructing a surface grid*. (The Cartesian grid approach would not require this step.) Given a smooth surface, the most widely used CAD representations of this surface are only guaranteed to be *geometrically* smooth – they are often not parametrically smooth. Thus, if grid lines are drawn on the surface, equally spaced in parameter space, the lines will not vary smoothly. Typically the parametric derivatives of the surface will not even be continuous. By relaxing the requirements of parametric smoothness, it is easier for the CAD system to represent the surface, but unfortunately such a representation causes major difficulties for the grid generation system. Furthermore, CAD programs often represent complicated surfaces by multiple patches and these patches may not join properly (there may be gaps between patches, or the patches may overlap). Grid generators must carefully examine the surfaces and fix such defects. This is a difficult task and one that in principle should not be necessary.

  Grid generators would like to have parametrically smooth surfaces so that the grid points vary smoothly over the surface. The smoothing of the surface parameterization typically involves solving an elliptic-like equation on the surface or, in the case of triangles, shifting vertices according to some averaging procedure. This step will also involve clustering of grid points, such as in regions of high curvature. Surface grid generation techniques are usually quite similar to volume grid generation methods.

- The third step is the generation of a volume grid. The procedure followed at this stage differs significantly between the various grid types, and will be described in the following sections.

3. Cartesian grid generation

Lately, there has been renewed interest in the Cartesian grid approach, due to its simplicity and ease of automatic grid generation. By combining the Cartesian approach with adaptive mesh refinement, several of the drawbacks

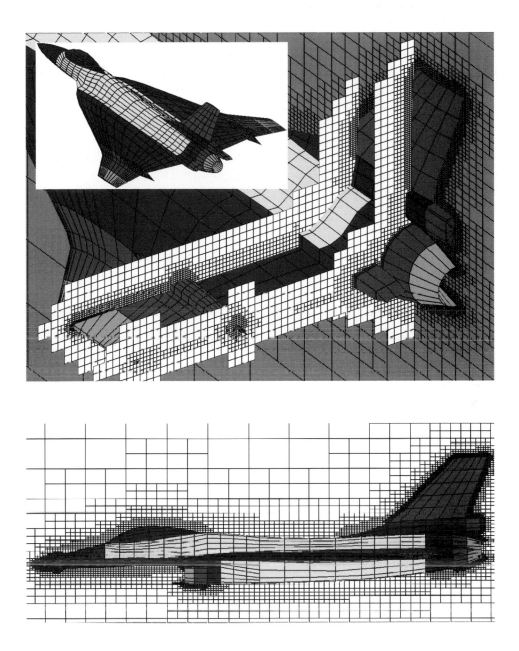

Fig. 5. Cartesian grid for the F16XL.

of the technique have been eased; see, for example, Berger and Melton (1994), and Coirier and Powell (1995). In the Cartesian grid approach, the region is covered by a rectangular grid. Domain boundaries cut out regions of the grid. The boundaries are not covered by boundary-fitted grids, but adaptive refinement can be used to improve surface resolution. Adaptively refined Cartesian grids combine elements of structured and unstructured grids and are perhaps best classified as hybrid grids. Cartesian grid solvers are faster and more efficient than more general unstructured solvers. Since the grids are all rectangular, much less geometrical information needs to be saved and there are significantly fewer operations required per grid point. Fig. 5 shows a Cartesian grid for an F16XL (Berger and Melton 1994).

The main drawback of the Cartesian-grid method lies in the representation of the boundary, where small cells are often formed. Without special treatment these small cells would force the time-step of a time-dependent solver to become prohibitively small. Typical applications only solve problems without boundary layers (Euler equations, for example, as opposed to Navier–Stokes equations). Since the boundary is not aligned with a grid line, in order to resolve a boundary layer it is necessary to refine the grid in two directions in two dimensions and three directions in three dimensions. In contrast, a three-dimensional boundary-fitted grid need only refine the grid in the direction normal to the boundary. This can be an important consideration, since the boundary layer grid spacing can be more than 10^3 times smaller than the spacing away from the boundary.

4. Multi-block-structured grid generation

In the multi-block-structured grid approach, the computational volume is divided into a set of non-overlapping logically rectangular *blocks*. A volume grid is created on each block; see Thompson (1988) and Spekreijse (1995). Usually, global smoothing is performed on the blocks to achieve some degree of continuity in the grid metrics at the block boundaries. Discontinuities in the grid spacing at block boundaries can result in poor solutions. Grid lines may or may not join across blocks; if not, the grid is sometimes called a *patched grid*. Patched grids require more general interpolation, but this can easily be made conservative.

The multi-block approach has been popular for many years for aerospace and other applications. It has improved flexibility over a single logically rectangular patch. High-quality grids can be created, and solvers are fast and efficient. Efficient numerical methods such as implicit methods and multigrid methods work well. Good-quality, highly stretched boundary-layer grids can be created. The main disadvantage with the method is that it is difficult to automate the decomposition of a region into non-overlapping blocks, especially in three dimensions. There is some difficulty with moving geometries

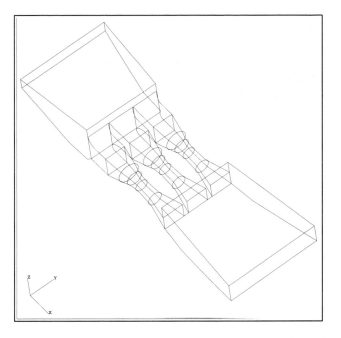

Fig. 6. Block decomposition of a hydroelectric power station.

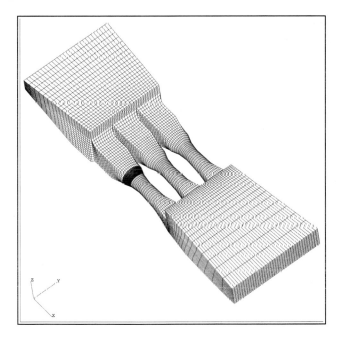

Fig. 7. Corresponding grid.

since the block decomposition may have to change. Generating a multi-block grid for a complicated three-dimensional region usually requires significant human intervention.

Figs 6 and 7 show a block structure grid for a hydroelectric power station (Spekreijse, Boerstoel, Vitagliano and Kuyvenhoven 1992). Fig. 8, showing a multi-block-structured grid for the space shuttle, is reproduced courtesy of Steven Alter, Lockheed Engineering and Sciences Company.

4.1. Structured component grid generation approaches

One of the most important parts of structured grid generation (whether multi-block or overlapping) is the creation of the individual blocks. The blocks will generally have some or all bounding surfaces specified, and the aim is to create a smooth volume-filling grid with appropriate grid spacing and orthogonality. The most common techniques fall into the following categories:

- algebraic
- elliptic and variational
- hyperbolic.

Algebraic grid generation methods create grids for the interior of a domain by algebraically combining the representations of the boundary surfaces. The transfinite interpolation procedure uses polynomials to interpolate the interior grid from the boundaries; see Thompson et al. (1985). For example, a two-dimensional grid bounded by the two curves $\mathbf{C}_1(s)$ and $\mathbf{C}_2(s)$, can be created using the simple shearing transformation

$$\mathbf{G}(r,s) = r\mathbf{C}_1(s) + (1-r)\mathbf{C}_2(s).$$

Whether the grid is useful depends strongly on the shape and parameterization of the curves. Algebraic methods are not as flexible as some of the other methods but their simplicity and speed of generation makes them popular.

Elliptic generation methods, pioneered by Thompson and co-workers, can handle more general cases. They can be used to construct high-quality grids on rather complicated domains; see, for example, Thompson (1987), Sorenson (1986), and Spekreijse (1995). A Poisson equation is solved to determine the location of the grid points. This equation commonly takes the form (in two dimensions):

$$\frac{\partial^2 r_i}{\partial x_1^2} + \frac{\partial^2 r_i}{\partial x_2^2} = P_i, \quad i = 1,2,$$

where $\{r_i\}$ are the unit square coordinates, $\{x_i\}$ are the physical domain coordinates and $\{P_i\}$ are the *control functions*. In practice, these equations are transformed so that $\{r_i\}$ are the independent variables,

$$\sum_{\mu,\nu} g_{\mu,\nu} \frac{\partial^2 x_i}{\partial r_\mu r_\nu} + \sum_\mu \frac{\partial x_i}{\partial r_\mu} P_\mu = 0, \quad i = 1,2.$$

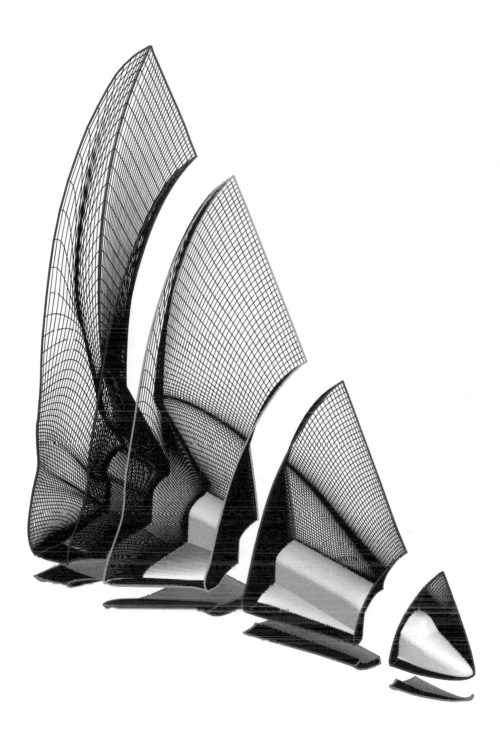

Fig. 8. Multi-block grid for the space shuttle.

Here,

$$g_{\mu,\nu} = \frac{\partial \mathbf{x}}{\partial r_\mu} \cdot \frac{\partial \mathbf{x}}{\partial r_\nu}$$

are the coefficients of the metric tensor. The equations are elliptic in nature and this means that the resulting grid has desirable smoothness properties. One of the keys to elliptic grid generation is the choice of control functions that determine the grid point spacing and grid orthogonality. The Poisson system that needs to be solved can be highly nonlinear and is difficult and time-consuming to solve. The solution to the system is generally not guaranteed to produce a single-valued grid, so care must be taken to prevent the grid from becoming multi-valued (folding).

The variational approach also produces an elliptic equation whose solution determines the locations of the grid points; see Brackbill and Saltzman (1982)[†] and Knupp and Steinberg (1993). The equations determining the grid point locations are derived by forming the Euler–Lagrange (variational) equations of a functional that measures properties of the grid such as orthogonality, cell area and smoothness. By weighting these different properties it is usually possible to obtain a grid with the desired features, although care must be taken to prevent folding grids.

Hyperbolic grid generation methods solve a hyperbolic set of equations to grow a grid from a boundary; see Starius (1977) and Chan and Steger (1992). Fig. 9 shows a grid generated in this way (Chan and Steger 1992).

Typically, the hyperbolic system is defined by requiring that the grid lines be orthogonal,

$$\frac{\partial \mathbf{x}}{\partial r_\mu} \cdot \frac{\partial \mathbf{x}}{\partial r_\nu} = 0, \quad \mu \neq \nu,$$

and that the cell area is specified

$$\left| \frac{\partial \mathbf{x}}{\partial r} \right| = \Delta.$$

Hyperbolic methods usually always add smoothing to prevent grid lines from crossing prematurely. The outer boundary of the grid is determined as the equations are solved, and thus this method is of limited use for block-structured grids. It is, however, an extremely useful technique in the context of overlapping grids. The method is much faster than an elliptic method since the grid is constructed by marching.

5. Overlapping grid generation

The overlapping (overlaid, overset or Chimera) grid approach is similar to the block-structured approach except that the component grids are allowed

[†] It doesn't hurt to cite your manager whenever possible.

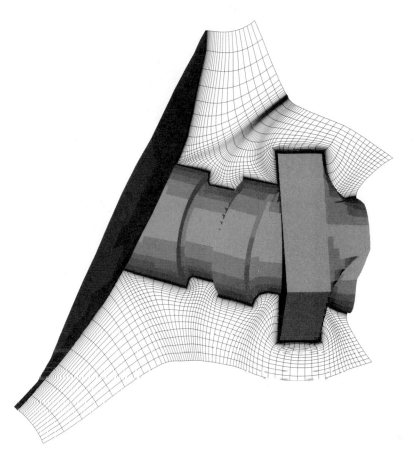

Fig. 9. Some sections of a three-dimensional grid for the liquid hydrogen feedline
of the space shuttle, created with hyperbolic grid generation methods.

to overlap, instead of aligning along block boundaries; see Steger and Benek
(1987), Chesshire and Henshaw (1990), Meakin (1995), Tu and Fuchs (1995).
This approach has added flexibility over the block-structured technique while
still retaining the efficiency of a set of logically rectangular grids. The great
strength of overlapping grids is that component grids can be created in a
manner that is relatively independent from the other component grids. New
features can be added to the composite grid in an incremental fashion and the
grid only changes locally. Fig. 10 shows part of a detailed overlapping grid
for the space shuttle (Gomez and Ma 1994). The method is also attractive
for moving geometries. Fig. 11 shows the overlapping grid used for a moving
grid computation (Meakin 1995).

Overlapping grids are not as flexible as unstructured grids. It is difficult
to get very many levels of coarser grids for a multigrid algorithm because the
coarsened grids do not overlap enough. Generally, the interpolation between

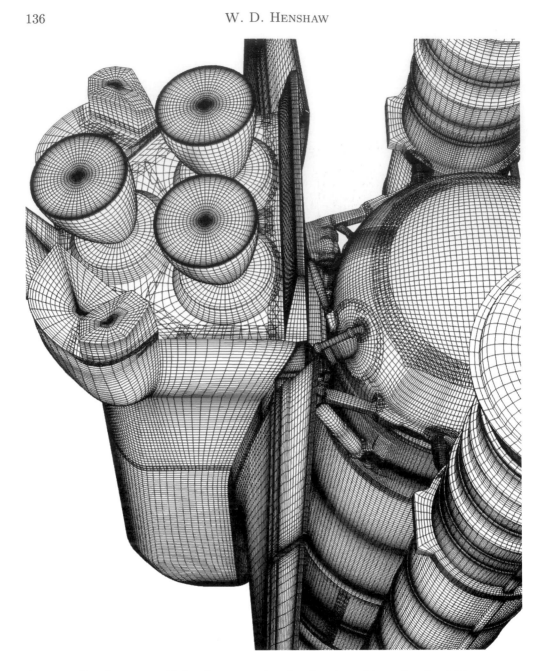

Fig. 10. Overlapping grid for the space shuttle; the three-dimensional grid has over 16 million grid points.

Fig. 11. Overlapping grid for the V-22 rotor and flapped wing, used in a moving grid computation.

component grids is not conservative (Chesshire and Henshaw 1994). In practice this rarely seems to be an issue. Generally, the grid generation proceeds in two steps. First, separate component grids are constructed for the various parts of the geometry, using algebraic, elliptic or hyperbolic methods. Then, given a set of component grids, the grid generation process of determining how the grids overlap can be entirely automatic. The process can fail, however, if there is insufficient overlap between components.

An approach similar to overlapping grids, but one that avoids using non-conservative interpolation, is the hybrid grid technique as shown in Fig. 12, reproduced courtesy of Dr. K.H. Kao at Nasa Lewis Research Center. The region is covered by overlapping blocks but the grid in the overlapping area is replaced by an unstructured grid of triangles.

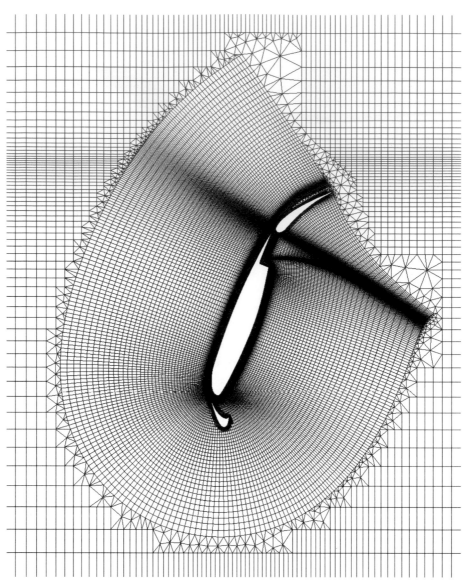

Fig. 12. A hybrid grid consisting of structured component grids joined with a region of triangles.

6. Unstructured grid generation

Unstructured grids have become very popular in recent years, due both to the influence of the finite-element method and to the increase in the power of computers. Unstructured grids and unstructured solvers have successfully demonstrated their capabilities to handle complex geometries in the demanding field of aerospace applications, an area dominated for many years by structured grids. The most flexible and automatic grid generation codes create unstructured grids. They are well suited to point-wise adaptive refinement and to moving mesh methods. See, for example, Shostko and Löhner (1995), Mavriplis (1995), Hasan, Probert, Morgan and Peraire (1995), George and Seveno (1994), Lo (1995), Johnson and Tezduyar (1995).

It is difficult to achieve good performance on unstructured grids; more memory is required and it is quite hard to apply certain fast algorithms such as implicit methods and multigrid. Attaining performance on vector, parallel and cache-based computer architectures is not easy for solvers using unstructured grids because these machines prefer that operations be performed on data that is stored locally in memory. On an unstructured grid, the data belonging to the neighbour of a point may be stored a long distance away. Moreover, triangular (and tetrahedral) meshes inherently require more elements and more computations per grid point; in three dimensions, there are some five to six times more tetrahedra per grid point than on a corresponding mesh of hexahedra. The creation of better-quality grids for hyperbolic problems and forming highly stretched elements in boundary layers continue to be active areas of research.

Fig. 13 shows a three-dimensional unstructured grid refined near the boundary, for use in a viscous flow computation. The figure has been provided by Professor Jaime Peraire.

Fig. 14, showing a cross-section of a three-dimensional grid for Yucca Mountain, is reproduced courtesy of Harold Trease, Los Alamos National Laboratory.

6.1. Un-structured grid generation approaches

Three popular methods for creating unstructured grids are

- Delaunay-based point insertion methods
- advancing front methods
- quadtree (octree) type methods.

Some of the most successful approaches use features of both the Delaunay method and the advancing front method, combining the efficiency of the former approach with the high element quality of the latter. Although quadrilateral (hexahedral) meshes are commonly used for structural problems, meshes for CFD tend to be based on triangles (tetrahedra), with perhaps

CONTENTS

PART I: THE HISTORY

1 The shape of the strongest column   150
2 Optimal partitioning of graphs   153
3 Multiple eigenvalues, optimality conditions, and
   algorithms   154
4 Interior-point   methods   and   polynomial-time
   algorithms   156
5 Polynomial-time approximations to NP-hard graph
   problems   157
6 Linear matrix inequalities in system and control
   theory   158
7 Non-Lipschitz eigenvalue optimization   158

PART II: THE MATHEMATICS

8 Conjugacy   160
9 Invariant norms   164
10 Functions of eigenvalues   166
11 Linear programming   170
12 Semidefinite programming   172
13 Strict complementarity and nondegeneracy   174
14 Primal-dual interior-point methods   176
15 Nonlinear semidefinite programming and eigenvalue optimization   180
16 Eigenvalues of nonsymmetric matrices   180
References   186

PART I: THE HISTORY

1. The shape of the strongest column

In 1773, Lagrange posed the following problem: determine the shape of
the strongest axially symmetric column with prescribed length, volume and
boundary conditions. The mathematical statement of this problem relies on
earlier work of J. Bernoulli and Euler. The latter, in 1744, established the
buckling load of such a column as the least eigenvalue of a self-adjoint fourth-
order differential operator. Consequently, Lagrange's problem requires the

maximization of this least eigenvalue, over all possible functions defining the cross-sectional area of the column.

Lagrange's problem, so easily stated, proved extraordinarily resistant to many attempts at its solution. Many authors made substantial contributions as well as serious errors. Lagrange must have the credit for posing the problem, yet several errors led to his incorrect conclusion that it is solved by the uniform column. The first to offer a correct solution was Clausen in 1851, in the case of clamped-free boundary conditions. The solution has the cigar shape shown in Fig. 1(a), where the cross-sectional area of the column is plotted as a function of its length. Clausen's paper is known primarily through later work of Pearson, who introduced many errors in an attempt to simplify the results.

Lagrange's problem then lay dormant for a century before it was taken up in a modern treatment by J. Keller in 1960. Keller established the solution, shown in Fig. 1(b), in the case of hinged-hinged boundary conditions. Then Tadjbakhsh and Keller (1962) offered solutions in the case of clamped-hinged and clamped-clamped boundary conditions. These are shown in Figs 1 (c) and (d) respectively. A conspicuous feature in both cases is the vanishing of the cross-sectional area at an internal point.

These solutions went unchallenged for fifteen years. Then Olhoff and Rasmussen (1977) claimed that the Tadjbaksh–Keller (TK) clamped-clamped solution was incorrect, because its solution procedure implicitly assumed that the least eigenvalue associated with the optimal solution is simple (that is, has multiplicity one). The solution offered by Olhoff and Rasmussen (OR), displayed in Fig. 1 (e), has a double least eigenvalue. However, no proof of the validity of this column was offered, nor were details of their numerical approximation procedure. Consequently, the issue remained quite controversial, with some authors defending the TK solution, and others, notably Masur and Seiranian, offering evidence for the OR solution. Recently, Cox and Overton (1992) gave the first proof of existence of a solution to the clamped-clamped problem, as well as the first proof that the OR solution indeed satisfies the Clarke (1983) first-order necessary conditions for optimality. In addition, Cox and Overton (1992) offered the first systematic numerical results using direct optimization techniques that take into account the possibility of a multiple eigenvalue. Both the theoretical contributions and the numerical techniques of Cox and Overton (1992) rely on the theory of convex analysis and its generalizations due to Rockafellar (1970) and Clarke (1983).

However, following in the footsteps of their illustrious predecessors in more ways than one, Cox and Overton also introduced a new error, claiming in an appendix that the TK clamped-hinged solution was also incorrect. Rather than believing their own numerical evidence, albeit uncertain given the vanishing of the cross-sectional area at an internal point and the corresponding absence of an existence proof (Cox and Overton 1992, p. 315), they placed

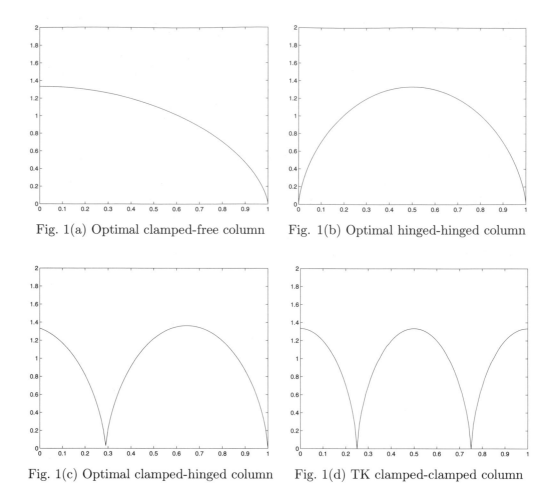

Fig. 1(a) Optimal clamped-free column   Fig. 1(b) Optimal hinged-hinged column

Fig. 1(c) Optimal clamped-hinged column   Fig. 1(d) TK clamped-clamped column

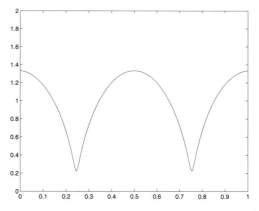

Fig. 1(e) Optimal clamped-clamped column

faith in a mathematical proof that contained a simple scaling error: the irony will doubtless be enjoyed by the readers of this journal. That the TK clamped-hinged solution is indeed correct (though the TK clamped-clamped solution is not) has now been established beyond doubt by Cox and Maddocks (1996). For more details, see the article by Cox in *Math. Intelligencer* (Cox 1992), accompanied by illustrations of the strongest columns on the cover, and also the follow-up discussion (Kirmser and Hu 1993, Cox 1993).

2. Optimal partitioning of graphs

Our next example of eigenvalue optimization could not be more different in character to the strongest column problem. Consider a nonnegative edge-weighting of the complete (undirected) graph on the vertex set $\{1, 2, \ldots, n\}$. We can associate any such weighting with an $n \times n$ symmetric matrix W with diagonal entries all zero and off-diagonal entries all nonnegative: entry W_{ij} is just the weight on the edge (i, j). Given integers $d_1 \geq d_2 \geq \cdots \geq d_k > 0$, with sum n, consider the problem of partitioning the vertex set into k subsets such that the ith subset contains exactly d_i vertices and the sum of weights of edges between subsets is minimized. Equivalently, the sum of the weights of edges whose endpoints are both inside the same subset is to be maximized. This problem is NP-hard. However, Donath and Hoffman (1973) suggested the clever idea of deriving bounds on the solution by means of eigenvalue optimization. (For other approaches to graph partitioning that exploit eigenvalues, though not necessarily eigenvalue optimization, see the early work of Fiedler (1973) and the recent survey paper of Pothen (1996).)

Denote the characteristic (column) vector for the ith subset by $x^i \in \mathbb{R}^n$: thus x_r^i is 1 if vertex r is in subset i, and is 0 otherwise. Let X be the $n \times k$ matrix $[x^1, x^2, \ldots, x^k]$. Then, by construction, $X^T X = \text{Diag}(d_1, d_2 \ldots, d_k)$, and we seek to maximize $\sum W_{ij} (XX^T)_{ij}$, or equivalently, the trace of $W X X^T$. Since for any matrices A and B we have $\text{tr } AB = \text{tr } BA$, we can write the partitioning problem as:

$$\max_{X \in \mathbb{R}^{n \times k}} \text{tr } X^T W X$$

$$\text{subject to } X^T X = \text{Diag}(d) \text{ and } X_{ij} \in \{0, 1\}. \tag{2.1}$$

Now let us replace the variable matrix X by making the normalized definitions $y^i = d_i^{1/2} x^i$ and $Y = [y^1, y^2, \ldots, y^k]^T$. With this change of variable, the optimization problem becomes

$$\max_{Y \in \mathbb{R}^{n \times k}} \sum_{i=1}^{k} d_i (y^i)^T W y^i$$

$$\text{subject to } Y^T Y = I \text{ and } \sqrt{d_i d_j} Y_{ij} \in \{0, 1\}. \tag{2.2}$$

The last constraint is the integrality constraint, which makes the problem difficult. So, let us *relax* the problem by dropping this constraint. As we shall see (in Lemma 10.2), the relaxed problem is solved by taking the columns of Y to be an orthonormal set of eigenvectors for the largest k eigenvalues of W: each y^i should be an eigenvector corresponding to $\lambda_i(W)$, the ith largest eigenvalue of W (counting multiplicities). The ordering is important since, by assumption, the d_i are given in descending order. Because the relaxation was obtained by dropping the integer constraint, the quantity $\sum_{i=1}^{k} d_i \lambda_i(W)$ is an upper bound for the optimal value of the problem (2.1).

Now we come to the key point: *a tighter upper bound can be obtained using eigenvalue optimization.* The diagonal elements of XX^T are all one, so we can replace the objective function of problem (2.1) by the trace of $(W+D)XX^T$ for any diagonal matrix D with zero trace. Equivalently, after the change of variables, we replace W by $W+D$ in the objective function of (2.2). Different choices of D give different relaxations when the integer constraint is dropped, and therefore different upper bounds. Thus D can be chosen to improve the upper bound, by *minimizing the weighted sum of the largest eigenvalues* of $W + D$, that is

$$G(D) = \sum_{i=1}^{k} d_i \lambda_i(W + D),$$

over all diagonal matrices D with zero trace.

Donath and Hoffman reasoned that since the function G is convex (as we shall see in Section 10), the task of minimizing G should be tractable. This turned out to be a more mathematically interesting and challenging problem than they anticipated at the time, as we shall now discuss.

3. Multiple eigenvalues, optimality conditions, and algorithms

Multiple eigenvalues had not been expected in the problem of Lagrange because, in all but the clamped-clamped case, the structure of the differential operator makes it impossible for the least eigenvalue, say $\hat{\lambda}_1$, to have multiplicity greater than one. If one considers more general eigenvalue optimization problems, however, it is clear that maximizing a least eigenvalue (equivalently minimizing a greatest eigenvalue) will potentially lead to coalescence of eigenvalues. Of course, *minimizing* a least eigenvalue has the opposite effect. The latter occurs, for example, in Rayleigh's problem of finding the shape of the two-dimensional drum with the least natural frequency. Mathematically, this means finding the shape of the domain that minimizes $\hat{\lambda}_1$, the least eigenvalue of the Laplacian. The least eigenvalue is necessarily simple, and the solution is a circle. An interesting variation is to find the shape that minimizes the ratio $\hat{\lambda}_1/\hat{\lambda}_2$. This was considered by Payne, Pólya and Weinberger (1956): they conjectured that the solution is a circle, but this was proved only recently (Ashbaugh and Benguria 1991). In this case, a double eigenvalue plays a role, because $\hat{\lambda}_2$ and $\hat{\lambda}_3$ coalesce at the solution. Eigenvalue optimization problems for plates (modelled by fourth-order differential operators in two dimensions) are also of interest, but these have received relatively little attention. All of these problems are difficult because they are infinite-dimensional and the operators depend on the variables in a complicated way. For the remainder of this article we confine our attention to matrix problems with linear dependence on the variables.

The Donath–Hoffman approach to graph partitioning requires minimizing a weighted sum of the largest eigenvalues of a matrix, the variables being simply the diagonal elements. This work led to a paper of Cullum, Donath and Wolfe (1975) that is remarkable for two significant contributions. The first was the development of an optimality condition using convex analysis, emphasizing the issue of multiple eigenvalues. Specifically, the authors recognized and addressed the fact that the sum-of-eigenvalues function, although convex, is not a differentiable function at points where the eigenvalues coalesce. The second contribution of Cullum et al. (1975) was the development of a convergent algorithm to find a minimizer. The significance of this work was not appreciated for some ten years or so. Then Fletcher (1985) revived interest in the problem, inspiring further analytical improvements by Overton and Womersley (1993) and Hirriart-Urruty and Ye (1995). These results are now largely subsumed by a more general but concise approach due to Lewis (1996a), presented in Part II of this survey. Specifically, rather general composite functions of the form $h \circ \lambda$ are considered, where λ is the eigenvalue map from symmetric matrix space to $\mathbb{R}^n$, and h is *any* convex function that is symmetric with respect to its arguments. A duality theory for this class of functions will be given in some detail, building on the fundamental results of convex analysis due to Rockafellar as well as key matrix theoretic results of von Neumann and others. Composite eigenvalue optimization includes semidefinite programming (SDP), a generalization of linear programming that has received much attention in the last few years.

The SDP problem is to minimize a linear function of a symmetric matrix variable subject to linear and positive semidefinite constraints on the matrix. Typically, SDPs have solutions with multiple zero eigenvalues. Semidefinite constraints have been considered in many contexts; two early papers are Bellman and Fan (1963) and Craven and Mond (1981). In fact, SDP was the variant of eigenvalue optimization that was primarily addressed by Fletcher (1985), introducing a new algorithmic approach and emphasizing the issues of multiple eigenvalues and quadratic convergence. This led to the computational work on minimizing a maximum eigenvalue due to Overton (1988, 1992) and the associated second-order convergence analysis (a complicated issue in the presence of multiple eigenvalues) given by Overton and Womersley (1995) and Shapiro and Fan (1995). However, since many eigenvalue optimization problems can be rephrased as equivalent SDPs, this work has now been largely overshadowed by the sudden advance of interior-point methods for SDP, to which we now turn.

4. Interior-point methods and polynomial-time algorithms

Linear programming (LP) was established as a discipline in the 1940s by Dantzig. The LP problem is to minimize a linear function subject to linear equality and inequality constraints on the variables, a problem which, remarkably, had largely escaped earlier attention, with the exception of some work on systems of linear inequalities by Fourier and Motzkin. As well as introducing the problem class, Dantzig gave an algorithm for solving LPs: the simplex method. Duality played a key role from the beginning, originating in a famous conversation between Dantzig and von Neumann at Princeton in 1947; see Dantzig (1991). The highly efficient simplex method went essentially unchallenged for 30 years, although it was known that, in the worst case, it required computation time exponential in the problem size. In 1979 Khaciyan showed that the ellipsoid method of Nemirovskii and Shor could be used to guarantee the solution of LPs in polynomial time. The ellipsoid method proved to be impractical, but it inspired the work of Karmarkar (1984), which established the interior-point framework as a practical, polynomial-time approach to solving LP. In the 10 years since, a profusion of interior-point methods for LP have been proposed, implemented and theoretically analysed; see the surveys by Lustig, Marsten and Shanno (1994), Gonzaga (1992) and Wright (1992). It is now generally accepted that the primal-dual interior-point method due to Monteiro and Adler (1989) and Kojima, Mizuno and Yoshise (1989) has substantial theoretical and practical advantages over the other interior-point methods, including Karmarkar's method.

As we already noted, the difference between LP and SDP is that, in the latter case, the variable is a symmetric matrix and the inequality constraint is a semidefinite matrix constraint. In the case that the matrix is constrained to be diagonal, SDP reduces to LP. There is no simplex method for SDP, because the feasible region is not polyhedral. In the late 1980s, Nesterov and Nemirovskii extended many of the interior-point methods and theoretical results from LP to a much broader class of convex programming problems, including SDP; see Nesterov and Nemirovskii (1994). Alizadeh (1991, 1995) and Karmarkar and Thakur (1992) also independently proposed such a generalization for SDP, a key component being the 'log determinant' barrier function. In the last three years there has been a burst of activity in the development of interior-point methods for SDP. Some of the most recent work, namely the derivation of a primal-dual interior-point method for SDP, will be discussed in Section 14. See Vandenberghe and Boyd (1996) for a survey article on SDP, including many applications not discussed here.

We now briefly discuss two important application areas that have successfully exploited the success of interior-point methods for SDP.

5. Polynomial-time approximations to NP-hard graph problems

The availability of polynomial-time algorithms for semidefinite programming has led to great interest by the combinatorial optimization community in provably good polynomial-time approximations to NP-hard problems. We consider one example.

As in Section 2, consider the complete graph with vertex set $\{1, 2, \ldots, n\}$ and edges (i, j) with associated nonnegative weights W_{ij}. The max-cut problem is to divide the vertices into two sets, V_1 and V_2, such that the weighted sum of edges crossing from one set to the other is maximized. This is *not* the same as the graph partitioning problem with $k = 2$ since the number of vertices in each set is not preassigned. The max-cut problem is NP-hard, although the min cut (max-flow) problem can be solved by standard fast algorithms. (The min-cut problem is trivial if one does not specify that V_1 and V_2 must be nonempty). The max-cut problem can be expressed as

$$\max_{x_1, x_2, \ldots, x_n \in \mathbb{R}} \left\{ \sum_{1 \leq i \leq j \leq n} W_{ij}(1 - x_i x_j) : |x_i| = 1 \text{ for all } i \right\}, \qquad (5.1)$$

where we adopt the convention that $x_i = 1$ means $i \in V_1$ and $x_i = -1$ means $i \in V_2$. Now consider the modified problem

$$\max_{x^1, x^2, \ldots, x^n \in \mathbb{R}^n} \left\{ \sum_{1 \leq i \leq j \leq n} W_{ij}(1 - (x^i)^T x^j) : \|x^i\| = 1 \text{ for all } i \right\}, \qquad (5.2)$$

where $\| \cdot \|$ denotes the Euclidean norm. If the vectors $x^1, \ldots, x^n$ solving problem (5.2) all happen to be parallel, then they can be associated with the scalar solutions $x_i = \pm 1$ to problem (5.1), and the max-cut problem is solved. Of course, this is very unlikely to occur. However, given any fixed optimal solution of problem (5.2), we can generate a cut for the graph by cutting the unit ball in half, and then assigning vertex i to set V_1 or V_2 according to which half of the ball contains the vector x^i. Goemans and Williamson (1996) recently established the surprising fact that, if one makes the division of the unit ball in the appropriate way, the resulting cut in the graph is an approximate solution of the max-cut problem with an objective value within a factor of 1.14 of the optimal value. Notice that problem (5.2) is equivalent to the SDP

$$\min_{Z \succeq 0}\{\operatorname{tr} WZ : Z_{ii} = 1, \text{ for each } i\},$$

the variable Z being a symmetric matrix associated with the vectors x^i by the equation $Z = X^T X$, where X is the matrix $[x^1, x^2, \ldots, x^n]$, and where $Z \succeq 0$ denotes the semidefinite constraint.

To summarize, the max-cut problem, which is NP-hard, is provably solvable within a factor of 1.14 in polynomial time, via the solution of a semidefinite program. For more on the max-cut problem, see the survey by Poljak and Tuza (1993) and the recent thesis of Helmberg (1994). For other applications of SDP and eigenvalue optimization to combinatorial optimization, see Grötschel, Lovász and Schriver (1988, Chapter 9), Mohar and Poljak (1993) and Rendl and Wolkowicz (1992).

6. Linear matrix inequalities in system and control theory

The title of this section is also the title of a recent book (Boyd, Ghaoui, Feron and Balakrishnan 1994). A *linear matrix inequality* (LMI) is generally understood to mean a positive semidefinite or definite constraint on a matrix depending affinely on parameters: as such, an LMI is simply the constraint of an SDP. However, the term is also sometimes used to describe more general matrix inequality constraints, especially bounds on the eigenvalues of a pencil (those scalars λ satisfying $\det(A - \lambda B) = 0$, where the matrices A and B are symmetric and depend affinely on parameters, and B is positive definite). The application of LMIs to control theory has its origins in the work of Lyapunov in the 1890s and Yakubovitch in the 1960s.

The impact of LMIs on system and control theory is hard to overstate: it is fair to say that the field has been revolutionized by the realization that optimization problems with LMI constraints can be effectively solved using interior-point methods. We give no further details here since the relevant material is available in Boyd et al. (1994).

7. Non-Lipschitz eigenvalue optimization

Up to this point we have discussed eigenvalue optimization for symmetric matrices and self-adjoint operators, which have real eigenvalues and orthonormal sets of eigenvectors. Eigenvalues of nonsymmetric matrices and operators also play many roles in applied mathematics, though it is well known that their potential sensitivity to perturbation requires caution. Stability issues arise in many applications, with instability generally associated with eigenvalues whose real parts are nonnegative. Indeed, the widespread use of symmetric linear matrix inequalities in system and control theory is, in part, motivated by stability issues for nonsymmetric matrices, via Lyapunov theory and its generalizations. It is therefore natural to consider direct application of

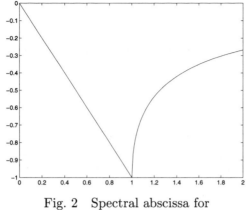

Fig. 2   Spectral abscissa for
the damped linear oscillator

optimization theory to functions of eigenvalues of nonsymmetric matrices. However, this is quite complicated because of the non-Lipschitz behaviour of the eigenvalues.

The damped linear oscillator provides a simple example of eigenvalue optimization in the nonsymmetric case. Consider the ordinary differential equation, for a given real b,

$$y''(t) + 2by'(t) + y(t) = 0. \tag{7.1}$$

Noting that the vector $z(t) = [y(t) \ \ y'(t)]^T$ satisfies the first-order system $z'(t) = A(b)z(t)$ where $A(b) = \begin{bmatrix} 0 & 1 \\ -1 & -2b \end{bmatrix}$, the initial value problem may be solved in terms of the eigenvalues and eigenvectors of $A(b)$. The effectiveness of the damping is measured by the spectral abscissa of $A(b)$ (that is, the largest real part of the eigenvalues of $A(b)$): we denote this function by $\alpha(b)$. Now $\alpha(b) = -b + \mathrm{Re}\sqrt{b^2 - 1}$, so, since the spectral abscissa achieves its minimum at $b = 1$, equation (7.1) is said to be over(under)damped if $b > 1$ ($b < 1$), and critically damped if $b = 1$. The function $\alpha(b)$ is plotted in Fig. 2. Note that α is not a Lipschitz function of b. The sharply different behaviour of the function α on the two sides of the minimizer occurs because, on one side, a double eigenvalue splits into a real pair, while on the other side, it splits into a complex conjugate pair. In both cases the changes in the eigenvalues are non-Lipschitz, but only in the former case do the real parts have non-Lipschitz behaviour. The optimal damping factor $b = 1$ yields a matrix $A(b)$ with an eigenvalue having algebraic multiplicity two, but geometric multiplicity one, and thus with a nontrivial Jordan block.

A similar phenomenon is well known from the analysis of the successive overrelaxation (SOR) iterative method for solving systems of linear equations;

see Ortega (1972). The critical value of the overrelaxation parameter is determined by an eigenvalue optimization problem in one variable. Over- and underrelaxation are well known to have very different consequences, again because of the presence of a nontrivial Jordan block at the minimizing point.

Of course, non-Lipschitz eigenvalue optimization problems also arise in more than one variable. Cox and Overton (1996) treat a generalization of the damped linear oscillator, namely the damped wave equation. Ringertz (1996) considers applications to stability issues for aircraft design.

Optimality conditions for non-Lipschitz eigenvalue optimization are rather complicated and beyond the scope of this article. For the present state of the art, see Burke and Overton (1994) and Overton and Womersley (1988).

Indeed, consider the following, far simpler question. Suppose A is a nonsymmetric matrix with multiple eigenvalues, and consider the eigenvalues of the perturbed matrix $A + \epsilon B$, where the matrix B is arbitrary and ϵ is a scalar perturbation parameter. How can we quantify the leading terms of the expansions of these eigenvalues in fractional powers of ϵ? When A has nontrivial Jordan structure, the behaviour of the eigenvalues under perturbation is quite complicated. Apparently, the only book that addresses this issue is Baumgärtel (1985), building on results of Lidskii and others published in the Russian literature in the 1960s, but remaining largely unknown in the West. See Moro, Burke and Overton (1995) for discussion of Lidskii's results and connections with the classical Newton diagram.

In the final section of this article we derive some apparently new variational results for functions of eigenvalues of nonsymmetric matrices. One special case amounts to a characterization of the spectral abscissa as the optimal value of a *symmetric* matrix eigenvalue optimization problem, a result well known to control theorists and one which may be viewed as a quantitative version of Lyapunov theory. Another special case implies the well known result that the spectral radius may be characterized as the infimum of all submultiplicative matrix norms. These results suggest a possible approach to non-Lipschitz eigenvalue optimization by means of symmetric eigenvalue optimization.

PART II: THE MATHEMATICS

8. Conjugacy

Convex analysis is an elegant and powerful tool for studying duality in optimization. Particularly for linearly constrained problems, it provides a concise and flexible framework. We begin by summarizing the relevant ideas.

Let E be a Euclidean space, by which we mean a finite-dimensional, real inner-product space. We could, of course, always identify E with $\mathbb{R}^n$, but a less concrete approach helps our future development. We call a real function f on E a *prenorm* if it is continuous, and satisfies

- *homogeneity:* $f(\alpha x) = |\alpha| f(x)$ for all real α and points x in E
- *positivity:* $f(x) > 0$ for all nonzero points x in E.

A norm is then just a prenorm satisfying the triangle inequality. For a prenorm f, we can define a real function f^D on E by

$$f^D(y) = \max\{\langle x, y \rangle : f(x) = 1\}.$$

The function f^D is actually a norm: we call it the *dual norm* of f.

Theorem 8.1. (von Neumann, 1937) A prenorm f is a norm if and only if $f = f^{DD}$.

The reader may consult Horn and Johnson (1985) for these ideas.

In optimization it is very convenient to consider *extended-real* functions $f : E \to [-\infty, +\infty]$. We call such a function *convex* (respectively *closed, polyhedral*) if its *epigraph* $\{(x, r) \in E \times \mathbb{R} : f(x) \le r\}$, is a convex (respectively closed, polyhedral) set. The *domain* of f is the set

$$\mathrm{dom}\, f = \{x \in E : f(x) < +\infty\};$$

if this set is nonempty and if f never takes the value $-\infty$, then f is called *proper*. For any extended-real function f we can define an extended-real function f^* on E by

$$f^*(y) = \sup\{\langle x, y \rangle - f(x) : x \in E\}.$$

The function f^* is always closed and convex: we call it the *(Fenchel) conjugate* of f. The basic reference for these and later convex-analytic ideas is Rockafellar (1970). Our definition of a closed function is slightly different from that of Rockafellar (1970): the definitions coincide for proper functions.

Theorem 8.2. (Fenchel–Hörmander, 1949) Suppose the extended-real function f is proper. Then f is closed and convex if and only if $f = f^{**}$. In this case, f^* is also proper.

The ideas of dual norms and conjugate functions are closely related: if f is a norm then a short calculation shows

$$(f^2/2)^* = (f^D)^2/2. \tag{8.1}$$

The first-order behaviour of a function $f : E \to (-\infty, +\infty]$ at a point x in its domain is fundamental to any study of optimality conditions and algorithms. For convex f this behaviour is encapsulated in the *subdifferential*

$$\partial f(x) = \{y \in E : \langle y, z - x \rangle \le f(z) - f(x) \text{ for all } z \text{ in } E\}.$$

Specifically, the directional derivative of f at x in a direction $w \in E$ is given by the formula

$$f'(x; w) = \sup\{\langle w, y\rangle : y \in \partial f(x)\}.$$

In particular, f is differentiable at x exactly when its subdifferential there is a singleton ($\partial f(x) = \{\nabla f(x)\}$). If $f(x)$ is infinite, we define $\partial f(x) = \emptyset$.

Immediately from its definition, we can relate the subdifferential to the conjugate:

$$y \in \partial f(x) \quad \Leftrightarrow \quad f(x) + f^*(y) = \langle x, y\rangle. \qquad (8.2)$$

Using the Fenchel–Hörmander Theorem (Theorem 8.2), we deduce that for a proper closed convex function f, the subdifferential map can be 'inverted':

$$y \in \partial f(x) \quad \Leftrightarrow \quad x \in \partial f^*(y). \qquad (8.3)$$

Example 8.3. (cones)   One benefit of convex analysis is the possibility of studying a subset K of E through its *indicator function*

$$\delta_K(x) = \begin{cases} 0 & \text{if } x \in K, \\ +\infty & \text{otherwise.} \end{cases}$$

This function is convex (closed) exactly when K is convex (closed). Suppose K is a *cone*: that is, $\mathbb{R}_+ K = K$. Then we deduce immediately that the function δ_K^* is just δ_{K^-}, the indicator function of the *polar* cone

$$K^- = \{y \in E : \langle x, y\rangle \leq 0 \text{ for all } x \text{ in } K\}.$$

The Fenchel–Hörmander Theorem (Theorem 8.2) then shows that a cone K is closed and convex exactly when $K^{--} = K$. From the subdifferential property (8.2) we deduce the 'complementarity' condition

$$y \in \partial\delta_K(x) \quad \Leftrightarrow \quad x \in K, \ y \in K^-, \text{ and } \langle x, y\rangle = 0. \qquad (8.4)$$

In particular, if the space E is $\mathbb{R}^n$ and the cone K is the nonnegative orthant $\mathbb{R}^n_+$, then the polar K^- is $-\mathbb{R}^n_+$, and for vectors x and y in $\mathbb{R}^n_+$ we deduce

$$y \in \partial\delta_{\mathbb{R}^n_+}(x) \quad \Leftrightarrow \quad x_j \geq 0, \ y_j \leq 0, \text{ and } x_j \text{ or } y_j = 0 \text{ for each } j. \qquad (8.5)$$

When f is a norm, the subdifferential property (8.2) has a simple analogue. For nonzero points x in E, an easy calculation shows

$$y \in \partial f(x) \quad \Leftrightarrow \quad f(x) = \langle x, y\rangle \text{ and } f^D(y) = 1, \qquad (8.6)$$

while $\partial f(0) = \{y \in E : f^D(y) \leq 1\}$.

The duality theory of linearly-constrained convex optimization is particularly transparent in this framework. We will always consider $\mathbb{R}^m$ as a Euclidean space of column vectors, with the standard inner product. Given a linear map $A : E \rightarrow \mathbb{R}^m$, we define the *adjoint* map $A^* : \mathbb{R}^m \rightarrow E$ by the property

$$y^T(Ax) = \langle A^*y, x\rangle \text{ for all points } x \text{ in } E \text{ and } y \text{ in } \mathbb{R}^m.$$

Suppose the function $f : E \to (-\infty, +\infty]$ is closed, convex and proper, fix a vector b in $\mathbb{R}^m$, and consider the pair of optimization problems,

$$\text{Primal:} \quad \rho \;=\; \inf\{f(x) : x \in E,\ Ax = b\};$$
$$\text{Dual:} \quad \delta \;=\; \sup\{y^T b - f^*(A^*y) : y \in \mathbb{R}^m\}.$$

The following result is derived from theory due to Rockafellar, dating from 1963 (Rockafellar 1970). We say the primal problem is *superconsistent* if there is a point $\hat{x}$ in int(dom f) satisfying $A\hat{x} = b$, and we say the dual problem is *superconsistent* if there is a point $\hat{y}$ in $\mathbb{R}^m$ with $A^*\hat{y}$ in int(dom f^*). By 'consistent' we mean the same properties with 'int' omitted.

Theorem 8.4. (Fenchel Duality)

(i) *Weak duality*: $\rho \geq \delta$.

(ii) *Dual attainment*: if the primal is superconsistent, then $\rho = \delta$, and δ is attained, if finite. Furthermore, if A is surjective, then, for any real α, the set

$$\{y \in \mathbb{R}^m : y^T b - f^*(A^*y) \geq \alpha\}$$

is compact.

(iii) *Primal attainment*: if the dual is superconsistent, then $\rho = \delta$, and ρ is attained if finite. Furthermore, for any real α, the set

$$\{x \in \mathbb{R}^n : f(x) \leq \alpha,\ Ax = b\}$$

is compact.

(iv) *Polyhedrality*: if f is polyhedral and either problem is consistent, then the other problem is attained, if finite, and $\rho = \delta$.

(v) *Complementary slackness*: suppose $\rho = \delta$. Then points $\bar{x}$ and $\bar{y}$ are optimal for the primal and the dual problems respectively, if and only if $A\bar{x} = b$ and $A^*\bar{y} \in \partial f(\bar{x})$.

The *complementary slackness* condition $A^*\bar{y} \in \partial f(\bar{x})$ is equivalent to $\bar{x} \in \partial f^*(A^*\bar{y})$, by the inversion formula (8.3). If in addition f^* is differentiable at $A^*\bar{y}$ then the primal solution $\bar{x}$ must therefore be $\nabla f^*(A^*y)$. In these circumstances we are thus able to recover a primal optimal solution by solving the dual problem.

A nice exercise is to apply the Fenchel Duality Theorem (Theorem 8.4) and Example 8.3 to the 'cone optimization problem'

$$\inf\{\langle c, x \rangle : Ax = b,\ x \in K\},$$

for a convex cone K and an element c of E. This model (*cf.* Nesterov and Nemirovskii 1994) subsumes both linear and semidefinite programming, which we discuss later.

9. Invariant norms

The theoretical foundations of eigenvalue optimization parallel the better-known theory of invariant matrix norms pioneered by von Neumann. A brief sketch of this theory's salient features is therefore illuminating. For clarity we consider only square, real matrices.

We consider the Euclidean space M_n of $n \times n$ real matrices, where the inner product is defined by $\langle X, Y \rangle = \operatorname{tr} X^T Y$. The singular values of a matrix X in M_n we denote $\sigma_1(X) \geq \sigma_2(X) \geq \cdots \geq \sigma_n(X)$. In this way we define the 'singular value map' $\sigma : M_n \to \mathbb{R}^n$.

We denote the groups of $n \times n$ permutation and orthogonal matrices by P_n and O_n respectively. We call a function f on $\mathbb{R}^n$ *symmetric* if, for any point x in $\mathbb{R}^n$ and any matrix Q in P_n, we have $f(Qx) = f(x)$. We say a norm ϕ on M_n is *(orthogonally) invariant* if, for any matrices X in M_n, and U and V in O_n, we have $\phi(UXV) = \phi(X)$.

For a vector x in $\mathbb{R}^n$, we denote the diagonal matrix with diagonal entries $x_1, x_2, \ldots, x_n$ by $\operatorname{Diag} x$. Clearly, for any invariant norm ϕ on M_n, the real function g on $\mathbb{R}^n$ defined by $g(x) = \phi(\operatorname{Diag} x)$ is a symmetric norm that is also *absolute*: $g((|x_1|, |x_2|, \ldots, |x_n|)^T) = g(x)$ for all vectors x in $\mathbb{R}^n$. Such norms are called *symmetric gauges*. The original norm ϕ is just the composite function $g \circ \sigma$. A beautiful result of von Neumann shows that this property characterizes invariant norms.

Theorem 9.1. (von Neumann, 1937) Invariant matrix norms are exactly those composite functions of the form $g \circ \sigma$, where g is a symmetric gauge.

For our purposes, almost more important than the result is the proof technique. Naturally, it relies heavily on the existence of an 'ordered singular value decomposition' for any matrix X:

$$X = U(\operatorname{Diag} \sigma(X))V \text{ for some orthogonal } U \text{ and } V.$$

If a second matrix Y satisfies $Y = U(\operatorname{Diag} \sigma(Y))V$, then we say X and Y have a *simultaneous ordered singular value decomposition*. Von Neumann's key step was the following result, of substantial independent interest.

Lemma 9.2. (von Neumann, 1937) Any $n \times n$ real matrices X and Y satisfy the inequality

$$\operatorname{tr} X^T Y \leq \sigma(X)^T \sigma(Y);$$

equality holds if and only if X and Y have a simultaneous ordered singular value decomposition.

Equipped with this (nontrivial) result, von Neumann's characterization (Theorem 9.1) follows from a beautifully transparent duality argument. For an absolute, symmetric prenorm g on $\mathbb{R}^n$, we first use Lemma 9.2 to deduce that the prenorm $g \circ \sigma$ satisfies

$$(g \circ \sigma)^D = g^D \circ \sigma. \tag{9.1}$$

Hence if g is actually a symmetric gauge, applying this formula twice and using Theorem 8.1, we deduce

$$(g \circ \sigma)^{DD} = (g^D \circ \sigma)^D = g^{DD} \circ \sigma = g \circ \sigma,$$

and, by Theorem 8.1, $g \circ \sigma$ must be a norm. The result is now easy to see.

Lemma 9.2 also greatly facilitates the calculation of subdifferentials. The following result, due to Ziętak (1993) (*cf.* Watson (1992)) follows immediately from the Lemma, the subdifferential characterization (8.6), and the duality formula (9.1) (*cf.* Lewis (1995a)).

Theorem 9.3. (Ziętak, 1994) If g is a symmetric gauge, then matrices X and Y satisfy $Y \in \partial(g \circ \sigma)(X)$ if and only if they have a simultaneous ordered singular value decomposition and satisfy $\sigma(Y) \in \partial g(\sigma(X))$.

Such techniques help reveal the intimate geometric connections between the two norms g and $g \circ \sigma$. For example, $g \circ \sigma$ is strict (respectively smooth) if and only if g is: see Arazy (1981) and Ziętak (1988). Furthermore, the facial structure of the unit ball of $g \circ \sigma$ can be derived from that of g (de Sá 1994a, 1994b, 1994c).

Example 9.4. (invariant approximation) Given a subspace of matrices and an invariant norm $g \circ \sigma$ (where g is a symmetric gauge), suppose we wish to approximate, in the norm $g \circ \sigma$, a given matrix by a matrix from the given subspace. We can rewrite this problem, for a suitable choice of matrices A_i and reals b_i (for $i = 1, 2, \ldots, m$), as

$$\inf_{X \in M_n} \{ (g(\sigma(X)))^2/2 : \operatorname{tr} A_i^T X = b_i \text{ for each } i \}. \tag{9.2}$$

By the Fenchel Duality Theorem (Theorem 8.4) and the dual norm equation (8.1), both this problem and its dual

$$\sup_{y \in \mathbb{R}^m} \left\{ \sum_i b_i y_i - \frac{1}{2} \left(g^D \left(\sigma \left(\sum_i y_i A_i \right) \right) \right)^2 \right\} \tag{9.3}$$

have optimal solutions, with equal optimal values: the form of the dual is a consequence of the duality formula (9.1). If the norm g is strict (that is, the unit sphere $\{ x : g(x) = 1 \}$ contains no line segments) then its dual norm g^D is smooth (see for example Deville, Godefroy and Zizler (1993, II.1.6)), whence so is $g^D \circ \sigma$: Ziętak's Theorem (Theorem 9.3) provides a simple formula for $\nabla(g^D \circ \sigma)$ in terms of ∇g^D. Then the dual problem (9.3) is an unconstrained, smooth, concave maximization, and if the vector $\overline{y}$ is a solution, then the unique primal optimal solution is given by

$$X = g^D(\sigma(\overline{Y}))U(\mathrm{Diag}\,\nabla g^D(\sigma(\overline{Y})))V,$$

where $\overline{Y} = \sum_i \bar{y}_i A_i$, and U and V are any orthogonal matrices for which $\overline{Y} = U(\mathrm{Diag}\,\sigma(\overline{Y}))V$.

10. Functions of eigenvalues

We turn next to our principal interest: variational properties of eigenvalues. Our development mimics that of the previous section. An invariant matrix function is simply an absolute, symmetric function of the singular values. Analogously, a function of a symmetric matrix X that is invariant under transformations $X \mapsto U^T X U$, for all orthogonal matrices U, must be a symmetric function of the eigenvalues of X.

We consider the Euclidean space S_n of $n \times n$ real symmetric matrices, where the inner product is defined by $\langle X, Y \rangle = \mathrm{tr}\,XY$. We denote the eigenvalues of a matrix X in S_n by $\lambda_1(X) \geq \lambda_2(X) \geq \cdots \geq \lambda_n(X)$. In this way we define the 'eigenvalue map' $\lambda : M_n \to \mathbb{R}^n$.

We say a function ψ on S_n is *weakly (orthogonally) invariant* if, for any matrices X in S_n and U in O_n, we have $\psi(U^T X U) = \psi(X)$. Clearly, for any weakly invariant convex function ψ on S_n, the extended-real function h on $\mathbb{R}^n$ defined by $h(x) = \psi(\mathrm{Diag}\,x)$ is symmetric and convex. Remarkably, just like von Neumann's Theorem (Theorem 9.1), this property is actually a characterization.

Theorem 10.1. (Davis, 1957) Functions on S_n that are weakly invariant and convex are exactly those composite functions of the form $h \circ \lambda$, where the function $h : \mathbb{R}^n \to [-\infty, +\infty]$ is symmetric and convex.

For proofs of this result, see Davis (1957), Martínez-Legaz (1995) and Lewis (1996c). A rather different characterization when the function h is differentiable may be found in Friedland (1981).

To pursue our analogy, we sketch a revealing, duality-based proof when the functions are closed. It begins with an analogue of von Neumann's Lemma (Lemma 9.2), for symmetric matrices. The inequality is actually an easy consequence of von Neumann's; the condition for equality is due to Theobald (1975). We say that two matrices X and Y in S_n have a *simultaneous ordered spectral decomposition* if there is an orthogonal matrix U with $X = U^T(\mathrm{Diag}\,\lambda(X))U$ and $Y = U^T(\mathrm{Diag}\,\lambda(Y))U$.

Lemma 10.2. (von Neumann–Theobald) Any $n \times n$ real symmetric matrices X and Y satisfy the inequality

$$\mathrm{tr}\,XY \leq \lambda(X)^T \lambda(Y);$$

equality holds if and only if X and Y have a simultaneous ordered spectral decomposition.

As in the singular value case, this inequality is the key tool. We first use it to prove that any extended-real symmetric function h satisfies

$$(h \circ \lambda)^* = h^* \circ \lambda. \tag{10.1}$$

Hence if h is also closed, proper and convex, then applying this formula twice and using the Fenchel–Hörmander Theorem (Theorem 8.2), we deduce

$$(h \circ \lambda)^{**} = (h^* \circ \lambda)^* = h^{**} \circ \lambda = h \circ \lambda,$$

and, by Theorem 8.2, $h \circ \lambda$ must be convex. Theorem 10.1 is now easy to see.

Very much as in the invariant norm case, the von Neumann–Theobald Lemma (Lemma 10.2) helps in the computation of subdifferentials. Using the Lemma, the subdifferential characterization (8.2), and the conjugacy formula (10.1), we obtain the following result (Lewis 1996a).

Theorem 10.3. (Lewis, 1996) If the function $h : \mathbb{R}^n \to (-\infty, \infty]$ is symmetric and convex, then matrices X and Y satisfy $Y \in \partial(h \circ \lambda)(X)$ if and only if they have a simultaneous ordered spectral decomposition and satisfy $\lambda(Y) \in \partial h(\lambda(X))$.

There are similar results for smooth and nonsmooth, nonconvex functions (Lewis 1996b, Tsing, Fan and Verriest 1994). Special versions of some of these ideas appeared independently in Barbara and Crouzeix (1994).

As in the invariant norm case, geometric/analytic properties of the two functions h and $h \circ \sigma$ are intimately related: strict convexity and smoothness are examples (Lewis 1996a). Furthermore, if the convex subset C of $\mathbb{R}^n$ is *symmetric* (that is, $PC = C$ for all matrices P in P_n), then by applying Davis's Theorem (Theorem 10.1), to the function $\delta_C \circ \lambda$ we see that the matrix set $\lambda^{-1}(C) = \{X \in S_n : \lambda(X) \in C\}$ is also convex: the extremal and facial structure of $\lambda^{-1}(C)$ may be deduced from that of C (Lewis 1996a, Lewis 1995b). Similar examples appear in Seeger (1996) and Martínez-Legaz (1995).

The parallel between the invariant norm case in the previous section and the development in this section is not accidental. There is a deeper, algebraic structure underlying both theorems (Lewis 1995c, Lewis 1996c).

Example 10.4. (semidefinite cone) Starting with the indicator function of the positive orthant, $\delta_{\mathbb{R}^n_+}$, the composite function $\delta_{\mathbb{R}^n_+} \circ \lambda$ is just the indicator function of the cone of positive semidefinite matrices. We denote this cone S_n^+, and for matrices X and Y in S_n, we write $X \succeq Y$ if $X - Y \in S_n^+$. The conjugacy formula (10.1) and Example 8.3 show Fejér's result that the positive semidefinite cone is 'self-dual' (that is, $(S_n^+)^- = -S_n^+$), since

$$\delta_{(S_n^+)^-} = \delta_{S_n^+}^* = (\delta_{\mathbb{R}^n_+} \circ \lambda)^* = \delta_{\mathbb{R}^n_+}^* \circ \lambda = \delta_{-\mathbb{R}^n_+} \circ \lambda = \delta_{-S_n^+}.$$

Furthermore, if matrices X and Y in S_n^+ satisfy $\operatorname{tr} XY = 0$, then in fact they must satisfy $XY = 0$. To see this, note that from the complementarity condition (8.4) and the self-duality of S_n^+, we deduce $-Y \in \partial \delta_{S_n^+}(X)$. By the subdifferential characterization, Theorem 10.3, X and $-Y$ have a simultaneous ordered spectral decomposition, and $\lambda(-Y) \in \partial \delta_{\mathbb{R}_+^n}(\lambda(X))$, whence (by relation (8.5)) $\lambda_j(X)\lambda_j(-Y) = 0$ for each j. Thus for some orthogonal matrix U,

$$
\begin{aligned}
-XY &= (U^T(\operatorname{Diag} \lambda(X))U)(U^T(\operatorname{Diag} \lambda(-Y))U) \\
&= U^T(\operatorname{Diag}[\lambda_j(X)\lambda_j(-Y)])U = 0.
\end{aligned}
$$

Example 10.5. (logarithmic barrier) For vectors x and y in $\mathbb{R}^n$, we write $x > y$ if $x_j > y_j$ for each index j. For matrices X and Y in S_n, we write $X \succ Y$ if $X - Y$ is positive definite. Define a symmetric closed convex function $h : \mathbb{R}^n \to (-\infty, +\infty]$ by

$$
h(x) = \begin{cases} -\sum_j \log x_j & \text{if } x > 0, \\ +\infty & \text{otherwise.} \end{cases} \tag{10.2}
$$

(Henceforth we will interpret $\log \alpha$ as $-\infty$ for any nonpositive real α.) The corresponding matrix function is

$$
(h \circ \lambda)(X) = \begin{cases} -\log \det X & \text{if } X \succ 0, \\ +\infty & \text{otherwise.} \end{cases} \tag{10.3}
$$

(Analogously, we henceforth interpret $\log \det X$ as $-\infty$ unless the symmetric matrix X is positive definite.) By Davis's Theorem (Theorem 10.1), this function is convex (and in fact essentially strictly convex, since h is; see Lewis (1996a). Using Theorem 10.3, a simple exercise shows, for positive definite X,

$$
\nabla(h \circ \lambda)(X) = -X^{-1}. \tag{10.4}
$$

Since $h^*(y) = -n + h(-y)$, we deduce from the conjugacy formula (10.1),

$$
(h \circ \lambda)^*(Y) = \begin{cases} -n - \log \det(-Y) & \text{if } 0 \succ Y, \\ +\infty & \text{otherwise.} \end{cases} \tag{10.5}
$$

In this example we see the intimate connection between the functions (10.2) and (10.3), two of the 'self-concordant barriers' fundamental to the development of Nesterov and Nemirovskii (1994). This connection suggests the following interesting question (Tunçel 1995): if the function h is a self-concordant barrier, is the same true of the matrix function $h \circ \lambda$?

Example 10.6. (BFGS updates – Fletcher, 1991) Given a matrix H in S_n which is positive definite, and vectors s and b in $\mathbb{R}^n$, we consider the primal problem

$$
\inf_{X \in S_n} \{ \operatorname{tr} H^{-1} X - \log \det X : Xs = b, \ X \succ 0 \}. \tag{10.6}
$$

Using the framework of the Fenchel Duality Theorem (Theorem 8.4), and formula (10.5), the dual problem is

$$\sup_{y \in \mathbb{R}^m} \{b^T y + \log \det(H^{-1} - (ys^T + sy^T)/2)\} + n. \qquad (10.7)$$

If $s^T b > 0$, standard quasi-Newton theory shows the primal problem (10.6) is superconsistent, and choosing $\hat{y} = 0$ shows the dual problem (10.7) is also superconsistent. Thus the primal and dual problems are both attained, by the Fenchel Duality Theorem, and routine calculation using the gradient formula (10.4) shows that the unique primal optimal solution is the 'BFGS update' of the 'Hessian approximation' H, subject to the 'secant equation' $Xs = b$ (Fletcher 1991, Lewis 1996a).

Example 10.7. (eigenvalue sums) For an integer k between 0 and n, define a symmetric closed convex function h on $\mathbb{R}^n$ by

$$h(x) = \text{sum of the } k \text{ largest } x_j. \qquad (10.8)$$

The corresponding matrix function is the sum of the k largest eigenvalues,

$$(h \circ \lambda)(X) = \sum_{j=1}^{k} \lambda_j(X).$$

A calculation shows the conjugate of h is the indicator function of the set

$$\left\{ z \in \mathbb{R}^n : \sum_{j=1}^{n} z_j = k, \ 0 \le z_j \le 1 \text{ for each } j \right\},$$

so by the conjugacy formula (10.1), the conjugate of $h \circ \lambda$ is the indicator function of the matrix set

$$H = \{Y \in S_n : \text{tr } Y = k, \ I \succeq Y \succeq 0\}. \qquad (10.9)$$

For given matrices $A^1, A^2, \ldots, A^m$ in S_n and a vector b in $\mathbb{R}^m$, consider the optimization problem

$$\inf_{X \in S_n} \left\{ \sum_{j=1}^{k} \lambda_j(X) : \text{tr } A^i X = b_i \text{ for each } i \right\};$$

cf. Fletcher (1985), Overton and Womersley (1993), Hirriart-Urruty and Ye (1995) and Pataki (1995). In the Fenchel Duality framework the dual problem is therefore

$$\sup_{y \in \mathbb{R}^m} \left\{ b^T y : \sum_{i=1}^{m} y_i A^i \in H \right\}, \qquad (10.10)$$

where the set H is given by equation (10.9).

Rather more generally, suppose the vector d in $\mathbb{R}^n$ has nonincreasing components. For any vector x in $\mathbb{R}^n$, let $\bar{x}$ denote the vector with components x_j rearranged into nonincreasing order. Then the function $h(x) = d^T\bar{x}$ is symmetric, closed and convex (since $h(x) = \max_{Q \in P_n}\{d^T Q x\}$). The corresponding matrix function is exactly the weighted sum of eigenvalues appearing in the graph partitioning problem in Section 2, namely $(h \circ \lambda)(W) = d^T \lambda(W)$.

11. Linear programming

An important area of eigenvalue optimization is semidefinite programming (SDP). Since the analogies with ordinary linear programming (LP) are very close, we begin by outlining the relevant classical theory.

For given vectors $c, a^1, a^2, \ldots, a^m$ in $\mathbb{R}^n$, and b in $\mathbb{R}^m$, the primal linear program we study is

$$\rho_0 = \inf_{x \in \mathbb{R}^n_+} \{c^T x : (a^j)^T x = b_j \text{ for each } j\}. \tag{11.1}$$

Using the framework of Theorem 8.4 (with objective function $f(x) = c^T x + \delta_{\mathbb{R}^n_+}(x)$), we obtain the dual problem

$$\delta_0 = \sup_{y \in \mathbb{R}^m} \left\{ b^T y : c \geq \sum_i y_i a^i \right\}. \tag{11.2}$$

By polyhedrality, we immediately see from the Fenchel Duality Theorem (Theorem 8.4) that if either the primal or dual problem is consistent, then $\rho_0 = \delta_0$, and both values are attained if finite. This is the classical linear programming duality theorem. The *complementary slackness* condition ((v) in Theorem 8.4) states that primal feasible $\bar{x}$ in $\mathbb{R}^n$ and dual feasible $\bar{y}$ in $\mathbb{R}^m$ are both optimal if and only if

$$\left(c - \sum_i \bar{y}_i a^i \right)^T \bar{x} = 0,$$

or, equivalently, $\bar{x}_j(c - \sum_i \bar{y}_i a^i)_j = 0$, for each index $j = 1, 2, \ldots, n$.

If we penalize the primal constraint $x \in \mathbb{R}^n_+$ using the logarithmic barrier (10.2) with a small positive parameter μ, we obtain the new primal problem

$$\rho_\mu = \inf_{x \in \mathbb{R}^n} \left\{ c^T x - \mu \sum_j \log x_j : (a^i)^T x = b_i \text{ for each } i \right\},$$

and the dual problem is

$$\delta_\mu = \sup_{y \in \mathbb{R}^m} \left\{ b^T y + \mu \sum_j \log \left(c_j - \sum_i y_i a^i_j \right) \right\} + n\mu(\log \mu - 1).$$

The Fenchel Duality Theorem now needs a regularity condition. We assume the following:

(i) *Primal superconsistency:* some vector $\hat{x} > 0$ in $\mathbb{R}^n$ satisfies

$$(a^i)^T \hat{x} = b_i, \text{ for each } i.$$

(ii) *Dual superconsistency:* some vector $\hat{y}$ in $\mathbb{R}^m$ satisfies $c > \sum_i \hat{y}_i a^i$.

(iii) *Independence:* the vectors $a^1, a^2, \ldots, a^m$ are linearly independent.

Assumptions (i) and (ii) guarantee $\rho_\mu = \delta_\mu$, by the Duality Theorem, and both values are attained. The primal objective is (essentially) strictly convex; assumption (iii) ensures the dual objective is too. Hence the primal and dual both have unique optimal solutions, $x = x^\mu$ in $\mathbb{R}^n$ and $y = y^\mu$ in $\mathbb{R}^m$ respectively, and by the complementary slackness condition, they are the unique solution of the system

$$(a^i)^T x = b_i, \text{ for each } i, \tag{11.3}$$

$$x_j \left(c_j - \sum_i y_i a_j^i \right) = \mu, \text{ for each } j, \tag{11.4}$$

$$x \geq 0 \text{ and } c \geq \sum_i y_i a^i. \tag{11.5}$$

Notice that when $\mu = 0$ these conditions reduce to the complementary slackness conditions for the original linear program.

The trajectory $\{(x^\mu, y^\mu) : \mu > 0\}$ is called the *central path*. From equations (11.3) and (11.4), we deduce the *duality gap*

$$c^T x^\mu - b^T y^\mu = n\mu. \tag{11.6}$$

Thus, using the *weak duality* inequality (Theorem 8.4(i)), we see that the feasible solutions x^μ and y^μ approach optimality:

$$\lim_{\mu \downarrow 0} c^T x^\mu = \rho_0 = \delta_0 = \lim_{\mu \downarrow 0} b^T y^\mu.$$

But our regularity assumptions (i), (ii) and (iii) then imply, using the Fenchel Duality compactness results, that the central path (x^μ, y^μ) stays bounded for small positive μ. Any limit point (x^0, y^0) must satisfy $c^T x^0 = b^T y^0$ (by the duality gap formula (11.6)), whence x^0 and y^0 are optimal for the primal and dual respectively. In fact, a more careful argument shows that the limit point (x^0, y^0) is unique: the vectors x^0 and y^0 are the 'analytic centres' of the optimal faces for the primal and dual problems respectively; see Megiddo (1989) and McLinden (1980). (Given a polytope $P = \{z \in L : z \geq 0\}$, where L is an affine subspace and P contains a point $z > 0$, the analytic centre of P is the unique minimizer of the logarithmic barrier $-\sum_j \log(z_j)$ over all $z \in P$.)

Condition (12.6) implies that, just like the solution pair $\overline{X}$ and $\overline{Z}$, X^μ and $C - \sum_i y_i^\mu A^i$ have a simultaneous ordered spectral decomposition. When $\mu = 0$, equation (12.6) reduces to the complementary slackness condition for the original semidefinite program.

Notice that, with the choices $C = \operatorname{Diag} c$ and $A^i = \operatorname{Diag} a^i$ for each i, the semidefinite theory developed in this section collapses to the linear theory of the previous section.

13. Strict complementarity and nondegeneracy

Let us go back to the linear programming problem and its dual, (11.1) and (11.2). We say a primal-dual solution $(\overline{x}, \overline{y})$ satisfies the *strict complementarity condition* if, for each j, exactly one of the two statements $\overline{x}_j = 0$ and $(c - \sum \overline{y}_i a^i)_j = 0$ holds. We say a strictly complementary solution is *nondegenerate* if the vector $\overline{x}$ has exactly m nonzero components and the corresponding m rows of the matrix $[a^1, a^2, \ldots, a^m]$ are linearly independent. It is well known that these conditions guarantee that $\overline{x}$ is the unique optimal solution of the primal problem (11.1) and that $\overline{y}$ is the unique optimal solution of the dual problem (11.2). Furthermore, these conditions hold 'generically' for a linear program: roughly speaking, this means that they hold with probability one, given randomly generated linear programs with associated nonempty feasible regions.

The situation is less clear in semidefinite programming. There is no difficulty with the idea of strict complementarity: we say a primal-dual solution $(\overline{X}, \overline{y})$ for (12.1) and (12.2) satisfies the *strict complementarity condition* if, for each index j, exactly one of the two statements $\lambda_j(\overline{X}) = 0$ and $\lambda_j(-\overline{Z}) = 0$ holds, where $\overline{Z} = C - \sum_i \overline{y}_i A^i$. Let r denote the rank of $\overline{X}$ and let s denote the rank of $\overline{Z}$; then strict complementarity holds if and only if $r + s = n$. Nondegeneracy conditions are more complicated and are discussed by Alizadeh, Haeberly and Overton (1996a) and Shapiro (1996). Assume that strict complementarity holds and let $\overline{U} = [\overline{U}_1\ \overline{U}_2]$ be the orthogonal matrix of eigenvectors which simultaneously diagonalizes $\overline{X}$ and $\overline{Z}$ (see equations (12.3)), with the first r columns (collected in $\overline{U}_1$) corresponding to nonzero eigenvalues of $\overline{X}$ and the last s columns (collected in $\overline{U}_2$) corresponding to nonzero eigenvalues of $\overline{Z}$. Then the appropriate nondegeneracy assumptions are the following two conditions, motivated by studying the primal and the dual separately: first, that the matrices

$$\begin{bmatrix} \overline{U}_1^T A^i \overline{U}_1 & \overline{U}_1^T A^i \overline{U}_2 \\ \overline{U}_2^T A^i \overline{U}_1 & 0 \end{bmatrix}, \quad \text{for } i = 1, 2, \ldots, m,$$

are linearly independent in the space S_n and second, that the matrices

$$\overline{U}_1^T A^i \overline{U}_1, \quad \text{for } i = 1, 2, \ldots, m,$$

Table 1. *Number of occurrences of rank(X) in 1000 randomly generated problems with $n = 10$ and various values of m.*

m	0	1	2	3	4	5	6	7	8	9	10
5	297	703									
10	0	494	506	0							
15		18	712	270	0						
20			100	813	87						
25				1	325	667	7				

span the space S_r. It is shown by Alizadeh et al. (1996a) that the strict complementarity and nondegeneracy conditions imply uniqueness of the primal and dual solutions, and also that the conditions are indeed generic properties of SDP, meaning roughly that they hold with probability one for an optimal solution pair, given random data with feasible solutions. An immediate consequence is the existence of generic bounds on the optimal solution matrix ranks r and s, and therefore on the multiplicity of the zero eigenvalues. Let $k^{\overline{2}}$ denote $k(k+1)/2$, and let $\sqrt[\overline{2}]{k} = \lfloor t \rfloor$, where t is the positive real root of $t^{\overline{2}} = k$. Then generic bounds on the rank of the primal optimal solution matrix $\overline{X}$ are given by

$$n - \sqrt[\overline{2}]{n^{\overline{2}} - m} \leq r \leq \sqrt[\overline{2}]{m}.$$

For further discussion of related issues, see Pataki (1995).

Experiments reported in Alizadeh et al. (1996a) show clearly that, given randomly generated data, the rank r is far more likely to lie in the centre of its range than near the end points. This is demonstrated by Table 1, which shows, for $n = 10$ and various choices of m, how many times the primal rank r occurred during 1000 runs with different random data. The zeros indicate possible values of r which did not occur, while the blanks indicate generically impossible values.

A natural question is: what is the underlying probability distribution for the primal solution rank r? We consider this to be a very interesting open question.

Table 1 also shows, incidentally, the reliability of the numerical method used to obtain the results: accurate solutions to 5000 different randomly generated problems were obtained without a single failure. (As with linear programming, it is easy to check the optimality of a solution pair, simply by checking primal and dual feasibility and the complementary slackness condition.) We now sketch the ideas behind the primal-dual interior-point method used to obtain these results.

14. Primal-dual interior-point methods

We begin again with the case of linear programming. The basic idea of the primal-dual interior-point method is to generate a sequence of iterates $(x^{(k)}, y^{(k)}) \in \mathbb{R}^n \times \mathbb{R}^m$ (for $k = 1, 2, \ldots$) approximating a sequence of points lying on the central path and converging to an optimal solution as $k \to \infty$. Briefly, this approximation is achieved by applying, at the kth iteration, one step of Newton's method to (11.3) and (11.4), a system of $n + m$ linear and quadratic equations in the $n+m$ variables x_j, y_i. Here μ is a positive number, fixed at the kth iteration to a value $\mu^{(k)}$, with $\mu^{(k)} \to 0$ as $k \to \infty$. (If we also introduce the equations $z_j = c_j - \sum_i y_i a_j^i$, substituting these into (11.4) to obtain $x_j z_j = \mu$, and treating z_j, $j = 1, \ldots, n$, as independent variables, Newton's method yields an equivalent iteration.) The Newton step is defined by the linear system

$$\begin{bmatrix} A^T & 0 \\ Z^{(k)} & -X^{(k)}A \end{bmatrix} \begin{bmatrix} \Delta x \\ \Delta y \end{bmatrix} = \begin{bmatrix} b - A^T x^{(k)} \\ (\mu^{(k)} I - X^{(k)} Z^{(k)})e \end{bmatrix}, \tag{14.1}$$

where A is the $n \times m$ matrix $[a^1, \ldots, a^m]$, $X^{(k)}$ and $Z^{(k)}$ are respectively the diagonal matrices $\mathrm{Diag}\, x^{(k)}$ and $\mathrm{Diag}(c - Ay^{(k)})$, I is the $n \times n$ identity matrix and e is the n-vector whose components are all one. Block Gauss elimination reduces this system to

$$\left(A^T (Z^{(k)})^{-1} X^{(k)} A\right) \Delta y = b - A^T (x^{(k)} - w^{(k)}) \tag{14.2}$$

$$\Delta x = w^{(k)} + (Z^{(k)})^{-1} X^{(k)} A \Delta y, \tag{14.3}$$

where $w^{(k)} = (\mu^{(k)} (Z^{(k)})^{-1} - X^{(k)})e$. New iterates are then obtained by $x^{(k+1)} = x^{(k)} + \alpha \Delta x$, $y^{(k+1)} = y^{(k)} + \beta \Delta y$, where steplengths α and β are chosen so that $x^{(k+1)} > 0$ and $c - Ay^{(k+1)} > 0$. A value $\mu^{(k+1)} < \mu^{(k)}$ is then chosen and the iterative step repeated. Different rules for reducing the parameter μ and choosing the steplengths α and β give different variants of the algorithm, with some specific rules known to guarantee a solution with prescribed accuracy in polynomial time. The original references for this method are Monteiro and Adler (1989) and Kojima et al. (1989).

We shall not discuss the global convergence theory. However, the following well known result is important for understanding the local convergence and numerical stability of the algorithm. It analyses the condition numbers of the two key matrices defining the algorithm at points on the central path.

Proposition 14.1 Suppose that $(\bar{x}, \bar{y})$ solves the LP primal-dual pair (11.1) and (11.2), with both the strict complementarity and nondegeneracy conditions holding. Then, using $X^\mu = \mathrm{Diag}\, x^\mu$ and $Z^\mu = \mathrm{Diag}(c - Ay^\mu)$, where (x^μ, y^μ) lies on the central path defined by (11.3), (11.4) and (11.5), the condition numbers of the matrices

$$\begin{bmatrix} A^T & 0 \\ Z^\mu & -X^\mu A \end{bmatrix} \quad \text{and} \quad A^T (Z^\mu)^{-1} X^\mu A \qquad (14.4)$$

are both bounded independent of μ as $\mu \downarrow 0$.

Proof. It is well known that the assumptions guarantee that $(\overline{x}, \overline{y})$ is the unique solution of the linear program and consequently, as discussed in Section 11, also the limit point of the central path (x^μ, y^μ) as $\mu \downarrow 0$. Without loss of generality, we may take

$$\overline{x} = \begin{bmatrix} \tilde{x} \\ 0 \end{bmatrix}, \qquad c - A\overline{y} = \begin{bmatrix} 0 \\ \tilde{z} \end{bmatrix}, \qquad A = \begin{bmatrix} A_1 \\ A_2 \end{bmatrix},$$

where all partitionings are from n rows into m and $n - m$ rows respectively, with $\tilde{x} > 0$, $\tilde{z} > 0$, and A_1 nonsingular. Then the first matrix in (14.4) converges to

$$\begin{bmatrix} A_1^T & A_2^T & 0 \\ 0 & 0 & (-\mathrm{Diag}\,\tilde{x})A_1 \\ 0 & \mathrm{Diag}\,\tilde{z} & 0 \end{bmatrix}$$

as $\mu \downarrow 0$. This matrix can be permuted into a block upper triangular matrix with nonsingular diagonal blocks A_1^T, $-\mathrm{Diag}(\tilde{x})A_1$, $\mathrm{Diag}\,\tilde{z}$. The second matrix does not have a limit, but $\mu A^T (Z^\mu)^{-1} X^\mu A$ has the limit

$$A_1^T (\mathrm{Diag}\,\tilde{x})^2 A_1$$

by virtue of (11.4), which completes the proof. $\square$

Let us refer to the first matrix in (14.4) as the *block Jacobian* matrix and to the second as the *Schur complement*. The consequence of the bounded condition number of the block Jacobian is that, given strict complementarity and nondegeneracy assumptions, the primal-dual interior-point method for linear programming has a quadratic rate of local convergence as long as the parameter μ is reduced sufficiently fast and the steplengths α and β are chosen sufficiently close to one. See Zhang, Tapia and Dennis (1992) for details. (Even without nondegeneracy assumptions, certain superlinear convergence properties still hold; see Zhang and Tapia (1993).) The consequence of the bounded condition number of the Schur complement is that, again under the given assumptions, there is no numerical difficulty with factorizing the matrix in (14.2) as $\mu \downarrow 0$. This is not necessarily the case in the absence of nondegeneracy assumptions, a fact that is a subject of some current interest (Wright 1995).

Now let us turn to semidefinite programming. As in linear programming, the essential idea of the primal-dual interior-point method is to generate a sequence of iterates $(X^{(k)}, y^{(k)}) \in S_n \times \mathbb{R}^m$ approximating a sequence of points on the central path, converging to a solution as $k \to \infty$. However, it is not clear in this case how to apply Newton's method. The key difficulty is that the left-hand side of (12.6) is *not* symmetric, so equations (12.5) and (12.6) do not map $S_n \times \mathbb{R}^m$ to itself; consequently, Newton's method is not directly applicable. The cleanest solution seems to be to replace (12.6) by

$$X\left(C - \sum_i y_i A^i\right) + \left(C - \sum_i y_i A^i\right) X = 2\mu I. \qquad (14.5)$$

That (12.6) implies (14.5) is immediate. That the converse holds for $X \succeq 0$ is easily seen by premultiplying (14.5) by U^T and postmultiplying by U, where the orthogonal matrix U diagonalizes X. Application of Newton's method to (12.5) and (14.5) leads to a very effective method for semidefinite programming called the $XZ + ZX$ method by Alizadeh, Haeberly and Overton (1996*b*); this method was used to generate the results shown in Table 1 in the previous section. On average, each problem was solved in less than 10 iterations, a property which is, in practice, almost independent of the problem dimension. Other variants of the primal-dual interior-point method given by Helmberg, Rendl, Vanderbei and Wolkowicz (1996), Kojima, Shindoh and Hara (1994), Nesterov and Todd (1996) and Vandenberghe and Boyd (1995) give similar performance, but the $XZ + ZX$ method is especially robust with respect to changes in the rules for reducing μ and choosing the steplengths α, β (Alizadeh et al. 1996*b*). It is proved by Alizadeh et al. (1996*b*) that, given the SDP strict complementarity and nondegeneracy assumptions stated in the previous section, the *first* part of Theorem 14.1, namely that the condition number of the block Jacobian is bounded, extends from LP to hold also for the $XZ + ZX$ method for SDP, but the *second does not*, that is, the condition number of the corresponding Schur complement matrix is unbounded for SDP, even with nondegeneracy assumptions. Consequently, the $XZ + ZX$ method *is* locally quadratically convergent, in contrast to other variants of the primal-dual interior-point method for SDP, given the nondegeneracy assumptions and appropriate μ reduction and steplength rules. However, under the same conditions, the method *is not* necessarily numerically stable as $\mu \downarrow 0$, since the condition number of the linear system which must be solved at each iteration is $O(1/\mu)$. Indeed, this was observed numerically by Alizadeh et al. (1996*b*): generally, it was possible to compute results accurate to only about the square root of the machine precision, given random data. The same difficulty applies to other variants of the primal-dual interior-point method as well as the $XZ + ZX$ method. By contrast, there is no difficulty solving modest-sized randomly generated linear programs to machine precision accuracy using the LP primal-dual interior-point method.

As in LP, if we use the substitution $Z = C - \sum_i y_i A^i$ in (12.6), introducing Z as an independent variable and $Z = C - \sum_i y_i A^i$ as an additional equation, Newton's method yields an equivalent iteration.

An appealing alternative primal-dual interior-point iteration for SDP is based on the following idea. Instead of treating the variable X (or X and Z) directly, recall from the discussion in Section 12 that for (X^μ, y^μ) to lie on the central path, X^μ and $Z = C - \sum_i y_i^\mu A^i$ must have a simultaneous ordered spectral decomposition. Therefore, consider the following set of variables: an orthogonal matrix U, which diagonalizes both X and $Z = C - \sum_i y_i A^i$, together with the eigenvalues of X and Z, say ξ_j and ζ_j, $j = 1, \ldots, n$. Equations (12.5) and (12.6) then reduce to

$$\mathrm{tr}\left(U^T A^i U (\mathrm{Diag}\, \xi_i)\right) = b_i, \text{ for each } i,$$

$$(\mathrm{Diag}\, \zeta_j) + \sum_i y_i U^T A^i U = U^T C U,$$

and

$$\xi_j \zeta_j = \mu, \text{ for each } j.$$

Borrowing a technique used by Friedland, Nocedal and Overton (1987), Overton (1988) and Overton and Womersley (1995), the orthogonal matrix U can be parametrized by $U = \exp(S) = I + S + \frac{1}{2}S^2 + \cdots$, where S is skew-symmetric, making the application of Newton's method straightforward. This leads to a method which, though it apparently has poor global convergence properties, is at present able to compute more accurate solutions than any other SDP interior-point method (Alizadeh et al. 1996b).

The eigenvalue optimization method of Overton (Overton 1988, Overton and Womersley 1995) is easily extended to apply to SDP. This method does not share the global convergence properties known for the interior-point methods. However, it can be used as an effective technique to obtain highly accurate solutions when an interior-point method reaches its limiting accuracy. The same presumably applies to Fletcher's method (Fletcher 1985), though this has not been tested. These methods are more difficult to describe because they use second derivatives, a complicated issue in the presence of multiple eigenvalues. They need second derivatives to achieve quadratic convergence because they are based on an appropriate form of Newton's method in the dual space only. These Newton methods use primal information to construct the second derivative of an appropriate Lagrangian function, but they are not primal-dual methods. A really remarkable property of the $XZ + ZX$ primal-dual interior-point method for SDP is that, in exact arithmetic, it generically achieves quadratic convergence with only first-order primal and dual information, even though the constraints are not polyhedral.

Primal-dual interior-point methods for LP have been used to solve very large problems; the best methods are generally thought to be superior to the simplex method, except for special problem classes. However, at present the implementation of interior-point methods for SDP has been limited to small problems, or problems with block-diagonal structure. If C and the A^i are block-diagonal with the same block structure, then, without loss of generality, the primal matrix X can be taken to have the same block structure: indeed, LP is a special case with block sizes all one. Consequently, the primal-dual interior-point methods for SDP can be implemented very efficiently if the block sizes are not large. However, if C and the matrices A^i have a more general sparse structure, then even if $C - \sum_i y_i^\mu A^i$ is sparse, the corresponding primal matrix $X^\mu = \mu(C - \sum_i y_i^\mu A^i)^{-1}$ is generally dense. For example, this is the case when C and the A^i are tridiagonal. In this situation, it is possible that an interior-point method based only on dual information is preferable to a primal-dual method. It may also be worth reconsidering some older and simpler first-order methods (Cullum et al. 1975, Overton 1992, Schramm and Zowe 1992).

15. Nonlinear semidefinite programming and eigenvalue optimization

The primal semidefinite program (12.1) permits only linear constraints; likewise the constraint in the dual program (12.2) is a semidefinite constraint on an affine matrix function $C - \sum_i y_i A^i$. In many applications, one finds eigenvalue optimization problems with nonlinear constraints, or with matrix functions depending nonlinearly on the variables. Such problems are, of course, substantially more difficult and a detailed discussion is beyond the scope of this article. However, we make two remarks.

First, although much of the duality theory described above fails to extend to the nonlinear case, some results are possible. Instead of subdifferentials, one may introduce the Clarke generalized gradient (Clarke 1983). A suitable chain rule yields first-order optimality conditions, though these are generally only necessary, not sufficient, conditions for optimality (Cox and Overton 1992, Lewis 1996b, Overton 1992). Second-order optimality conditions may also be derived (Shapiro 1996).

Second, some of the essential ideas of interior-point methods can be extended to nonlinear, nonconvex problems. Specifically, the logarithmic barrier function remains a very useful tool (Ringertz 1995). Whether primal-dual methods have an important role to play in the nonlinear case is not clear. However, the main idea remains valid, namely the application of Newton's method to a perturbed form of the optimality conditions, which, as in the linear case, involve a complementarity condition.

16. Eigenvalues of nonsymmetric matrices

The eigenvalues of a real symmetric matrix, which we described by the function $\lambda : S_n \to \mathbb{R}^n$, are Lipschitz functions of the matrix elements. Our development in Section 10 and our analysis of semidefinite programming depend heavily on the symmetry of the matrices. A completely parallel theory holds for complex Hermitian matrices. However, the eigenvalues of a real nonsymmetric or a general complex matrix are, in general, non-Lipschitz functions of the matrix elements.

In this section we give some apparently new variational results for functions of eigenvalues of nonsymmetric matrices. One special case characterizes the spectral abscissa of a nonsymmetric matrix, in a quantitative version of Lyapunov theory, while another special case yields a well known characterization of the spectral radius.

We can order the complex numbers $\mathbb{C}$ *lexicographically*: in this order, one complex number, z, dominates another, w if either $\operatorname{Re} z > \operatorname{Re} w$, or $\operatorname{Re} z = \operatorname{Re} w$ and $\operatorname{Im} z > \operatorname{Im} w$. For a matrix X in the vector space of $n \times n$ complex matrices, $M_n(\mathbb{C})$, let us denote the eigenvalues of X by $\lambda_1(X), \lambda_2(X), \ldots, \lambda_n(X)$, counted by multiplicity and ordered lexicographically. In this way we can extend the eigenvalue function λ to the space $M_n(\mathbb{C})$. If the matrices X and Z in $M_n(\mathbb{C})$ are *similar* (that is, some matrix L satisfies $Z = LXL^{-1}$), then we write $X \sim Z$.

Proposition 16.1 If the function $F : M_n(\mathbb{C}) \to [-\infty, +\infty]$ satisfies

$$F(X) \geq F(\operatorname{Diag} \lambda(X)) \quad \text{for all } X \text{ in } M_n(\mathbb{C}), \tag{16.1}$$

and if, for some matrix Y in $M_n(\mathbb{C})$, the function F is upper semicontinuous at $\operatorname{Diag} \lambda(Y)$, then

$$F(\operatorname{Diag} \lambda(Y)) - \inf_{Z \sim Y} F(Z). \tag{16.2}$$

Proof. If Z is similar to Y, then $\lambda(Z) = \lambda(Y)$, whence by inequality (16.1) we obtain $F(Z) \geq F(\operatorname{Diag} \lambda(Z)) = F(\operatorname{Diag} \lambda(Y))$. Thus $F(\operatorname{Diag} \lambda(Y)) < \inf_{Z \sim Y} F(Z)$.

On the other hand, by Schur's Theorem (Horn and Johnson 1985, Theorem 2.3.1), there is a unitary matrix Q and an upper triangular matrix T with main diagonal $\lambda(Y)$, satisfying $QYQ^* - T$. For positive real t, let D_t denote the matrix $\operatorname{Diag}(t, t^2, \ldots, t^n)$. As t approaches $+\infty$, we have

$$(D_t Q)Y(D_t Q)^{-1} - D_t T D_t^{-1} \to \operatorname{Diag} \lambda(Y),$$

and since F is upper semicontinuous at $\operatorname{Diag} \lambda(Y)$, we deduce

$$\inf_{Z \sim Y} F(Z) \leq \limsup_{t \to +\infty} F((D_t Q)Y(D_t Q)^{-1}) \leq F(\operatorname{Diag} \lambda(Y)).$$

Equation (16.2) follows. $\square$

The key technique in this proof, using diagonal similarity transformations to reduce the strictly upper triangular part of the Schur triangular form, is well known: see, for example, Horn and Johnson (1985, Lemma 5.6.10). Notice that if Y is not diagonalizable, the infimum in (16.2) may not be attained.

The following two propositions begin to look reminiscent of the material in Sections 9 and 10. Indeed, the complex versions of von Neumann's Lemma (Lemma 9.2) and the von Neumann–Theobald Lemma (Lemma 10.2) may be used to prove the propositions (although we quote intermediate results). For a vector z in $\mathbb{C}^n$, we write $\operatorname{Re} z$ and $|z|$ for the vectors with entries $\operatorname{Re} z_j$ and $|z_j|$ respectively.

Proposition 16.2 If the function $h : \mathbb{R}^n \to [-\infty, +\infty]$ is symmetric and convex, then any matrix X in $M_n(\mathbb{C})$ satisfies the inequality

$$h\left(\tfrac{1}{2}\lambda(X + X^*)\right) \geq h(\operatorname{Re}(\lambda(X))). \tag{16.3}$$

Proof. Since h is 'Schur convex' (see Marshall and Olkin (1979)), it suffices to show that the inequalities

$$\tfrac{1}{2}\sum_{j=1}^{k} \lambda_j(X + X^*) \geq \sum_{j=1}^{k} \operatorname{Re}\lambda_j(X)$$

hold for each index $k = 1, 2, \ldots, n$, with equality for $k = n$. This is exactly Horn and Johnson (1991, (3.3.33)). $\square$

Proposition 16.3 If g is a symmetric gauge on $\mathbb{R}^n$, then any matrix X in $M_n(\mathbb{C})$ satisfies the inequality

$$g(\sigma(X)) \geq g(|\lambda(X)|). \tag{16.4}$$

Proof. By Horn and Johnson (1985, Theorem 7.4.45), it suffices to show that the inequalities

$$\sum_{j=1}^{k} \sigma_j(X) \geq \sum_{j=1}^{k} |\lambda_{\pi(j)}(X)|$$

hold for each index $k = 1, 2, \ldots, n$, where π is any permutation for which $|\lambda_{\pi(j)}(X)|$ is nonincreasing in j. But this is precisely a result of Weyl (Horn and Johnson 1991, Theorem 3.3.13). $\square$

The analogy between the previous two propositions is clear if we recall that the components of $\sigma(X)$ are just $(\lambda_i(X^*X))^{1/2}$.

Putting together the three previous propositions, we arrive at the main result of this section.

Theorem 16.4

(a) Suppose the function $h : \mathbb{R}^n \to [-\infty, +\infty]$ is convex and symmetric. If the matrix Y in $M_n(\mathbb{C})$ has $\operatorname{Re}\lambda(Y)$ in $\operatorname{int}(\operatorname{dom} h)$, then it satisfies

$$h(\operatorname{Re}\lambda(Y)) = \inf_{Z \sim Y} h\left(\tfrac{1}{2}\lambda(Z + Z^*)\right).$$

(b) Suppose g is a symmetric gauge on $\mathbb{R}^n$. Then any matrix Y in $M_n(\mathbb{C})$ satisfies

$$g(|\lambda(Y)|) = \inf_{Z \sim Y} g(\sigma(Y)).$$

Proof.

(a) We choose $F(X) = h(\frac{1}{2}\lambda(X+X^*))$ in Proposition 16.1. Inequality (16.1) follows from Proposition 16.2, and since the convex function h must be continuous on the interior of its domain (Rockafellar 1970, Theorem 10.1) and λ is continuous, it follows that F is continuous at $\mathrm{Diag}\,\lambda(Y)$.

(b) We choose $F(X) = g(\sigma(X))$ in Proposition 16.1.

$\square$

Example 16.5. (spectral abscissa)   The *spectral abscissa* of a matrix Y in $M_n(\mathbb{C})$ is $\mathrm{Re}\,\lambda_1(Y)$. Applying Theorem 16.4(a) with the function h defined by $h(x) = \max_j x_j$, we obtain, for any matrix Y in $M_n(\mathbb{C})$,

$$\text{spectral abscissa of } Y = \tfrac{1}{2} \inf_{Z \sim Y} \lambda_1(Z + Z^*), \qquad (16.5)$$

or equivalently

$$\text{spectral abscissa of } Y = \tfrac{1}{2} \inf_{L:\det L \neq 0} \lambda_1(LYL^{-1} + L^{-*}Y^*L^*). \qquad (16.6)$$

We can interpret the spectral abscissa characterization (16.5) as a quantitative version of the Lyapunov Stability Theorem. We say a matrix A is *positive stable* if all its eigenvalues have strictly positive real part.

Corollary 16.6. (Lyapunov, 1947)   For any matrix A in $M_n(\mathbb{C})$, the following statements are equivalent.

(a) The matrix A is positive stable.
(b) There is a matrix B similar to A for which $B + B^*$ is positive definite.
(c) There is a positive definite matrix W for which $WA + A^*W$ is positive definite.

Proof.   The matrix A is positive stable exactly when the spectral abscissa of $-A$ is strictly negative. The equivalence of parts (a) and (b) now follows from the characterization (16.5). The equivalence of parts (b) and (c) follows by observing that W is positive definite if and only if $W = L^*L$ for some invertible L. $\square$

We can give a third form of the spectral abscissa characterization (16.5), (16.6) using the notation of generalized eigenvalue problems. For Hermitian matrices H and W, let $\lambda_1(H, W)$ denote the largest real μ for which there is a nonzero vector x in $\mathbb{C}^n$ satisfying $Hx = \mu W x$. With this notation, we have, as an immediate consequence of (16.6),

$$\text{spectral abscissa of } Y = \tfrac{1}{2} \inf_{W \succ 0} \lambda_1(WY + Y^*W, W). \qquad (16.7)$$

This result is apparently well known in the control theory community, although we are not aware of a standard reference.

Other quantitative results related to the Lyapunov Theorem may be stated by making different choices for the function h in Theorem 16.4(a). Two interesting examples follow.

Example 16.7. (products of eigenvalue real parts)   Letting the function h be the logarithmic barrier function (10.2), we have, for a positive stable matrix Y,

$$\prod_{j=1}^{n} \operatorname{Re} \lambda_j(Y) = \sup\{\det \tfrac{1}{2}(Z + Z^*) : Z \sim Y, \ Z + Z^* \succ 0\}.$$

Example 16.8. (sums of eigenvalue real parts)   Choosing the function $h(x)$ to be the sum of the k largest x_j (see (10.8)), we obtain

$$\operatorname{Re} \sum_{j=1}^{k} \lambda_j(Y) = \tfrac{1}{2} \inf_{Z \sim Y} \sum_{j=1}^{k} \lambda_j(Z + Z^*).$$

Let us now turn to Theorem 16.4(b):

Example 16.9. (spectral radius)   The *spectral radius* of a matrix X in $M_n(\mathbb{C})$ is $\max_j |\lambda_j(X)|$. Applying Theorem 16.4(b) with the symmetric gauge $g(\cdot) = \|\cdot\|_\infty$, we obtain, for any matrix Y in $M_n(\mathbb{C})$,

$$\text{spectral radius of } Y = \inf_{Z \sim Y} \sigma_1(Z) \tag{16.8}$$

(recalling that σ_1 denotes the largest singular value), or equivalently

$$\text{spectral radius of } Y = \inf_{L:\det L \neq 0} \sigma_1(LYL^{-1}). \tag{16.9}$$

A norm f on $M_n(\mathbb{C})$ is *submultiplicative* if $f(AB) \leq f(A)f(B)$ for all $A, B \in M_n(\mathbb{C})$. Clearly, the function $f(Y) = \sigma_1(LYL^{-1})$ is a submultiplicative matrix norm. Furthermore, it is easy to check that the spectral radius of any matrix Y cannot exceed the value of a submultiplicative matrix norm of Y: to prove this, choose $A = Y$ and B such that every column of B is the eigenvector of Y corresponding to the spectral radius. Consequently, equation (16.9) proves the well known fact that the spectral radius is the infimum of all submultiplicative matrix norms (Horn and Johnson 1985, Lemma 5.6.10).

More generally, we have the following example.

Example 16.10. (sums of eigenvalue moduli)   Choosing $g(x)$ to be the sum of the k largest $|x_j|$, we obtain

$$\text{sum of } k \text{ largest } |\lambda_j(Y)| = \inf_{Z \sim Y} \sum_{j=1}^{k} \sigma_j(Z).$$

Theorem 16.4 suggests a simple approach to nonsymmetric eigenvalue optimization which, to some extent, avoids the technical difficulties associated with the non-Lipschitz nature of the problem. Given a function $A : \mathbb{R}^m \to M_n(\mathbb{C})$ and a symmetric convex function $h : \mathbb{R}^n \to [-\infty, +\infty]$, consider the optimization problem

$$\inf_{w \in \mathbb{R}^m} \{h(\operatorname{Re} \lambda(A(w)))\}.$$

Using Theorem 16.4(a) we can rephrase this as

$$\inf_{w \in \mathbb{R}^m,\ L \in M_n(\mathbb{C})} \{(h \circ \lambda)(\tfrac{1}{2}(Z + Z^*)) : Z = LA(w)L^{-1},\ L \text{ invertible}\}.$$

Likewise, given a function $f : \mathbb{R}^m \to \mathbb{R}$, the problem

$$\inf_{w \in \mathbb{R}^m} \{f(w) : \operatorname{Re} \lambda(A(w)) \geq 0\}$$

can be rewritten as

$$\inf_{w \in \mathbb{R}^m,\ L \in M_n(\mathbb{C})} \{f(w) : LA(w)L^{-1} + L^{-*}A(w)^*L^* \succeq 0,\ L \text{ invertible}\}.$$

At the expense of introducing the extra variable matrix L, we have reduced these problems to symmetric eigenvalue optimization. Indeed, this idea (using an equivalent Lyapunov formulation based on (16.7)) is exploited in the application of linear matrix inequalities to system and control theory (Boyd et al. 1994); for applications to structural mechanics, see Ringertz (1996).

A similar technique could be applied to the problem

$$\inf_{w \in \mathbb{R}^m} \{g(|\lambda(A(w))|)\},$$

for a symmetric gauge g, this time using Theorem 16.4(b).

However, we caution that there are at least two difficulties with this approach. The first is the expense of introducing so many extra variables (the entire matrix L) into the optimization problem. The second is that the infimum is not likely to be achieved for many interesting applications, a fact that is likely to cause serious difficulties with ill-conditioning.

Acknowledgements. We thank Farid Alizadeh, Steve Cox and Levent Tunçel for various helpful comments. We especially thank Jim Burke and Jean-Pierre Haeberly for contributing unpublished key observations, giving parallel characterizations for the spectral radius and the spectral abscissa, which led to the results of the final section.

REFERENCES

F. Alizadeh (1991), Combinatorial Optimization with Interior Point Methods and Semidefinite Matrices, PhD thesis, University of Minnesota.

F. Alizadeh (1995), 'Interior point methods in semidefinite programming with applications to combinatorial optimization', *SIAM J. Optim.* **5**, 13–51.

F. Alizadeh, J.-P. A. Haeberly and M. L. Overton (1996a), 'Complementarity and nondegeneracy in semidefinite programming', *Mathematical Programming (Series B)*. To appear.

F. Alizadeh, J.-P. A. Haeberly and M. L. Overton (1996b), Primal-dual interior-point methods for semidefinite programming: convergence rates, stability, and numerical results. In preparation.

J. Arazy (1981), 'On the geometry of the unit ball of unitary matrix spaces', *Integral Equations and Operator Theory* **4**, 151–171.

M. S. Ashbaugh and R. D. Benguria (1991), 'Proof of the Payne–Pólya–Weinberger conjecture', *Bulletin of the American Math Society* **25**, 19–29.

A. Barbara and J.-P. Crouzeix (1994), 'Concave gauge functions and applications', *ZOR – Mathematical Methods of Operations Research* **40**, 43–74.

H. Baumgärtel (1985), *Analytic Perturbation Theory for Matrices and Operators*, Birkhäuser, Boston.

R. Bellman and K. Fan (1963), On systems of linear inequalities in Hermitian matrix variables, in *Convexity*, American Mathematical Society, Providence, pp. 1–11. Proceedings of Symposia in Pure Mathematics VII.

S. Boyd, L. E. Ghaoui, E. Feron and V. Balakrishnan (1994), *Linear Matrix Inequalities in System and Control Theory*, SIAM, Philadelphia.

J. V. Burke and M. L. Overton (1994), 'Differential properties of the spectral abscissa and the spectral radius for analytic matrix–valued mappings', *Nonlinear Analysis, Theory, Methods and Applications* **23**, 467–488.

F. H. Clarke (1983), *Optimization and Nonsmooth Analysis*, Wiley, New York. Reprinted by SIAM, Philadelphia, 1990.

S. J. Cox (1992), 'The shape of the ideal column', *Mathematical Intelligencer* **14**, 16–31.

S. J. Cox (1993), 'Steven Cox responds', *Mathematical Intelligencer* **15**, 68.

S. J. Cox and J. H. Maddocks (1996), Optimal design of the nonlinear elastica. In preparation.

S. J. Cox and M. L. Overton (1992), 'The optimal design of columns against buckling', *SIAM J. Math. Anal.* **23**, 287–325.

S. J. Cox and M. L. Overton (1996), 'Perturbing the critically damped wave equation', *SIAM J. Math. Anal.* To appear.

B. D. Craven and B. Mond (1981), 'Linear programming with matrix variables', *Linear Algebra and its Applications* **38**, 73–80.

J. Cullum, W. E. Donath and P. Wolfe (1975), 'The minimization of certain nondifferentiable sums of eigenvalues of symmetric matrices', *Mathematical Programming Study* **3**, 35–55.

G. B. Dantzig (1991), Linear programming, in *History of Mathematical Programming: A Collection of Personal Reminiscences* (J. K. Lenstra, A. H. G. Rinnooy Kan and A. Schrijver, eds), North-Holland, Amsterdam, pp. 19–31.

C. Davis (1957), 'All convex invariant functions of Hermitian matrices', *Archiv der Mathematik* **8**, 276–278.

E. M. de Sá (1994*a*), 'Exposed faces and duality for symmetric and unitarily invariant norms', *Linear Algebra and its Applications* **197,198**, 429–450.

E. M. de Sá (1994*b*), 'Faces and traces of the unit ball of a symmetric gauge function', *Linear Algebra and its Applications* **197,198**, 349–395.

E. M. de Sá (1994*c*), 'Faces of the unit ball of a unitarily invariant norm', *Linear Algebra and its Applications* **197,198**, 451–493.

R. Deville, G. Godefroy and V. Zizler (1993), *Smoothness and Renormings in Banach Spaces*, Longman, Harlow, UK.

W. E. Donath and A. J. Hoffman (1973), 'Lower bounds for the partitionings of graphs', *IBM Journal of Research and Development* **17**, 420–425.

M. Fiedler (1973), 'Algebraic connectivity of graphs', *Czech. Math. Journal* **23**, 298–305.

R. Fletcher (1985), 'Semi-definite constraints in optimization', *SIAM J. Control Optim.* **23**, 493–513.

R. Fletcher (1991), 'A new variational result for quasi-Newton formulae', *SIAM J. Optim.* **1**, 18–21.

S. Friedland (1981), 'Convex spectral functions', *Linear and multilinear algebra* **9**, 299–316.

S. Friedland, J. Nocedal and M. L. Overton (1987), 'The formulation and analysis of numerical methods for inverse eigenvalue problems', *SIAM J. Numer. Anal.* **24**, 634–667.

M. X. Goemans and D. P. Williamson (1996), 'Improved Approximation Algorithms for Maximum Cut and Satisfiability Problems Using Semidefinite Programming', *J. Assoc. Comput. Mach.* To appear.

C. C. Gonzaga (1992), 'Path-following methods for linear programming', *SIAM Review* **34**, 167–224.

M. Grötschel, L. Lovász and A. Schriver (1988), *Geometric Algorithms and Combinatorial Optimization*, Springer, New York.

C. Helmberg (1994), An Interior Point Method for Semidefinite Programming and Max-Cut Bounds, PhD thesis, Technische Universität Graz, Austria.

C. Helmberg, F. Rendl, R. Vanderbei and H. Wolkowicz (1996), 'An interior-point method for semidefinite programming', *SIAM J. Optim.* To appear.

J.-B. Hirriart-Urruty and D. Ye (1995), 'Sensitivity analysis of all eigenvalues of a symmetric matrix', *Numer. Math.* **70**, 45–72.

R. A. Horn and C. R. Johnson (1985), *Matrix Analysis*, Cambridge University Press, UK.

R. A. Horn and C. R. Johnson (1991), *Topics in Matrix Analysis*, Cambridge University Press, UK.

N. Karmarkar (1984), 'A new polynomial time algorithm for linear programming', *Combinatorica* **4**, 373–395.

N. Karmarkar and S. Thakur (1992), An interior-point approach to a tensor optimization problem with application to upper bounds in integer quadratic optimization problems, in *Integer Programming and Combinatorial Optimization* (E. Balas, G. Cornuejils and R. Kannan, eds), pp. 406–419. Proceedings of a conference held at Carnegie Mellon University by the Math. Programming Society.

P. G. Kirmser and K.-K. Hu (1993), 'The shape of the ideal column reconsidered', *Mathematical Intelligencer* **15**, 62–67.

M. Kojima, S. Mizuno and A. Yoshise (1989), A primal-dual interior point method for linear programming, in *Progress in Mathematical Programming, Interior-Point and Related Methods* (N. Megiddo, ed.), Springer, New York, pp. 29–47.

M. Kojima, S. Shindoh and S. Hara (1994), Interior-point methods for the monotone linear complementarity problem in symmetric matrices. Technical Report B-282, Department of Information Sciences, Tokyo Inst. of Technology.

A. S. Lewis (1995*a*), 'The convex analysis of unitarily invariant functions', *J. Convex Anal.* **2**, 173–183.

A. S. Lewis (1995*b*), Eigenvalue-constrained faces. Technical Report CORR 95-22, University of Waterloo. Submitted to Linear Algebra and its Applications.

A. S. Lewis (1995*c*), Von Neumann's lemma and a Chevalley-type theorem for convex functions on Cartan subspaces. Technical Report CORR 95-19, University of Waterloo. Submitted to Transactions of the American Mathematical Society.

A. S. Lewis (1996*a*), 'Convex analysis on the Hermitian matrices', *SIAM J. Optim.* **6**, 164–177.

A. S. Lewis (1996*b*), 'Derivatives of spectral functions', *Mathematics of Operations Research.* To appear.

A. S. Lewis (1996*c*), 'Group invariance and convex matrix analysis', *SIAM J. Matrix Anal. Appl.* To appear.

I. J. Lustig, R. E. Marsten and D. F. Shanno (1994), 'Interior point methods for linear programming: Computational state of the art', *ORSA Journal on Computing* **6**, 1–14.

A. W. Marshall and I. Olkin (1979), *Inequalities: Theory of Majorization and its Applications*, Academic Press, New York.

J.-E. Martínez-Legaz (1995), On convex and quasi-convex spectral functions. Technical report, Universitat Autònoma de Barcelona.

L. McLinden (1980), 'An analogue of Moreau's proximation theorem with applications to nonlinear complementarity problems', *Pacific Journal of Mathematics* pp. 101–161.

N. Megiddo (1989), Pathways to the optimal set in linear programming, in *Progress in Mathematical Programming, Interior-Point and Related Methods* (N. Megiddo, ed.), Springer, New York, pp. 131–158.

B. Mohar and S. Poljak (1993), Eigenvalues in combinatorial optimization, in *Combinatorial and graph-theoretic problems in linear algebra* (R. Brualdi, S. Friedland and V. Klee, eds), Vol. 50, Springer. IMA Volumes in Mathematics and its Applications.

R. Monteiro and I. Adler (1989), 'Interior path-following primal-dual algorithms: Part I: Linear programming', *Mathematical Programming* **44**, 27–41.

J. Moro, J. V. Burke and M. L. Overton (1995), On the Lidskii–Vishik–Lyusternik perturbation theory for eigenvalues of matrices with arbitrary Jordan structure. Technical Report 710, NYU Comp. Sci. Dept. Submitted to *SIAM J. Matr. Anal. Appl.*

Y. Nesterov and A. Nemirovskii (1994), *Interior Point Polynomial Methods in Convex Programming*, SIAM, Philadelphia.

Y. Nesterov and M. Todd (1996), 'Self-scaled barriers and interior-point methods for convex programming', *Mathematics of Operations Research*. To appear.

N. Olhoff and S. Rasmussen (1977), 'On single and bimodal optimum buckling loads of clamped columns', *Int. J. Solids Struct.* **9**, 605–614.

J. M. Ortega (1972), *Numerical Analysis: A Second Course*, Academic Press, New York and London. Reprinted by SIAM, Philadelphia, 1990.

M. L. Overton (1988), 'On minimizing the maximum eigenvalue of a symmetric matrix', *SIAM J. Matrix Anal. Appl.* **9**, 256–268.

M. L. Overton (1992), 'Large-scale optimization of eigenvalues', *SIAM J. Optim.* **2**, 88–120.

M. L. Overton and R. S. Womersley (1988), 'On minimizing the spectral radius of a nonsymmetric matrix function – optimality conditions and duality theory', *SIAM J. Matrix Anal. Appl.* **9**, 473–498.

M. L. Overton and R. S. Womersley (1993), 'Optimality conditions and duality theory for minimizing sums of the largest eigenvalues of symmetric matrices', *Mathematical Programming* **62**, 321–357.

M. L. Overton and R. S. Womersley (1995), 'Second derivatives for optimizing eigenvalues of symmetric matrices', *SIAM J. Matrix Anal. Appl.* **16**, 697–718.

G. Pataki (1995), On the rank of extreme matrices in semidefinite programs and the multiplicity of optimal eigenvalues. Technical Report MSRR-604 (revised), Graduate School of Industrial Administration, Carnegie–Mellon University. Submitted to *Math. of Operations Research*.

L. E. Payne, G. Pólya and H. F. Weinberger (1956), 'On the ratio of consecutive eigenvalues', *Journal of Mathematics and Physics* **35**, 289–298.

S. Poljak and Z. Tuza (1993), Max-cut problem – a survey. Technical report, Charles University, Prague.

A. Pothen (1996), Graph partitioning algorithms in scientific computing, in *Parallel Numerical Algorithms* (D. Keyes, A. Sameh and V. Venkatakrishnan, eds), Kluwer. To appear.

F. Rendl and H. Wolkowicz (1992), 'Applications of parametric programming and eigenvalue maximization to the quadratic assignment problem', *Mathematical Programming* **53**, 63–78.

U. T. Ringertz (1995), 'An algorithm for optimization of nonlinear shell structures', *International Journal for Numerical Methods in Engineering* **38**, 299–314.

U. T. Ringertz (1996), Eigenvalues in optimum structural design, in *Proceedings of an IMA workshop on Large-Scale Optimization* (A. R. Conn, L. T. Biegler, T. F. Coleman and F. Santosa, eds). To appear.

R. T. Rockafellar (1970), *Convex Analysis*, Princeton University Press, Princeton University.

H. Schramm and J. Zowe (1992), 'A version of the bundle idea for minimizing a nonsmooth function: conceptual idea, convergence analysis, numerical results', *SIAM J. Optim.* **2**, 121–152.

A. Seeger (1996), 'Convex analysis of spectrally-defined matrix functions', *SIAM J. Optim.* To appear.

A. Shapiro (1996), 'First and second order analysis of nonlinear semidefinite programs', *Mathematical Programming (Series B)*. To appear.

A. Shapiro and M. K. H. Fan (1995), 'On eigenvalue optimization', *SIAM J. Optim.* **3**, 552–568.

I. Tadjbakhsh and J. B. Keller (1962), 'Strongest columns and isoperimetric inequalities for eigenvalues', *J. Appl. Mech.* **29**, 159–164.

C. M. Theobald (1975), 'An inequality for the trace of the product of two symmetric matrices', *Mathematical Proceedings of the Cambridge Philosophical Society* **77**, 265–266.

N.-K. Tsing, M. K. H. Fan and E. I. Verriest (1994), 'On analyticity of functions involving eigenvalues', *Linear Algebra and its Applications* **207**, 159–180.

L. Tunçel (1995). Private communication.

L. Vandenberghe and S. Boyd (1995), 'Primal-dual potential reduction method for problems involving matrix inequalities', *Mathematical Programming (Series B)* **69**, 205–236.

L. Vandenberghe and S. Boyd (1996), 'Semidefinite programming', *SIAM Review* **38**, 49–95.

G. A. Watson (1992), 'Characterization of the subdifferential of some matrix norms', *Linear Algebra and its Applications* **170**, 33–45.

M. H. Wright (1992), Interior methods for constrained optimization, in *Acta Numerica*, Cambridge University Press, UK, pp. 341–407.

S. J. Wright (1995), 'Stability of linear equations solvers in interior-point methods', *SIAM J. Matrix Anal. Appl.* **16**, 1287–1307.

Y. Zhang and R. A. Tapia (1993), 'A superlinearly convergent polynomial primal-dual interior-point algorithm for linear programming', *SIAM J. Optim.* **3**, 118–133.

Y. Zhang, R. A. Tapia and J. E. Dennis (1992), 'On the superlinear and quadratic convergence of primal-dual interior point linear programming algorithms', *SIAM J. Optim.* **2**, 304–324.

K. Ziętak (1988), 'On the characterization of the extremal points of the unit sphere of matrices', *Linear Algebra and its Applications* **106**, 57–75.

K. Ziętak (1993), 'Subdifferentials, faces and dual matrices', *Linear Algebra and its Applications* **185**, 125–141.

Acta Numerica (1996), *pp.* 191–257

On the computation of crystalline microstructure

Mitchell Luskin *

School of Mathematics
University of Minnesota
Minneapolis, MN 55455, USA
E-mail: luskin@math.umn.edu

Microstructure is a feature of crystals with multiple symmetry-related energy-minimizing states. Continuum models have been developed explaining microstructure as the mixture of these symmetry-related states on a fine scale to minimize energy. This article is a review of numerical methods and the numerical analysis for the computation of crystalline microstructure.

CONTENTS

1	Introduction	191
2	Continuum theory for martensitic crystals	195
3	Microstructure	201
4	Finite element methods	227
5	Approximation of microstructure	236
6	Numerical analysis of microstructure	242
7	Relaxation	246
8	Acknowledgments	249
	References	249

1. Introduction

Advances in the understanding of material microstructure are playing an important role in the development of many new technologies that depend on material properties such as shape memory, magnetostriction, and ferroelectricity. Microstructure occurs in many materials as the fine-scale spatial oscillation

* This work was supported in part by the NSF through grants DMS 91-11572 and DMS 95-05077, by the AFOSR through grant AFOSR-91-0301, by the ARO through grant DAAL03-92-G-0003, by the Institute for Mathematics and its Applications, and by a grant from the Minnesota Supercomputer Institute.

Fig. 1. Photomicrograph of an austenitic-martensitic phase boundary (see Section 3.9) for a single crystal of Cu-14 at.% Al-3.9 at.% Ni from the laboratory of C. Chu and R. James. The martensitic phase is laminated or 'twinned'. (Field of view: 1.25 mm × 0.86 mm.)

between symmetry-related states. In this article, we survey the recent development of numerical methods and their analysis to compute microstructure in materials. We will be mainly concerned here with the microstructure of martensitic crystals where lattice structure oscillates between 'twinned' states (see Fig. 1 and Fig. 2).

During the past several years a geometrically nonlinear continuum theory for the equilibria of martensitic crystals based on elastic energy minimization has been developed (Ericksen 1986, 1987a, 1987b, Ball and James 1987, James and Kinderlehrer 1989, Ball and James 1992). The invariance of the energy density with respect to symmetry-related states implies that the elastic energy density is non-convex and must have multiple energy wells. For a large class of boundary conditions, the gradients of energy-minimizing sequences of deformations must oscillate between the energy wells to allow the energy to converge to the lowest possible value. Even though the deformation gradients of such energy-minimizing sequences do not converge pointwise, certain kinds of averages of the deformation gradients converge for a large class of boundary conditions. This convergence has been studied intensively using

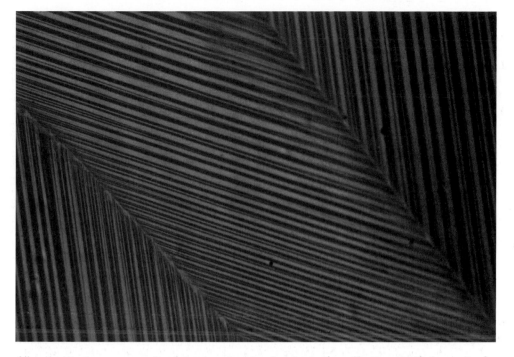

Fig. 2. Photomicrograph of a second-order laminate (see Section 3.10) for a single crystal of Cu-14 at.% Al-3.9 at.% Ni from the laboratory of C. Chu and R. James. (Field of view: 1.25 mm × 0.86 mm.)

the Young measure (Tartar 1984, Kinderlehrer and Pedregal 1991, Ball and James 1992) and the H-measure (Tartar 1990, Kohn 1991).

A geometrically linear theory for the equilibria of martensitic crystals was developed by Eshelby (1961), Khachaturyan (1967, 1983), Khachaturyan and Shatalov (1969), and Roitburd (1969, 1978). This theory is nonlinear, though, because the energy density has local minima at multiple stress-free strains. The relationship between the geometrically linear theory and the geometrically nonlinear theory has been explored by Kohn (1991), Ball and James (1992), and Bhattacharya (1993). Most of the results for the geometrically nonlinear theory that we discuss in this article have related counterparts for the geometrically linear theory.

These theories have presented a major challenge to the development and analysis of numerical methods, since they have features very unlike those of the physical theories usually approximated by numerical methods. The presence of microstructure has motivated the development of numerical methods that can capture macroscopic information without resolving the microstructure on the physical length scale (which can vary from nanometres to millimetres).

Although much progress has been made in the analysis of global minima of models for the energy of martensitic crystals, such crystals typically exhibit hysteretic behaviour and are usually observed in local minima or in meta-stable states (Burkart and Read 1953, Basinski and Christian 1954, Ball, Chu and James 1994, Ball, Chu and James 1995). Since the analytic study of these local minima is difficult, the computational approach offers an important tool for the exploration of meta-stable states. Thus, a further computational challenge is presented by the multitude of local minima which the numerical models necessarily inherit from the continuum models (Ball, Holmes, James, Pego and Swart 1991), as well as the local minima that occur from representing the same microstructure on the different scales possible for a given grid.

Early three-dimensional computations and numerical algorithms for a geometrically nonlinear model of microstructure in martensitic crystals have been given by Collins and Luskin (1989) for the In-20.7 at.% Tl alloy, and Silling (1989) has reported computations for a two-dimensional model exhibiting microstructure. Later computational results and numerical algorithms for equilibrium problems are given by Collins, Luskin and Riordan (1993) and Collins (1993a). Computations and numerical algorithms for geometrically linear models of martensitic crystals have been given by Wen, Khachaturyan and Morris Jr. (1981), Wang, Chen and Khachaturyan (1994), Kartha, Castán, Krumhansl and Sethna (1994) and Kartha, Krumhansl, Sethna and Wickham (1995).

A theory for the numerical analysis of microstructure was proposed by Collins, Kinderlehrer and Luskin (1991a) and Collins and Luskin (1991b) and extended in Chipot (1991), Chipot and Collins (1992), Gremaud (1994) and Chipot, Collins and Kinderlehrer (1995). This theory has been used to give an analysis of the convergence of numerical methods for three-dimensional, physical models of microstructure in ferromagnetic crystals (Luskin and Ma 1992) and in martensitic crystals with an orthorhombic to monoclinic transformation (Luskin 1996a, Luskin 1996b) and a cubic to tetragonal transformation (Li and Luskin 1996).

The theory for the numerical analysis of microstructure gives error estimates for the local mixture, rather than the pointwise values, of the deformation gradients; so the representations of the same microstructure on different scales are shown to yield almost identical macroscopic properties. These estimates show that many macroscopic properties converge as the length scale of the underlying microstructure converges to zero, which gives a justification for computing microstructure on a length scale that can be orders of magnitude larger than the physical length scale.

The relaxed energy density for a given deformation gradient $F \in \mathbb{R}^{3 \times 3}$ is given by the *infimum* of the average energy of deformations defined on a smooth domain and constrained to be equal to an Fx on the boundary. Under appropriate conditions, the *infimum* of the relaxed energy is attained by deformations that are the limit of energy-minimizing (for the original energy) sequences of deformations (Ekeland and Temam 1974, Dacorogna 1989). Although an explicit formula or practical computational algorithm for the relaxed energy density (the quasi-convex envelope) is generally not known for the non-convex energies used to model martensitic crystals, representations of the polyconvex and rank-one convex envelopes have been given, which can be numerically approximated to give lower and upper bounds for the relaxed energy density (Dacorogna 1989). These representations, especially that given by Kohn and Strang (1986), have been used in Nicolaides and Walkington (1993), Roubíček (1994), Carstensen and Plecháč (1995), Roubíček (1996a), Pedregal (1996), Pedregal (1995) and Kružík (1995).

The computation of the dynamics of the development and propagation of microstructure is important for the modelling and control of materials with microstructure. Swart and Holmes (1992) have studied the 'viscoelastodynamics' of a scalar, two-dimensional model, and Klouček and Luskin (1994a) and Klouček and Luskin (1994b) have computed the viscoelastodynamics of a three-dimensional model for the In-20.7 at.% Tl alloy.

This article focuses on computational methods for continuum theories for single martensitic crystals. Our bibliography contains references to many topics that we do not consider in detail in the text, such as homogenization, polycrystals, surface energy, and dynamics. We refer the reader to Luskin and Ma (1992, 1993) and Ma (1993) for recent developments in numerical methods and numerical analysis for the computation of the microstructure in the magnetization of ferromagnetic crystals.

2. Continuum theory for martensitic crystals

We give here a brief outline of the geometrically nonlinear continuum theory for martensitic crystals (Ericksen 1986, 1987a, 1987b; Ball and James 1987, 1992). The crystallographic background for the topics treated in this section will be given in the forthcoming book by Pitteri and Zanzotto (1996a). Martensitic crystals have a high-temperature phase known as *austenite*, and a low-temperature, less symmetric phase known as *martensite*. The austenitic phase exists in one variant, but the martensitic phase exists in several symmetry-related variants and can form a microstructure by the fine-scale mixing of the variants.

2.1. The elastic energy and admissible deformations

We use the austenitic phase at the transformation temperature as the reference configuration $\Omega \subseteq \mathbb{R}^3$ of the crystal. We assume that Ω is either a polyhedron or a smooth, bounded domain. We denote deformations by functions $y(x) : \Omega \to \mathbb{R}^3$, and we denote the corresponding deformation gradients by $F(x) = \nabla y(x)$.

We shall denote the elastic energy per unit volume at temperature θ and deformation gradient $F \in \mathbb{R}^{3 \times 3}$ by $\phi(F, \theta)$, which shall always be assumed to be continuous and to satisfy the growth condition

$$\phi(F, \theta) \geq C_1 \|F\|^p - C_0 \qquad \text{for all } F \in \mathbb{R}^{3 \times 3}, \tag{2.1}$$

where C_0 and C_1 are positive constants independent of $F \in \mathbb{R}^{3 \times 3}$, where we assume $p > 3$ to ensure that deformations with finite energy are continuous (see (2.4) below), and where we are using the matrix norm

$$\|F\|^2 \equiv \sum_{i,j=1}^{3} F_{ij}^2 \qquad \text{for } F \in \mathbb{R}^{3 \times 3}.$$

It is not realistic to consider deformations with arbitrarily large deformation gradients $F(x) = \nabla y(x)$ within the theory of elasticity (we can expect non-elastic behaviour such as fracture and plasticity to occur at large deformation gradients), so our use of the growth condition (2.1) can be viewed as a mathematical convenience. Also, we will be concerned only with temperatures in a neighbourhood (θ_L, θ_U) of the transformation temperature θ_T, so we need only assume that the growth condition (2.1) is valid uniformly for $\theta \in (\theta_L, \theta_U)$.

We expect that observed deformations $\hat{y}(x)$ are local minima of the total elastic energy

$$\mathcal{E}(\hat{y}) = \int_\Omega \phi(\nabla \hat{y}(x), \theta) \, dx \tag{2.2}$$

among all deformations satisfying appropriate boundary conditions and having finite energy. However, we will see that there generally do not exist energy-minimizing deformations to (2.2) for the non-convex energy densities ϕ that we use to model martensitic crystals, and so we must consider energy-minimizing sequences.

Since $p > 3$ in the growth condition (2.1), we have that the deformations with finite energy are uniformly continuous (Adams 1975), so we can denote the set of deformations of finite energy by

$$W^\phi = \left\{ y \in C(\bar{\Omega}; \mathbb{R}^3) : \int_\Omega \phi(\nabla y(x), \theta) \, dx < \infty \right\}. \tag{2.3}$$

We note that

$$W^\phi \subset W^{1,p}(\Omega; \mathbb{R}^3) \subset C(\bar{\Omega}; \mathbb{R}^3), \tag{2.4}$$

where $W^{1,p}(\Omega; \mathbb{R}^3)$ is the Sobolev space of measurable deformations $y : \Omega \to \mathbb{R}^3$ such that (Adams 1975)

$$\int_\Omega \left[|y(x)|^p + \|\nabla y(x)\|^p \right] dx < \infty.$$

In what follows (and above in the definition of $\mathcal{E}$ and W^ϕ), we shall often

suppress the explicit dependence on temperature where we do not think that
there is a danger of misunderstanding.

To model an unconstrained crystal, we define the admissible set of deformations $\mathcal{A}$ to be the set of deformations of finite energy

$$\mathcal{A} = W^\phi,$$

and we consider energy-minimizing sequences of deformations for the problem

$$\inf_{y \in \mathcal{A}} \mathcal{E}(y). \tag{2.5}$$

For a crystal that is constrained on the entire boundary by the condition

$$y(x) = y_0(x), \qquad \text{for all } x \in \partial\Omega, \tag{2.6}$$

for some $y_0(x) \in W^\phi$, we consider energy-minimizing sequences of deformations for the problem (2.5), where the set $\mathcal{A}$ of admissible deformations
consists of all deformations of finite energy constrained on the boundary by
(2.6), that is,

$$\mathcal{A} = \left\{ y \in W^\phi : y(x) = y_0(x), \text{ for } x \in \partial\Omega \right\}.$$

Our model and analysis can also accommodate more general boundary conditions, such as the inclusion of boundary loads.

Admissible deformations should be orientation-preserving isomorphisms,
that is, $\det \nabla y(x) > 0$ for all $x \in \Omega$. However, we shall not explicitly impose
this constraint since we have found that computed solutions have always
satisfied this condition.

2.2. Frame indifference and crystal symmetry

The elastic energy density ϕ is required to be frame-indifferent, that is,

$$\phi(RF, \theta) = \phi(F, \theta) \qquad \text{for all } R \in \mathrm{SO}(3) \text{ and } F \in \mathbb{R}^{3\times3}, \tag{2.7}$$

where $\mathrm{SO}(3)$ denotes the set of orthogonal matrices with determinant equal
to 1. We assume that the energy density inherits the symmetry of the more
symmetric high temperature phase of the crystal when the domain of the
energy density is suitably restricted (Ericksen 1980, Pitteri 1984), so

$$\phi(R_i F R_i^T, \theta) = \phi(F, \theta) \qquad \text{for all } R_i \in \mathcal{G}, \tag{2.8}$$

where $\mathcal{G} = \{R_1, \ldots, R_L\}$ is the symmetry group of the austenite.

2.3. Local minima of the energy density

Near the transformation temperature, we will assume that the energy density
$\phi(F, \theta)$ has local minima at the deformation gradients that describe the austenitic and the martensitic phases, and is therefore non-convex. The reference

configuration has been taken to be the austenitic phase at the transformation temperature, so the identity deformation gradient I describes the austenitic phase, and by the frame indifference of the energy density (2.7), every $R \in SO(3)$ should then be a local minimum of the energy density $\phi(F, \theta)$. We note that for simplicity we have neglected the thermal expansion of the austenite in the above conditions, since the deformations describing the austenitic phase are taken to be independent of temperature.

We shall assume that the energy density $\phi(F, \theta)$ for the temperature θ near the transformation temperature θ_T also has local minima at the set of *variants*

$$\left\{ R_i U_1 R_i^T : R_i \in \mathcal{G} \right\} = \{U_1, \dots, U_M\} \tag{2.9}$$

which describe the martensitic phase. Here the $U_i = U_i(\theta)$ are deformation gradients for an unstressed crystal in the low-temperature, martensitic phase. It follows from the symmetry of the energy density (2.8) that

$$\phi(U_1, \theta) = \cdots = \phi(U_M, \theta). \tag{2.10}$$

Since M (defined in (2.9)) is equal to the number of cosets of the subgroup

$$\mathcal{H} = \left\{ R_i \in \mathcal{G} : R_i U_1 R_i^T = U_1 \right\}$$

in $\mathcal{G}$, we have by Lagrange's Theorem (Herstein 1975) that

$$M = \frac{|\mathcal{G}|}{|\mathcal{H}|}.$$

It follows from (2.10) and the frame indifference of the energy density (2.7) that $\phi(F, \theta)$ has local minima at the energy wells of each variant given by

$$\mathcal{U}_\rangle = SO(3)U_i = \{ RU_i : R \in SO(3) \} . \tag{2.11}$$

If we denote the union of the energy wells by

$$\mathcal{U} = \mathcal{U}_1 \cup \cdots \cup \mathcal{U}_M,$$

then it follows from the frame indifference (2.7) of the energy density and (2.10) that

$$\phi(U, \theta) = \phi(U_1, \theta) = \cdots = \phi(U_M, \theta) \qquad \text{for all } U \in \mathcal{U}.$$

Also, since admissible deformations are required to be orientation-preserving isomorphisms, we shall always assume that $\det U_1 > 0$, so by (2.9) and (2.11) we have that

$$\det U = \det U_1 > 0 \qquad \text{for all } U \in \mathcal{U}. \tag{2.12}$$

2.4. The orthorhombic to monoclinic transformation

We next present two examples of martensitic phase transformations. First, we describe the symmetry group $\mathcal{G}$ and the corresponding martensitic variants $\{R_i U_1 R_i^T : R_i \in \mathcal{G}\}$ for one of the orthorhombic to monoclinic transformations (Ball and James 1992). The symmetry group of the orthorhombic (high-temperature) phase is composed of the rotations of π radians about an orthogonal set of axes, so

$$\mathcal{G} = \{\, I, \, -I + 2e_1 \otimes e_1, \, -I + 2e_2 \otimes e_2, \, -I + 2e_3 \otimes e_3 \,\},$$

where $\{e_1, e_2, e_3\}$ is an orthonormal basis of $\mathbb{R}^3$. We recall that $v \otimes w \in \mathbb{R}^{3\times3}$ for $v, w \in \mathbb{R}^3$ is the tensor product defined by $(v \otimes w)_{kl} = v_k w_l$, or, equivalently, $(v \otimes w)u = (w \cdot u)v$ for $u \in \mathbb{R}^3$.

The variants of the monoclinic (low-temperature) phase can then be given by

$$U_1 = (I - \eta e_2 \otimes e_1)\, D \qquad \text{and} \qquad U_2 = (I + \eta e_2 \otimes e_1)\, D, \qquad (2.13)$$

where $\eta > 0$, and where $D \in \mathbb{R}^{3\times3}$ is the positive diagonal matrix

$$D = d_1 e_1 \otimes e_1 + d_2 e_2 \otimes e_2 + d_3 e_3 \otimes e_3$$

for $d_1, d_2, d_3 > 0$. We note that

$$\left\{ R_i U_1 R_i^T : R_i \in \mathcal{G} \right\} = \{U_1, U_2\}.$$

2.5. The cubic to tetragonal transformation

For the more common cubic (high-temperature) to tetragonal (low-temperature) transformation, the group $\mathcal{G}$ is the symmetry group of the cube

$$\mathcal{G} = \{\, R_1, \ldots, R_{24} \,\}, \qquad (2.14)$$

which is given by the group of matrices

$$R_i = (-1)^{\upsilon(1)} e_{\pi(1)} \otimes e_1 + (-1)^{\upsilon(2)} e_{\pi(2)} \otimes e_2 + (-1)^{\upsilon(3)} e_{\pi(3)} \otimes e_3,$$

where $\upsilon : \{1, 2, 3\} \to \{0, 1\}$; $\pi : \{1, 2, 3\} \to \{1, 2, 3\}$ is a permutation; and $\det R_i = 1$. We also assume as above that $\{e_1, e_2, e_3\}$ is an orthonormal basis of $\mathbb{R}^3$. The variants of the tetragonal phase can be taken to be

$$U_1 = \nu_1 I + (\nu_2 - \nu_1)e_1 \otimes e_1, \qquad U_2 = \nu_1 I + (\nu_2 - \nu_1)e_2 \otimes e_2,$$
$$U_3 = \nu_1 I + (\nu_2 - \nu_1)e_3 \otimes e_3 \qquad (2.15)$$

where $0 < \nu_1$, $0 < \nu_2$, and $\nu_1 \neq \nu_2$. For this transformation,

$$\left\{ R_i U_1 R_i^T : R_i \in \mathcal{G} \right\} = \{U_1, U_2, U_3\}.$$

2.6. Global minima of the energy density

The reference configuration has been chosen so that $F = I$ is the deformation gradient for the high-temperature phase at the transformation temperature $\theta = \theta_T$. The elastic energy density should then predict that the high-temperature phase (represented by $F \in \mathrm{SO}(3)$) is a global minimum for $\theta > \theta_T$ and the low-temperature phase (represented by $U \in \mathcal{U} = \mathcal{U}_1 \cup \cdots \cup \mathcal{U}_M$) is a global minimum for $\theta < \theta_T$. Thus, the elastic energy density should satisfy the conditions that, for $\theta > \theta_T$,

$$\phi(F, \theta) > \phi(R, \theta) \quad \text{for all } F \notin \mathrm{SO}(3), \ R \in \mathrm{SO}(3); \qquad (2.16)$$

for $\theta = \theta_T$,

$$\phi(F, \theta_T) > \phi(R, \theta_T) = \phi(U, \theta_T) \quad \text{for all } F \notin \mathrm{SO}(3) \cup \mathcal{U}, \ R \in \mathrm{SO}(3), \ U \in \mathcal{U};$$
$$(2.17)$$

and for $\theta < \theta_T$,

$$\phi(F, \theta) > \phi(U, \theta) \quad \text{for all } F \notin \mathcal{U}, \ U \in \mathcal{U}. \qquad (2.18)$$

2.7. The Ericksen–James energy density for the cubic to tetragonal transformation

The development of a computational model for martensitic crystals requires the construction of an energy density $\phi(F, \theta)$ that is frame-indifferent (2.7), has the symmetry group of the crystal (2.8), satisfies the qualitative properties of the first-order phase transition (2.16)–(2.18), and matches available experimental data such as the linear elastic moduli of the pure phases and the dependence of the transformation temperature on stress. The following such energy density for the cubic to tetragonal transformation was developed by Ericksen and James (Ericksen 1986, Ericksen 1987a, Collins and Luskin 1989):

$$\begin{aligned}
\phi(F, \theta) = {} & \frac{b(\theta)}{6} \left[\left(\frac{3C_{11}}{\mathrm{tr}\, C} - 1 \right)^2 + \left(\frac{3C_{22}}{\mathrm{tr}\, C} - 1 \right)^2 + \left(\frac{3C_{33}}{\mathrm{tr}\, C} - 1 \right)^2 \right] \\
& + \frac{c(\theta)}{2} \left(\frac{3C_{11}}{\mathrm{tr}\, C} - 1 \right) \left(\frac{3C_{22}}{\mathrm{tr}\, C} - 1 \right) \left(\frac{3C_{33}}{\mathrm{tr}\, C} - 1 \right) \\
& + \frac{d(\theta)}{36} \left[\left(\frac{3C_{11}}{\mathrm{tr}\, C} - 1 \right)^2 + \left(\frac{3C_{22}}{\mathrm{tr}\, C} - 1 \right)^2 + \left(\frac{3C_{33}}{\mathrm{tr}\, C} - 1 \right)^2 \right]^2 \\
& + \frac{e}{2} (C_{12}^2 + C_{13}^2 + C_{23}^2 + C_{21}^2 + C_{31}^2 + C_{32}^2) + f(\mathrm{tr}\, C - 3)^2,
\end{aligned} \qquad (2.19)$$

where $C = F^T F$ is the right Cauchy–Green strain and $\mathrm{tr}\, C$ is the trace of C. The energy density (2.19) is frame-indifferent since it is a function of the right Cauchy–Green strain C. Ericksen has also shown that it has the cubic symmetry group, and that the coefficients b, c, d, e, and f can be chosen so that the energy density satisfies the qualitative conditions for the first-order phase transition with

$$\nu_1 = \sqrt{1 - \epsilon} \qquad \nu_2 = \sqrt{1 + 2\epsilon}$$

for $0 < \epsilon < 1$ (Ericksen 1986).

Ericksen and James have also determined moduli that fit experimental data for the In-20.7 at% Tl alloy. These moduli are given by (θ in °C and moduli in gigapascals)

$$b = 0.38 + (1.22 \times 10^{-3})(\theta - \theta_T), \qquad c = -29.23, \qquad d = 562.13$$
$$e = 3.26, \qquad f = 5.25,$$

where the transformation temperature is $\theta_T = 70°C$. The size of In-20.7 at% Tl crystals used in laboratory experiments is typically on the order of several centimetres in diameter.

An easy calculation establishes that

$$\phi(U(\epsilon), \theta) = b(\theta)\epsilon^2 + c\epsilon^3 + d\epsilon^4,$$

where $U(\epsilon) = \text{diag}(\sqrt{1 + 2\epsilon}, \sqrt{1 - \epsilon}, \sqrt{1 - \epsilon})$. Thus $\phi(U(\epsilon), \theta)$ has a local minimum in ϵ corresponding to the austenitic phase at $\epsilon(\theta) = 0$, for all temperatures satisfying $b(\theta) > 0$ (or for $\theta > -240°C$). Further, there is a local minimum at

$$\epsilon(\theta) = \frac{-3c + \sqrt{9c^2 - 32db(\theta)}}{8d}$$

corresponding to the martensitic phase for $\theta < \theta^*$, where $\theta^* = 108.92°C$ satisfies $9c^2 - 32db(\theta^*) - 0$. Thus, $\epsilon(\theta_T) = 0.026$.

3. Microstructure

In this section we will describe some examples and properties of microstructures.

3.1. Interfaces and the rank-one property

We first give a necessary and sufficient condition for the existence of a continuous deformation with a planar interface separating two regions with constant deformation gradients $F_0 \in \mathbb{R}^{3\times3}$ and $F_1 \in \mathbb{R}^{3\times3}$.

Lemma 1   Let $n \in \mathbb{R}^3$, $|n| = 1$, and $s \in \mathbb{R}$. There exists a continuous deformation $w(x) \in C(\mathbb{R}^3; \mathbb{R}^3)$ such that

$$\nabla w(x) - \begin{cases} F_0 & \text{for all } x \text{ such that } x \cdot n < s, \\ F_1 & \text{for all } x \text{ such that } x \cdot n > s, \end{cases} \tag{3.1}$$

if and only if there exists $a \in \mathbb{R}^3$ such that

$$F_1 = F_0 + a \otimes n. \tag{3.2}$$

Proof.   If $w(x) \in C(\mathbb{R}^3; \mathbb{R}^3)$ satisfies (3.1), then the equality of the directional derivatives of $w(x)$ in directions orthogonal to the normal of the interface implies that

$$F_1 v = F_0 v \qquad \text{for all } v \in \mathbb{R}^3 \text{ such that } v \cdot n = 0.$$

Thus, we have that (3.2) holds with

$$a = F_1 n - F_0 n.$$

Conversely, if (3.2) holds, then the deformation

$$w(x) = \begin{cases} F_0 x & \text{for all } x \text{ such that } x \cdot n < s, \\ F_1 x - sa & \text{for all } x \text{ such that } x \cdot n > s, \end{cases}$$

is continuous and satisfies (3.1). □

Lemma 1 can be strengthened to state that if $w(x)$ is a continuous deformation whose gradient takes constant values $F_0 \in \mathbb{R}^{3 \times 3}$ and $F_1 \in \mathbb{R}^{3 \times 3}$, with $F_0 \neq F_1$ in two regions separated by a smooth interface, then the interface is planar and (3.2) holds for some $a, n \in \mathbb{R}^3$, $|n| = 1$, with n a normal to the planar interface. A more general result for a deformation with a gradient taking two values can be found in Ball and James (1987). We also note that the condition $|n| = 1$ above is not essential since we can always rewrite $a \otimes n$ by $|n|a \otimes \frac{n}{|n|}$ when $n \neq 0$.

The above lemma motivates the following definition.

Definition 1   We say that $F_0 \in \mathbb{R}^{3 \times 3}$ and $F_1 \in \mathbb{R}^{3 \times 3}$ are *rank-one connected* if there exist $a \in \mathbb{R}^3$ and $n \in \mathbb{R}^3$, $|n| = 1$, such that

$$F_1 = F_0 + a \otimes n. \tag{3.3}$$

3.2. Laminated microstructure

More generally, if F_0 and F_1 are rank-one connected as in (3.3), then we can construct a continuous deformation having parallel planar interfaces

$$S_i = \{ x \in \Omega : x \cdot n = s_i \}$$

for $s_1 < \cdots < s_m$ with the same normal n separating the layers in which the deformation gradient alternates between F_0 and F_1 by

$$w(x) = F_0 x + \left[\int_0^{x \cdot n} \chi(s) \, ds \right] a, \tag{3.4}$$

where $\chi(s) : \mathbb{R} \to \mathbb{R}$ is the characteristic function

$$\chi(s) = \begin{cases} 0 & \text{if } x \in (s_{2l}, s_{2l+1}) \text{ for } 0 \leq 2l \leq m \text{ where } l \in \mathbb{Z}, \\ 1 & \text{if } x \in (s_{2l+1}, s_{2l+2}) \text{ for } 1 \leq 2l + 1 \leq m \text{ where } l \in \mathbb{Z}, \end{cases}$$

where we take $s_0 = -\infty$ and $s_{m+1} = \infty$. This deformation satisfies the property that

$$\nabla w(x) = F_0 + \chi(x \cdot n)a \otimes n = \begin{cases} F_0 & \text{for all } x \text{ such that } \chi(x \cdot n) = 0, \\ F_1 & \text{for all } x \text{ such that } \chi(x \cdot n) = 1. \end{cases}$$

Deformations $w(x)$ of the form (3.4) with layer thickness $s_{i+1} - s_i$ small for $i = 1, \ldots, m$ are the simplest examples of *microstructure*.

We define $\phi_{\min}(\theta)$ to be the lowest attainable energy at the temperature θ, that is,

$$\phi_{\min}(\theta) = \min_{F \in \mathbb{R}^{3 \times 3}} \phi(F, \theta). \tag{3.5}$$

For $\theta \leq \theta_T$ and F_0, $F_1 \in \mathcal{U}$, we have that

$$\phi_{\min}(\theta) = \phi(F_0, \theta) = \phi(F_1, \theta).$$

Thus, if $\theta \leq \theta_T$ and F_0, $F_1 \in \mathcal{U}$, then the deformation $w(x)$ defined by (3.4) attains the minimum energy, since

$$\mathcal{E}(w) = \int_\Omega \phi(\nabla w(x), \theta) \, \mathrm{d}x = \phi_{\min}(\theta) \mathrm{meas}\, \Omega.$$

Furthermore, if $\theta > \theta_T$ and F_0, $F_1 \in \mathcal{U}$, then the deformations $w(x)$ defined by (3.4) are equilibria, since every $F \in \mathcal{U}$ is a local minimum of $\phi(F, \theta)$ (see Section 2.3), and hence

$$\mathcal{E}(w + z) - \mathcal{E}(w) = \int_\Omega [\phi(\nabla w(x) + \nabla z(x), \theta) - \phi(\nabla w(x), \theta)] \, \mathrm{d}x \geq 0$$

for all $z \in W^{1,\infty}(\Omega; \mathbb{R}^3)$ such that ess $\sup_{x \in \Omega} \|\nabla z(x)\|$ is sufficiently small.

3.3. Surface energy

The *surface energy* $\mathcal{S}$ associated with all the interfaces S_i can be modelled by

$$\mathcal{S}(w) = \alpha \sum_{i=1}^m \mathrm{area}\, S_i, \tag{3.6}$$

where $\alpha > 0$ is the surface energy density and m is the number of interfaces. For $\theta \leq \theta_T$ and F_0, $F_1 \in \mathcal{U}$, the total energy is the sum of the bulk energy and the surface energy given by

$$\mathcal{E}(w) + \mathcal{S}(w) = \phi_{\min}(\theta) \, \mathrm{meas}\, \Omega + \alpha \sum_{i=1}^m \mathrm{area}\, S_i, \tag{3.7}$$

which is minimized when the deformation w does not have any interfaces, that is, when $w(x) = F_0 x$ or $w(x) = F_1 x$. So, how do we explain the presence of interfaces in martensitic crystals? We will see later in this section that the constraint of boundary conditions or the constraint of continuity between austenitic and martensitic regions can make deformations with closely spaced interfaces energetically advantageous. The presence of interfaces can also be explained by the meta-stability of such deformations (Abeyaratne, Chu and James 1994, Ball et al. 1995).

For analytical and computational purposes, the surface energy is usually

transformation (2.15) each $F_0 \in \mathcal{U}_i$ is not rank-one connected to any $F_1 \in \mathcal{U}_i$ with $F_0 \neq F_1$, but that every $F_0 \in \mathcal{U}_i$ is rank-one connected to two distinct $F_1 \in \mathcal{U}_j$ for all $j \neq i$, $j \in \{1, \cdots, M\}$.

Lemma 3 If $F_0 \in \mathcal{U}_i$ for some $i \in \{1, \cdots, M\}$, then there does not exist $F_1 \in \mathcal{U}_i$ with $F_0 \neq F_1$, such that F_0 and F_1 are rank-one connected.

Proof. If $F_0 = R_0 U_i \in \mathcal{U}_i$ and $F_1 = R_1 U_i \in \mathcal{U}_i$ are rank-one connected where R_0, $R_1 \in SO(3)$, then

$$R_1 U_i = R_0 U_i + a \otimes n$$

for $a \in \mathbb{R}^3$ and $n \in \mathbb{R}^3$, $|n| = 1$. So,

$$R_1 = R_0 + a \otimes U_i^{-T} n. \tag{3.11}$$

It then follows from (3.11) and Lemma 2 that $R_1 = R_0$ which proves the lemma. $\square$

The following lemma will allow us to reduce the problem of determining the rank-one connections for the orthorhombic to monoclinic transformation (2.13) and the cubic to tetragonal transformation (2.15) to a two-dimensional problem.

Lemma 4 Suppose that U_1, $U_2 \in \mathcal{U}$ satisfy the conditions

$$U_1 e_3 = U_1^T e_3 = U_2 e_3 = U_2^T e_3 = \hat{\nu} e_3 \tag{3.12}$$

for $\hat{\nu} \neq 0$. If there exists $R \in SO(3)$, $a \in \mathbb{R}^3$, and $n \in \mathbb{R}^3$ with $|n| = 1$, such that

$$R U_2 = U_1 + a \otimes n, \tag{3.13}$$

then $a \cdot e_3 = n \cdot e_3 = 0$ and $R = R(\sigma e_3)$ is the rotation matrix of angle σ about the axis e_3, which satisfies

$$R(\sigma e_3) U_2 v = U_1 v \tag{3.14}$$

for $v \in \mathbb{R}^3$ satisfying $v \cdot n = v \cdot e_3 = 0$, $v \neq 0$.

Conversely, if (3.14) holds for some $v \in \mathbb{R}^3$ satisfying $v \cdot e_3 = 0$, $v \neq 0$, then (3.13) holds for $R = R(\sigma e_3)$, $n \in \mathbb{R}^3$ satisfying $n \cdot e_3 = n \cdot v = 0$, $|n| = 1$, and $a = (R U_2 - U_1) n$.

Proof. We suppose that (3.13) holds. It then follows that

$$R U_2 = U_1 + a \otimes n = (I + a \otimes U_1^{-T} n) U_1, \tag{3.15}$$

so, since $\det U_1 = \det U_2 \neq 0$ by (2.12), we have that

$$\det U_1 = \det(R U_2) = \det(I + a \otimes U_1^{-T} n) \det U_1 = \left(1 + a \cdot U_1^{-T} n\right) \det U_1.$$

Hence, it follows that

$$a \cdot U_1^{-T} n = 0. \tag{3.16}$$

We have by (3.15) that

$$RU_2 U_1^{-1} = I + a \otimes U_1^{-T} n, \tag{3.17}$$

so we have for

$$C \equiv \left(RU_2 U_1^{-1}\right)^T \left(RU_2 U_1^{-1}\right) = \left(U_2 U_1^{-1}\right)^T \left(U_2 U_1^{-1}\right) \tag{3.18}$$

that

$$C = \left(I + U_1^{-T} n \otimes a\right)\left(I + a \otimes U_1^{-T} n\right). \tag{3.19}$$

Now it follows from (3.12) and (3.18) that

$$C e_3 = e_3 \tag{3.20}$$

and from (3.19) that

$$C e_3 = e_3 + (U_1^{-T} n \cdot e_3) a + \left[a \cdot e_3 + |a|^2 (U_1^{-T} n \cdot e_3)\right] U_1^{-T} n. \tag{3.21}$$

Since a and $U_1^{-T} n$ are linearly independent by (3.16), it follows from (3.20) and (3.21) that

$$U_1^{-T} n \cdot e_3 = 0 \qquad \text{and} \qquad a \cdot e_3 = 0. \tag{3.22}$$

Next, we have by (3.12) and (3.22) that

$$n \cdot e_3 - n \cdot \left(\hat{\nu} U_1^{-1} c_3\right) = \hat{\nu} U_1^{-T} n \cdot e_3 = 0. \tag{3.23}$$

We then obtain from (3.12), (3.13), and (3.23) that

$$\hat{\nu} R e_3 = RU_2 e_3 = (U_1 + a \otimes n) e_3 = U_1 e_3 = \hat{\nu} e_3. \tag{3.24}$$

We have that $R e_3 = e_3$ by (3.24), so we can conclude that

$$R = R(\sigma e_3),$$

where $R(\sigma e_3)$ is a rotation matrix of angle σ about the axis e_3. The result (3.14) now follows by (3.13) for $v \in \mathbb{R}^3$ satisfying $v \cdot n = v \cdot e_3 = 0$, $v \neq 0$.

Conversely, if (3.14) holds, then it is easy to check that (3.13) holds for $R = R(\sigma e_3)$, $n \in \mathbb{R}^3$ satisfying $n \cdot e_3 - n \cdot v = 0$ with $|n| = 1$, and $a = (RU_2 - U_1)n$. $\square$

Lemma 5 We consider the orthorhombic–monoclinic transformation (2.13). If $F_0 \in \mathcal{U}_i$ for $i \in \{1, 2\}$, then for $j \neq i$, $j \in \{1, 2\}$, there exist two distinct $F_1 \in \mathcal{U}_j$ such that F_0 and F_1 are rank-one connected.

Proof. Without loss of generality we may assume that $F_0 = U_1$, and we show that there exist two distinct $R \in SO(3)$ such that

$$RU_2 = U_1 + a \otimes n \tag{3.25}$$

for some $a, n \in \mathbb{R}^3$, $|n| = 1$. Since (3.12) holds with $\hat{\nu} = 1$, by Lemma 4 it is sufficient to determine all $\sigma \in \mathbb{R}$ and $v \in \text{Span}\{e_1, e_2\}$, $v \neq 0$, such that

$$R(\sigma e_3)U_2 v = U_1 v. \tag{3.26}$$

Now there exist $\sigma \in \mathbb{R}$ and $v \in \text{Span}\{e_1, e_2\}$, $v \neq 0$, such that

$$R(\sigma e_3)U_2 v = U_1 v$$

if and only if there exists $v \in \text{Span}\{e_1, e_2\}$, $v \neq 0$, such that

$$|U_1 v| = |U_2 v|. \tag{3.27}$$

For $v = v_1 e_1 + v_2 e_2$ where v_1, $v_2 \in \mathbb{R}$, we have that

$$|U_1 v| = |U_2 v| \tag{3.28}$$

if and only if

$$v_1 v_2 = 0.$$

The solution to (3.28) given by $v_1 = 0$ or $v = e_2$ corresponds to the obvious solution

$$U_2 = U_1 + 2\eta d_1 e_2 \otimes e_1$$

to (3.25) given by $n = e_1$ and $\sigma = 0$. The solution to (3.28) given by $v_2 = 0$ or $v = e_1$ corresponds to the solution to (3.25) given by

$$n = e_2, \qquad \cos \sigma = \frac{U_1 v \cdot U_2 v}{|U_1 v||U_2 v|} = \frac{d_1^2 - \eta^2 d_2^2}{d_1^2 + \eta^2 d_2^2} \qquad \text{for } -\pi < \sigma < 0.$$

We note that solutions v and $-v$ to (3.28) give the same solutions to (3.25). $\square$

Lemma 6   We consider the cubic to tetragonal transformation (2.15). If $F_0 \in \mathcal{U}_i$ for some $i \in \{1, 2, 3\}$, then for any $j \neq i$, $j \in \{1, 2, 3\}$, there exist two distinct $F_1 \in \mathcal{U}_j$ such that F_0 and F_1 are rank-one connected.

Proof.   Without loss of generality we again assume that $F_0 = U_1$ and $j = 2$, and we show that there exist two distinct $R \in SO(3)$ such that

$$RU_2 = U_1 + a \otimes n \tag{3.29}$$

for some a, $n \in \mathbb{R}^3$, $|n| = 1$. Since (3.12) holds with $\hat{\nu} = \nu_1$, by Lemma 4 it is sufficient to determine all $\sigma \in \mathbb{R}$ and $v \in \text{Span}\{e_1, e_2\}$, $v \neq 0$, such that

$$R(\sigma e_3)U_2 v = U_1 v. \tag{3.30}$$

Again, there exist $\sigma \in \mathbb{R}$ and $v \in \text{Span}\{e_1, e_2\}$, $v \neq 0$, such that

$$R(\sigma e_3)U_2 v = U_1 v$$

if and only if there exists $v \in \text{Span}\{e_1, e_2\}$, $v \neq 0$, such that

$$|U_1 v| = |U_2 v|. \tag{3.31}$$

For $v = v_1 e_1 + v_2 e_2$ where v_1, $v_2 \in \mathbb{R}$, we have that

$$|U_1 v| = |U_2 v| \tag{3.32}$$

if and only if

$$v_1^2 = v_2^2.$$

We have for the solution to (3.32) given by $v = e_1 - e_2$ the corresponding solution to (3.29) given by

$$n = \frac{1}{\sqrt{2}}(e_1 + e_2)$$

and

$$\cos \sigma = \frac{U_1 v \cdot U_2 v}{|U_1 v||U_2 v|} = \frac{2 v_1 v_2}{v_1^2 + v_2^2}$$

where

$$0 < \sigma < \frac{\pi}{2} \quad \text{if} \quad v_2 > v_1,$$

$$-\frac{\pi}{2} < \sigma < 0 \quad \text{if} \quad v_1 > v_2.$$

We also have for the solution $v = e_1 + e_2$ of (3.32) the corresponding solution to (3.29) given by

$$n - \frac{1}{\sqrt{2}}(e_1 - e_2)$$

and

$$\cos \sigma = \frac{U_1 v \cdot U_2 v}{|U_1 v||U_2 v|} = \frac{2 v_1 v_2}{v_1^2 + v_2^2}$$

where

$$0 < \sigma < \frac{\pi}{2} \quad \text{if} \quad v_1 > v_2,$$

$$-\frac{\pi}{2} < \sigma < 0 \quad \text{if} \quad v_2 > v_1.$$

Thus, the solutions to (3.29) give two distinct families of parallel interfaces corresponding to

$$n = \frac{1}{\sqrt{2}}(e_1 + e_2) \quad \text{and} \quad n = \frac{1}{\sqrt{2}}(e_1 - e_2).$$

It follows from symmetry that there are four additional distinct families of parallel interfaces corresponding to

$$n = \frac{1}{\sqrt{2}}(e_1 + e_3), \qquad n = \frac{1}{\sqrt{2}}(e_1 - e_3),$$

and

$$n = \frac{1}{\sqrt{2}}(e_2 + e_3), \qquad n = \frac{1}{\sqrt{2}}(e_2 - e_3).$$

□

The homogeneous austenitic phase can be separated from the homogeneous martensitic phase by a planar interface with normal n if and only if there exist a rotation $R \in SO(3)$ and vectors $a \in \mathbb{R}^3$ and $n \in \mathbb{R}^3$, $|n| = 1$, such that

$$RU_i = I + a \otimes n$$

for some $i \in \{1, \ldots, M\}$, where U_i is one of the variants defined by (2.9). The following theorem gives a necessary and sufficient condition for (3.33) to have a solution.

Lemma 7 We consider the cubic to tetragonal transformation (2.15). If $\nu_1 \neq 1$, then there does not exist a rotation $R \in SO(3)$ and vectors $a \in \mathbb{R}^3$ and $n \in \mathbb{R}^3$, $|n| = 1$, such that

$$RU_i = I + a \otimes n \tag{3.33}$$

for any $i \in \{1, 2, 3\}$. If $\nu_1 = 1$, then

$$U_i = I + (\nu_2 - 1)e_i \otimes e_i \tag{3.34}$$

for any $i \in \{1, 2, 3\}$.

Proof. We first assume that $\nu_1 \neq 1$ and $\nu_2 = 1$ and that there exist a rotation $R \in SO(3)$ and vectors $a \in \mathbb{R}^3$ and $n \in \mathbb{R}^3$, $|n| = 1$, such that (3.33) holds for some $i \in \{1, 2, 3\}$. We have that $|RU_i v| = |v|$ if and only if v lies in the one-dimensional subspace spanned by e_i. However, $|(I + a \otimes n)v| = |v|$ for all v in the two-dimensional subspace for which $v \cdot n = 0$, which contradicts (3.33).

We next assume that $\nu_1 \neq 1$ and $\nu_2 \neq 1$. By multiplying (3.33) by its transpose, we have

$$U_i^2 = (RU_i)^T RU_i = (I + n \otimes a)(I + a \otimes n), \tag{3.35}$$

since $U_i^T = U_i$ and $R^T = R^{-1}$ because $R \in SO(3)$. Further, a is nonzero, because otherwise (3.33) implies that $U_i \in SO(3)$. Now $a \times n$ is an eigenvector of $(I + n \otimes a)(I + a \otimes n)$ with eigenvalue 1, so we have reached a contradiction, since 1 is not an eigenvalue of U_i^2 in this case.

The proof of the result (3.34) follows directly from the definition of the U_i given in (2.15). □

Bhattacharya (1992) has shown that martensitic crystals exhibiting the shape-memory phenomenon that is important for many technological applications can be expected to have a transformation that is approximately volume preserving, that is, $\det U_i = \det I$ or $\nu_1^2 \nu_2 = 1$. Hence, we do not expect to

observe the homogeneous austenitic phase separated from the homogeneous
martensitic phase by a smooth interface in martensitic crystals exhibiting the
shape-memory phenomena. We shall see in Section 3.9 that if

$$\nu_1 < 1 < \nu_2 \quad \text{and} \quad \frac{1}{\nu_1^2} + \frac{1}{\nu_2^2} < 2,$$

$$\text{or} \quad \nu_2 < 1 < \nu_1 \quad \text{and} \quad \nu_1^2 + \nu_2^2 < 2,$$

then the homogeneous austenitic phase can be separated by a planar interface
from a martensitic phase that is composed of a fine-scale laminate of two
martensitic variants.

3.5. Boundary constraints and fine-scale laminates

We can construct energy-minimizing deformations w with arbitrarily fine-
scale oscillations from energy-minimizing deformation gradients $F_0 \in \mathcal{U}$ and
$F_1 \in \mathcal{U}$ that are rank-one connected as in (3.3). To construct a laminated microstructure having deformation gradient F_0 for volume fraction $1 - \lambda$ (where
$0 < \lambda < 1$) and having deformation gradient F_1 for volume fraction λ, we
construct the continuous deformation $w_\gamma(x)$ with layer thickness $\gamma > 0$ by

$$w_\gamma(x) = \gamma w \left(\frac{x}{\gamma} \right), \tag{3.36}$$

where

$$w(x) = F_0 x + \left[\int_0^{x \cdot n} \chi(s) \, ds \right] a$$

and where $\chi(s) : \mathbb{R} \to \mathbb{R}$ is the characteristic function with period 1 defined
by

$$\chi(s) = \begin{cases} 0 & \text{for all} \quad 0 \le s \le 1 - \lambda, \\ 1 & \text{for all} \quad 1 - \lambda < s < 1. \end{cases}$$

Now

$$\left| w_\gamma(x) - F_\lambda(x) \right| - \gamma \left| w(x/\gamma) - F_\lambda(x/\gamma) \right| - \gamma \left| \int_0^{x \cdot n/\gamma} (\chi(s) - \lambda) \, ds \, a \right|$$
$$\le \lambda(1 - \lambda)|a|\gamma \tag{3.37}$$

where

$$F_\lambda = (1 - \lambda)F_0 + \lambda F_1 = F_0 + \lambda a \otimes n.$$

We also have

$$\nabla w_\gamma(x) = F_0 + \chi \left(\frac{x \cdot n}{\gamma} \right) a \otimes n, \quad \text{for almost all } x \in \Omega,$$

so

$$\nabla w_\gamma(x) = \begin{cases} F_0 & \text{if } j\gamma < x \cdot n < (j+1-\lambda)\gamma & \text{for some } j \in \mathbb{Z}, \\ F_1 & \text{if } (j+1-\lambda)\gamma < x \cdot n < (j+1)\gamma & \text{for some } j \in \mathbb{Z}. \end{cases}$$

$$(3.38)$$

The deformations $w_\gamma(x)$ converge uniformly to $F_\lambda x$ as $\gamma \to 0$ by (3.37), but the deformation gradients oscillate between F_0 in layers of thickness $(1-\lambda)\gamma$ and F_1 in layers of thickness $\lambda\gamma$. In the laboratory, we do not observe laminates with arbitrarily small layer thickness γ. Laminates with arbitrarily small layer thickness exist in our model because we neglect surface energy. However, even with the inclusion of surface energy in the total energy, the constraint of boundary conditions makes the formation of layers of finite thickness with a deformation gradient oscillating between F_0 and F_1 energetically advantageous.

The *infimum* of the energy with respect to deformations constrained by the boundary condition

$$y(x) = Fx \qquad \text{for all } x \in \partial\Omega \tag{3.39}$$

for a fixed $F \in \mathbb{R}^{3\times3}$ has been the subject of much research, since it gives the minimum energy attainable by a microstructure with average deformation gradient F. The value of this *infimum* is called the *relaxation* of ϕ at F and is discussed further in Section 7 and in more detail in Ekeland and Temam (1974) and Dacorogna (1989). For the boundary condition (3.39), we denote the set of admissible deformations by

$$W_F^\phi = \left\{ v \in W^\phi : v(x) = Fx \text{ for } x \in \partial\Omega \right\}.$$

We know from (3.5) that

$$\inf_{z \in W_F^\phi} \mathcal{E}(z) \geq \phi_{\min}(\theta)\text{meas}(\Omega)$$

for all $F \in \mathbb{R}^{3\times3}$. The following theorem shows that the *infimum* of the total energy over deformations constrained by the boundary condition

$$y(x) = F_\lambda x = [(1-\lambda)F_0 + \lambda F_1]\, x \qquad \text{for all } x \in \partial\Omega,$$

where $F_0 \in \mathcal{U}$ and $F_1 \in \mathcal{U}$ are rank-one connected as in (3.3) and $\theta \leq \theta_T$, is equal to the lowest energy attainable for deformations that are not constrained on the boundary. The proof of the following theorem also shows that an energy-minimizing sequence can be constructed which is equal to the laminate $w_\gamma(x)$ except for a boundary layer whose thickness converges to zero as $\gamma \to 0$.

Theorem 1 If $F_0 \in \mathcal{U}$ and $F_1 \in \mathcal{U}$ are rank-one connected as in (3.3) and $\theta \leq \theta_T$, then there exist deformations $\hat{w}_\gamma \in W_{F_\lambda}^\phi$ defined for $\gamma > 0$ such that

$$\det\left(\nabla \hat{w}_\gamma(x)\right) > 0, \qquad \text{for almost all } x \in \Omega$$

and

$$\lim_{\gamma \to 0} \mathcal{E}(\hat{w}_\gamma) = \phi_{\min}(\theta)\text{meas}(\Omega).$$

Proof. The deformation $\hat{w}_\gamma(x)$ that we construct is equal to $w_\gamma(x)$ as defined in (3.36) in the subset

$$\Omega_\gamma^1 \equiv \left\{\, x \in \Omega : \text{dist}(x, \partial\Omega) > \upsilon\gamma \,\right\},$$

where $\upsilon > 0$ is a constant to be determined to ensure that $\det\left(\nabla \hat{w}_\gamma(x)\right) > 0$; $\hat{w}_\gamma(x)$ is equal to $F_\lambda x$ on $\partial\Omega$, and it interpolates between $w_\gamma(x)$ and $F_\lambda x$ on $\Omega \setminus \Omega_\gamma^1$. To construct the interpolation, we define the scalar-valued function $\psi_\gamma(x) : \Omega \to \mathbb{R}$ by

$$\psi_\gamma(x) = \begin{cases} 1 & \text{for all } x \in \Omega_\gamma^1, \\ (\upsilon\gamma)^{-1}\text{dist}(x, \partial\Omega) & \text{for all } x \in \Omega \setminus \Omega_\gamma^1. \end{cases}$$

The function $\psi_\gamma(x)$ is easily seen to satisfy the following properties:

$$\begin{aligned} 0 \leq \psi_\gamma(x) \leq 1 & \qquad \text{for all } x \in \Omega, \\ \psi_\gamma(x) = 1 & \qquad \text{for all } x \in \Omega_\gamma^1, \\ \psi_\gamma(x) = 0 & \qquad \text{for all } x \in \partial\Omega, \\ |\nabla\psi_\gamma(x)| \leq (\upsilon\gamma)^{-1} & \qquad \text{for almost all } x \in \Omega. \end{aligned} \qquad (3.40)$$

We define the deformation $\hat{w}_\gamma(x) : \Omega \to \mathbb{R}^3$ by

$$\hat{w}_\gamma(x) \equiv \psi_\gamma(x)w_\gamma(x) + (1 - \psi_\gamma(x))F_\lambda x \qquad \text{for all } x \in \Omega, \qquad (3.41)$$

so we have for $x \in \Omega$ that

$$\nabla \hat{w}_\gamma(x) = (w_\gamma(x) - F_\lambda x) \otimes \nabla\psi_\gamma(x) + \psi_\gamma(x)\nabla w_\gamma(x) + (1 - \psi_\gamma(x))F_\lambda.$$

It then follows from (3.37), (3.38), and (3.40) that

$$\begin{aligned} |\hat{w}_\gamma(x) - F_\lambda x| = \psi_\gamma(x)\,|w_\gamma(x) - F_\lambda x| \leq \lambda(1-\lambda)|a|\gamma, & \quad x \in \Omega, \\ \nabla \hat{w}_\gamma(x) = \nabla w_\gamma(x) \subset \{F_0, F_1\} \subset \mathcal{U}, & \quad x \in \Omega_\gamma^1, \\ \|\nabla \hat{w}_\gamma(x)\| \leq C, & \quad \text{almost all } x \subset \Omega, \\ \hat{w}_\gamma(x) = F_\lambda x, & \quad x \in \partial\Omega, \qquad (3.42) \end{aligned}$$

where $C > 0$ above and in what follows denotes a generic constant that is independent of γ.

Since ϕ is continuous, it is bounded on bounded sets in $\mathbb{R}^{3\times3}$. Thus, it follows from (3.5) and (3.42) that for $\theta \leq \theta_T$

$$\left| \int_\Omega \left[\phi\left(\nabla \hat{w}_\gamma(x), \theta\right) - \phi_{\min}(\theta) \right] dx \right| = \int_\Omega \left[\phi\left(\nabla \hat{w}_\gamma(x), \theta\right) - \phi_{\min}(\theta) \right] dx$$

$$= \int_{\Omega \setminus \Omega_\gamma^1} \left[\phi\left(\nabla \hat{w}_\gamma(x), \theta\right) - \phi_{\min}(\theta) \right] dx + \int_{\Omega_\gamma^1} \left[\phi\left(\nabla \hat{w}_\gamma(x), \theta\right) - \phi_{\min}(\theta) \right] dx$$

$$= \int_{\Omega \setminus \Omega_\gamma^1} \left[\phi\left(\nabla \hat{w}_\gamma(x), \theta\right) - \phi_{\min}(\theta) \right] dx$$

$$\leq C\gamma \tag{3.43}$$

since $\mathrm{meas}\,(\Omega \setminus \Omega_\gamma^1) \leq C\gamma$.

We next show that

$$\det\left(\nabla \hat{w}_\gamma(x)\right) > 0, \qquad \text{for almost all } x \in \Omega, \tag{3.44}$$

for all $v > 0$ sufficiently large. Since F_0 and F_1 are rank-one connected as in (3.3), we have for any ξ satisfying $0 \leq \xi \leq 1$ that

$$F_\xi = (1 - \xi)F_0 + \xi F_1 = F_0 + \xi a \otimes n = \left(I + \xi a \otimes F_0^{-T} n \right) F_0. \tag{3.45}$$

Hence, we have by (3.45) that

$$\det F_\xi = \left(1 + \xi a \cdot F_0^{-T} n \right) \det F_0 \tag{3.46}$$

for all $0 \leq \xi \leq 1$. Since $F_0, F_1 \in \mathcal{U}$, it follows from (2.12) that

$$\det F_0 = \det F_1 > 0,$$

so we have from (3.46) that

$$a \cdot F_0^{-T} n = 0, \tag{3.47}$$

and

$$\det F_\xi = \det F_0 \tag{3.48}$$

for all $0 \leq \xi \leq 1$.

Now, by (3.40) and (3.42),

$$\left\| \left(w_\gamma(x) - F_\lambda x \right) \otimes \nabla \psi_\gamma(x) \right\| \leq C v^{-1} \qquad \text{for almost all } x \in \Omega, \tag{3.49}$$

and

$$\psi_\gamma(x) \nabla w_\gamma(x) + \left(1 - \psi_\gamma(x)\right) F_\lambda = F_{\xi(x)} \tag{3.50}$$

where

$$\xi(x) = \begin{cases} (1 - \psi_\gamma(x))\lambda & \text{if } \nabla w_\gamma(x) = F_0, \\ \psi_\gamma(x) + (1 - \psi_\gamma(x))\lambda & \text{if } \nabla w_\gamma(x) = F_1. \end{cases}$$

So, (3.44) follows from (3.49), (3.48), and (3.50) for $v > 0$ sufficiently large. $\square$

The results of Kohn and Müller (1992a, 1992b and 1994) for scalar problems with strain gradient surface energies of the form (3.8) show that we can expect the energy-minimizing deformations to have layers that branch in the neighbourhood of the boundary to form infinitesimally small layers, so that the deformation is compatible with the boundary conditions. However, these layers are usually several orders of magnitude smaller than our numerical grid, so the effect of the surface energy is often negligible on macroscopic properties. Our results in Section 6 show that we can approximate the macroscopic properties of energy-minimizing microstructures for the energy (2.2) by solutions obtained on a grid of finite mesh size.

There are affine boundary conditions

$$y(x) = Fx \qquad \text{for all } x \in \partial\Omega$$

for which energy minimization requires a construction more complicated than first-order laminates of the form $w_\gamma(x)$. Higher-order laminates than the first-order laminates $w_\gamma(x)$ are commonly observed (Arlt 1990) and can be constructed from layers of compatible laminates (Bhattacharya 1991, Pedregal 1993, Kohn 1991, Bhattacharya 1992). We shall give a construction of a second-order laminate in Section 3.10. Furthermore, Šverák (1992) has given an energy density for which the *infimum* may only be attained by a microstructure that is not even one of these higher-order laminates, although it is not yet known whether such a property holds for the energy densities used to model martensitic crystals.

3.6. The Young measure and macroscopic densities

The Young measure is a useful device for calculating macroscopic densities from microscopic densities and for describing the pointwise volume fractions of the mixture of the gradient of sequences of energy-minimizing deformations (Tartar 1984, Chipot and Kinderlehrer 1988, Kinderlehrer and Pedregal 1991, Ball and James 1992). We will give a description of the Young measure following most closely the viewpoint of Ball (1989).

We suppose that $\{y_k\} \subset W^\phi$ is a sequence of deformations having uniformly bounded energy $\mathcal{E}(y_k) < C$, and enjoying the property that, for any $f \in C(\mathbb{R}^{3\times3}, \mathbb{R})$ such that $f(F) = o(\|F\|)\|F\|^p$ as $\|F\| \to \infty$, there exists $\tilde{f} \in L^1(\Omega, \mathbb{R})$ so that

$$\lim_{k \to \infty} \int_\omega f(\nabla y_k(x)) \, dx = \int_\omega \tilde{f}(x) \, dx \qquad (3.51)$$

for every measurable set $\omega \subset \Omega$. It can then be shown that there exists a family μ_x of probability measures on $\mathbb{R}^{3\times3}$, depending measurably on $x \in \Omega$, such that $\tilde{f}(x)$ is given by the formula

$$\tilde{f}(x) = \int_{\mathbb{R}^{3\times3}} f(F) \, d\mu_x(F). \qquad (3.52)$$

The family of probability measures μ_x is the *Young measure* associated with the sequence y_k. In the above, we note that it follows from the growth condition (2.1) that

$$\int_\Omega \|\nabla v(x)\|^p \, \mathrm{d}x \le C_1^{-1} \int_\Omega \phi(\nabla v(x), \theta) \, \mathrm{d}x + C_1^{-1} C_0 \operatorname{meas} \Omega \qquad \text{for all } v \in W^\phi.$$

If a sequence of deformations $y_k \in W^\phi$ with uniformly bounded energy has a Young measure and if for some $y \in W^\phi$ we have that

$$\nabla y_k(x) \to \nabla y(x) \qquad \text{for almost all } x \in \Omega,$$

then we have by (3.51) and the Lebesgue dominated convergence theorem (Rudin 1987) that

$$\lim_{k \to \infty} \int_\omega f(\nabla y_k(x)) \, \mathrm{d}x = \int_\omega f(\nabla y(x)) \, \mathrm{d}x$$

for all measurable sets $\omega \subset \Omega$ and for all deformations $f \in C_c(\mathbb{R}^{3 \times 3}; \mathbb{R})$ where $C_c(\mathbb{R}^{3 \times 3}; \mathbb{R})$ denotes the set of continuous deformations $f(F) \in C(\mathbb{R}^{3 \times 3}; \mathbb{R})$ with compact support. Thus, it follows from the representation (3.52) that

$$\mu_x = \delta_{\nabla y(x)} \qquad \text{for almost all } x \in \Omega.$$

It can be shown by a compactness argument that every sequence has at least one subsequence with the property that, for every $f \in C(\mathbb{R}^{3 \times 3}, \mathbb{R})$ such that $f(F) = \mathrm{o}(\|F\|)\|F\|^p$ as $\|F\| \to \infty$, there exists a $\tilde{f} \in L^1(\Omega, \mathbb{R})$ such that (3.51) holds. Thus, every bounded sequence of deformations in W^ϕ contains a subsequence with a Young measure.

The thermodynamic properties of materials, such as energy density and stress, depend nonlinearly on the deformation gradient and can be described by densities $f(F) \in C(\mathbb{R}^{3 \times 3}; \mathbb{R})$ (the dependence of f on temperature is suppressed in this paragraph). We can identify $f(\nabla y_k(x))$ with a *microscopic* density and $\tilde{f}(x)$ with the corresponding *macroscopic* density. We observe that the microscopic density $f(\nabla y_k(x))$ can be oscillatory, while the corresponding macroscopic density $\tilde{f}(x)$ is smooth. For example, we have for the energy-minimizing sequence $\hat{w}_{\gamma_k}(x)$ defined by (3.41) that the macroscopic density

$$\tilde{f}(x) = (1 - \lambda)f(F_0) + \lambda f(F_1)$$

is constant for every $f \in C(\mathbb{R}^{3 \times 3}, \mathbb{R})$ such that $f(F) = \mathrm{o}(\|F\|)\|F\|^p$ as $\|F\| \to \infty$ even though $f(\nabla \hat{w}_{\gamma_k}(x))$ is oscillatory.

For any deformation $y \in W^\phi$, $x \in \Omega$, and $R > 0$, we can define a probability measure on the Borel sets $\Upsilon \subset \mathbb{R}^{3 \times 3}$ by

$$\mu_{x,R,\nabla y}(\Upsilon) = \frac{\operatorname{meas}\{\tilde{x} \in B_R(x) : \nabla y(\tilde{x}) \in \Upsilon\}}{\operatorname{meas} B_R(x)} \tag{3.53}$$

where

$$B_R(x) = \{\, \tilde{x} \in \Omega : |\tilde{x} - x| < R \,\}\,.$$

The probability measure $\mu_{x,R,\nabla y}(\Upsilon)$ gives the volume fraction for which $\nabla y(\tilde{x}) \in \Upsilon$, where $\tilde{x} \in B_R(x)$. We can easily check that

$$\int_{\mathbb{R}^{3\times3}} f(F)\,\mathrm{d}\mu_{x,R,\nabla y}(F) = \frac{1}{\text{meas } B_R(x)} \int_{B_R(x)} f\left(\nabla y(\tilde{x})\right)\,\mathrm{d}\tilde{x} \qquad (3.54)$$

for $f(F) \in C_c(\mathbb{R}^{3\times3}; \mathbb{R})$, so

$$\mu_{x,R,\nabla y} = \frac{1}{\text{meas } B_R(x)} \int_{B_R(x)} \delta_{\nabla y(\tilde{x})}\,\mathrm{d}\tilde{x}.$$

If y_k is a bounded sequence of deformations in W^ϕ with Young measure μ_x, so that (3.51) holds for every $f(F) \in C_c(\mathbb{R}^{3\times3}; \mathbb{R})$ for $\tilde{f}$ given by (3.52); then it follows from (3.54) that

$$\lim_{k\to\infty} \int_{\mathbb{R}^{3\times3}} f(F)\,\mathrm{d}\mu_{x,R,\nabla y_k}(F) =$$

$$\text{meas } B_R(x)^{-1} \int_{B_R(x)} \int_{\mathbb{R}^{3\times3}} f(F)\,\mathrm{d}\mu_{\tilde{x}}(F)\,\mathrm{d}\tilde{x} = \int_{\mathbb{R}^{3\times3}} f(F)\,\mathrm{d}\mu_{x,R}(F)$$

$$(3.55)$$

where

$$\mu_{x,R} \equiv \frac{1}{\text{meas } B_R(x)} \int_{B_R(x)} \mu_{\tilde{x}}\,\mathrm{d}\tilde{x}. \qquad (3.56)$$

The result (3.55) can be restated as

$$\mu_{x,R,\nabla y_k} \overset{*}{\rightharpoonup} \mu_{x,R} \qquad \text{as } n \to \infty,$$

where the limit is understood to be in the sense of measures (weak-* convergence). It further follows from (3.56) and the Lebesgue differentiation theorem (Ball 1989) that

$$\mu_{x,R} \overset{*}{\rightharpoonup} \mu_x \qquad \text{as } R \to 0, \text{ for almost all } x \in \Omega.$$

We can thus characterize the Young measure by the result that

$$\lim_{R\to 0} \lim_{k\to\infty} \mu_{x,R,\nabla y_k} = \mu_x.$$

3.7. Computation of the Young measure for a first-order laminate

We next compute the Young measure of the sequence of first-order laminates constructed in Section 3.5. For the energy-minimizing sequence of first-order laminates $\hat{w}_\gamma$ defined by (3.41), we have that if $\Upsilon \subset \mathbb{R}^{3\times3}$ is an open set with smooth boundary, such that

$$F_0 \notin \Upsilon, \qquad F_1 \notin \Upsilon;$$

then we have by the above construction that

$$\mu_{x,R,\nabla\hat{w}_\gamma}(\Upsilon) \leq \min\left\{\frac{\gamma}{R}, 1\right\}.$$

(In fact, if $B_R(x) \subset \Omega_\gamma^1$, then we have that $\mu_{x,R,\nabla\hat{w}_\gamma}(\Upsilon) = 0$.) Also, if $\Upsilon \subset \mathbb{R}^{3\times3}$ is an open set with smooth boundary, such that

$$F_0 \in \Upsilon, \qquad F_1 \notin \Upsilon,$$

then

$$\left|\mu_{x,R,\nabla\hat{w}_\gamma}(\Upsilon) - (1-\lambda)\right| \leq \min\left\{\frac{\gamma}{R}, 1\right\};$$

and if $\Upsilon \subset \mathbb{R}^{3\times3}$ is an open set with smooth boundary such that

$$F_1 \in \Upsilon, \qquad F_0 \notin \Upsilon,$$

then we have that

$$\left|\mu_{x,R,\nabla\hat{w}_\gamma}(\Upsilon) - \lambda\right| \leq \min\left\{\frac{\gamma}{R}, 1\right\}.$$

Thus, we can conclude that for any open set $\Upsilon \subset \mathbb{R}^{3\times3}$ with smooth boundary $\Upsilon \subset \mathbb{R}^{3\times3}$, we have that

$$\left|\mu_{x,R,\nabla\hat{w}_\gamma}(\Upsilon) - [(1-\lambda)\delta_{F_0}(\Upsilon) + \lambda\delta_{F_1}(\Upsilon)]\right| \leq \min\left\{\frac{\gamma}{R}, 1\right\} \qquad (3.57)$$

where $\delta_F(\Upsilon)$ is the Dirac measure of unit mass at $F \in \mathbb{R}^{3\times3}$.

It follows from (3.57) that we have for any sequence $\gamma_k \to 0$ that

$$\mu_{x,R} = \lim_{\gamma_k \to 0} \mu_{x,R,\nabla\hat{w}_{\gamma_k}} = (1-\lambda)\delta_{F_0} + \lambda\delta_{F_1}.$$

Hence, we have that the Young measure for the energy-minimizing sequence $\hat{w}_{\gamma_k}(x)$ defined by (3.41) satisfies

$$\mu_x = \lim_{R \to 0} \mu_{x,R} = (1-\lambda)\delta_{F_0} + \lambda\delta_{F_1}.$$

We note that in this special case the Young measure μ_x is independent of $x \in \Omega$, although in general the Young measure depends on $x \in \Omega$.

3.8. The failure of the direct method of the calculus of variations to give an energy-minimizing deformation

The direct method of the calculus of variations is widely used to construct energy-minimizers to variational problems (2.5) by taking the limit of energy-minimizing sequences of deformations (Dacorogna 1989). On the other hand, if $(1-\lambda)F_0 + \lambda F_1 \notin \mathcal{U}$, then we cannot use this technique to construct an energy-minimizing deformation for our models of martensitic crystals, since we have by (3.43) and (2.18) that

$$\lim_{\gamma \to 0} \int_\Omega \phi\left(\nabla \hat{w}_\gamma(x)\right) \, dx = \phi_{\min} \text{meas}\,(\Omega) < \phi((1-\lambda)F_0 + \lambda F_1)\text{meas}\,(\Omega)$$

$$= \int_\Omega \phi\left(\nabla(F_\lambda x)\right) \, dx = \int_\Omega \phi\left(\nabla\left(\lim_{\gamma \to 0} \hat{w}_\gamma(x)\right)\right) \, dx.$$

This result, together with the fact that $\nabla \hat{w}_\gamma$ converges weakly to F_λ, shows that the functional $\mathcal{E}(y)$ is not *weakly lower semi-continuous* (Dacorogna 1989).

The following lemmas show that $(1-\lambda)F_0 + \lambda F_1 \notin \mathcal{U}$ for $0 < \lambda < 1$ in the orthorhombic to monoclinic case (2.13) and the cubic to tetragonal case (2.15).

Lemma 8   If $F_0 \in \mathcal{U}$ and $F_1 \in \mathcal{U}$ with $F_0 \neq F_1$ are rank-one connected and

$$\left\{ R_i U_1 R_i^T : R_i \in \mathcal{G} \right\} = \{ U_1, U_2 \},$$

then

$$(1-\lambda)F_0 + \lambda F_1 \notin \mathcal{U}$$

for $0 < \lambda < 1$.

Proof.   We prove the result by contradiction, so we assume that $F_0 \in \mathcal{U}$ and $F_1 \in \mathcal{U}$ are rank-one connected and that

$$(1-\lambda)F_0 + \lambda F_1 \in \mathcal{U} \tag{3.58}$$

for some $0 < \lambda < 1$. It follows by Lemma 3 that we may assume that

$$F_0 = R_0 U_1 \quad \text{and} \quad F_1 = R_1 U_2 \tag{3.59}$$

for R_0, $R_1 \in SO(3)$ and that we may assume by (3.58) that

$$(1-\lambda)F_0 + \lambda F_1 = Q U_1 \tag{3.60}$$

for $Q \in SO(3)$. Since $F_0 \in \mathcal{U}$ and $F_1 \in \mathcal{U}$ are rank-one connected, we have by (3.3) that there exist $a \in \mathbb{R}^3$ and $n \in \mathbb{R}^3$, $|n| = 1$, such that

$$(1-\lambda)F_0 + \lambda F_1 = F_0 + \lambda a \otimes n. \tag{3.61}$$

It follows from (3.59)–(3.61) that

$$Q U_1 = R_0 U_1 + \lambda a \otimes n, \tag{3.62}$$

so it follows from Lemma 3 that $Q = R_0$. Since $0 < \lambda < 1$, it follows from (3.62) that $a = 0$ and $F_0 = F_1$, which is a contradiction with the hypothesis of the lemma. $\square$

Lemma 9   For the cubic to tetragonal transformation (2.15), if $F_0 \in \mathcal{U}$ and $F_1 \in \mathcal{U}$ are rank-one connected, then

$$(1-\lambda)F_0 + \lambda F_1 \notin \mathcal{U}$$

for $0 < \lambda < 1$.

Proof. If $F_0 \in \mathcal{U}$ and $F_1 \in \mathcal{U}$ are rank-one connected, then it follows from Lemma 3 that we may assume without loss of generality that

$$F_1 = RU_2, \qquad F_0 = U_1,$$

for $R \in SO(3)$, and by Lemma 6 that

$$RU_2 = U_1 + a \otimes n \tag{3.63}$$

where

$$n = \frac{1}{\sqrt{2}}(e_1 + e_2) \quad \text{or} \quad n = \frac{1}{\sqrt{2}}(e_1 - e_2). \tag{3.64}$$

We suppose that $(1 - \lambda)F_0 + \lambda F_1 \in \mathcal{U}$. It then follows from the proof of Lemma 8 that

$$(1 - \lambda)F_0 + \lambda F_1 \notin \mathcal{U}_1 \cup \mathcal{U}_2,$$

so we conclude that

$$(1 - \lambda)F_0 + \lambda F_1 = QU_3 \tag{3.65}$$

for $Q \in SO(3)$. We next obtain from (3.63) and (3.65) that

$$U_1 + \lambda a \otimes n = QU_3. \tag{3.66}$$

We have thus reached a contradiction with (3.64) since Lemma 6 implies the relation

$$n = \frac{1}{\sqrt{2}}(e_1 \pm e_3)$$

for any solution to (3.66). $\square$

The following result shows that for the orthorhombic to monoclinic case (2.13) and for the cubic to tetragonal case (2.15) there does not exist an energy-minimizing deformation (Ball and James 1992).

Theorem 2   For the orthorhombic to monoclinic case (2.13) and for the cubic to tetragonal case (2.15) there does not exist a deformation $y(x) \in W_{F_\lambda}^\phi$ such that

$$\mathcal{E}(y) = \inf_{z \in W_{F_\lambda}^\phi} \mathcal{E}(z). \tag{3.67}$$

Proof.   We give a proof that covers both the orthorhombic to monoclinic case (2.13) and the cubic to tetragonal case (2.15). We assume that (3.67) holds, so by Theorem 1 (which holds for both the orthorhombic to monoclinic case (2.13) and the cubic to tetragonal case (2.15)) and (3.67) we have that

$$\mathcal{E}(y) = \int_\Omega \phi(\nabla y, \theta)\, dx = \phi_{\min}(\theta)\text{meas}\,\Omega. \tag{3.68}$$

Since (3.67) holds, we can conclude from Theorem 7 in Section 6 (which also holds for both the orthorhombic to monoclinic case (2.13) and the cubic to tetragonal case (2.15)) that for all $x \in \Omega$ and $R > 0$ we have that

$$\text{meas} \left\{ \tilde{x} \in B_R(x) : \nabla y(x) = F_0 \right\} = (1 - \lambda) \, \text{meas} \, B_R(x),$$
$$\text{meas} \left\{ \tilde{x} \in B_R(x) : \nabla y(x) = F_1 \right\} = \lambda \, \text{meas} \, B_R(x). \tag{3.69}$$

It then follows from (3.69) that

$$\frac{1}{\text{meas} \, B_R(x)} \int_{B_R(x)} \nabla y(\tilde{x}) \, d\tilde{x} = (1 - \lambda) F_0 + \lambda F_1 = F_\lambda. \tag{3.70}$$

Now y is an element of W^ϕ, so the Lebesgue differentiation theorem (Rudin 1987) implies that

$$\lim_{R \to 0} \frac{1}{\text{meas} \, B_R(x)} \int_{B_R(x)} \nabla y(\tilde{x}) \, d\tilde{x} = \nabla y(x), \tag{3.71}$$

for almost all $x \in \Omega$. Hence, we can conclude from (3.70), (3.71), and (3.68) that

$$\phi((1 - \lambda) F_0 + \lambda F_1, \theta) = \phi_{\min}(\theta),$$

which is a contradiction, since $(1 - \lambda) F_0 + \lambda F_1 \notin \mathcal{U}$ by Lemma 8. $\square$

3.9. The austenitic martensitic interface

Microstructure is observed in phase transformations between the austenitic and the martensitic phases (see Fig. 1). A phase boundary is observed to separate a homogeneous austenitic region from a microstructured martensitic region (Basinski and Christian 1954, Burkart and Read 1953). Ball and James (1987) have shown that this phenomenon can be explained by the geometrically nonlinear continuum theory and Chu and James (1995) have used this theory to explain the austenitic–martensitic phase boundary presented in Fig. 1. The kinematic condition that the martensitic phase be compatible with the austenitic phase imposes a boundary condition similar to that of (3.39).

For the cubic to tetragonal case (2.15), Ball and James (1987) have shown that if

$$\nu_1 < 1 < \nu_2 \text{ and } \tfrac{1}{\nu_1^2} + \tfrac{1}{\nu_2^2} < 2,$$
$$\text{or } \nu_2 < 1 < \nu_1 \text{ and } \nu_1^2 + \nu_2^2 < 2,$$

then the continuum theory predicts that there are fine-scale mixtures of any two variants of the martensite that can be separated from a homogeneous austenitic phase by a planar interface. For example, we can construct the mixture w_γ using $F_0 = U_1$ and $F_1 = RU_2$ where RU_2 and U_1 are as defined in (3.29). By (3.37), $w_\gamma \to F_\lambda x$ uniformly as $\gamma \to 0$. It turns out that for the volume fraction $0 < \lambda^* < 1/2$ given by

$$\lambda^* = \frac{1}{2}\left[1 - \left(2(\nu_2^2 - 1)(\nu_1^2 - 1)(\nu_1^2 + \nu_2^2)(\nu_2^2 - \nu_1^2)^{-2} + 1\right)^{1/2}\right],$$

there exists a continuous deformation with deformation $F_{\lambda^*}x$ on one side of a planar interface with normal m, and the homogeneous austenitic deformation Qx, where $Q \in SO(3)$, on the opposite side. Here we have used the fact that there is a $Q \in SO(3)$ and corresponding $b, m \in \mathbb{R}^3$, with $|m| = 1$, such that

$$F_{\lambda^*} = (1 - \lambda^*)U_1 + \lambda^* RU_2 = Q(I + b \otimes m) \qquad (3.72)$$

where in the orthonormal basis $\{e_1, e_2, e_3\}$

$$b = (1 + \chi^2 + \tau^2)(-\zeta(\chi + \tau), \zeta(\chi - \tau), \beta),$$
$$m = (1 + \chi^2 + \tau^2)^{-1}(-(\chi + \tau), (\chi - \tau), 1),$$

with

$$\chi = \tfrac{1}{2}\left[(\nu_2^2 + \nu_1^2 - 2)(1 - \nu_1^2)^{-1}\right]^{1/2},$$
$$\tau = \tfrac{1}{2}\left[(2\nu_1^2\nu_2^2 - \nu_1^2 - \nu_2^2)(1 - \nu_1^2)^{-1}\right]^{1/2},$$
$$\zeta = (1 - \nu_1^2)(1 + \nu_2)^{-1},$$
$$\beta = \nu_2(\nu_1^2 - 1)(1 + \nu_2)^{-1}.$$

All of the remaining austenitic–martensitic interfaces can be obtained from (3.72) by symmetry considerations, and we obtain that there are 24 distinct ways a parallel, planar interface can separate the homogeneous austenitic phase from a microstructured martensitic phase.

We say (3.72) represents an austenitic–martensitic interface because Ball and James (1987) have constructed an energy-minimizing sequence u_γ of continuous deformations such that

$$\int_\Omega \phi(\nabla u_\gamma(x), \theta_T) \, dx \to \phi_{\min}(\theta_T) \text{ meas } \Omega \qquad (3.73)$$

and $u_\gamma(x) \to u(x)$ uniformly as $\gamma \to 0$, where

$$u(x) = \begin{cases} Qx & \text{for } x \cdot m < 0, \\ F_{\lambda^*}x & \text{for } x \cdot m > 0. \end{cases}$$

We note that

$$\phi_{\min}(\theta_T) = \phi(Q, \theta_T) = \phi(RU_2, \theta_T) = \phi(U_1, \theta_T).$$

We can construct $u_\gamma(x)$ by

$$u_\gamma(x) = \begin{cases} Qx & \text{for } x \cdot m < 0, \\ \psi_\gamma(x)w_\gamma(x) + (1 - \psi_\gamma(x))F_{\lambda^*}x & \text{for } x \cdot m > 0, \end{cases}$$

where $w_\gamma(x)$ is the first-order laminate defined by (3.36), and where

$$\psi_\gamma(x) = \begin{cases} \gamma^{-1}x \cdot m & \text{if } 0 < x \cdot m < \gamma, \\ 1 & \text{if } x \cdot m > \gamma. \end{cases}$$

It is easy to check that $u_\gamma(x)$ satisfies the scaling $u_\gamma(x) = \gamma u_1(\gamma^{-1}x)$ for $\gamma > 0$ and $x \in \mathbb{R}^3$. We also note that we can ensure that $\det \nabla u_\gamma(x) > 0$ almost everywhere by replacing $\psi_\gamma(x)$ by $\psi_\gamma(v^{-1}x)$ in the definition of $u_\gamma(x)$ if the constant $v > 0$ is sufficiently large (*cf.* Theorem 1).

Then $u_\gamma(x)$ satisfies

$$\nabla u_\gamma(x) = \begin{cases} Q & \text{if } x \cdot m < 0, \\ U_1 & \text{if } x \cdot m > \gamma \text{ and } j\gamma \le x \cdot n \le (j+1-\lambda^*)\gamma \\ & \quad \text{for some } j \in \mathbb{Z}, \\ RU_2 & \text{if } x \cdot m > \gamma \text{ and } (j+1-\lambda^*)\gamma < x \cdot n < (j+1)\gamma \\ & \quad \text{for some } j \in \mathbb{Z}, \end{cases}$$

and

$$\|\nabla u_\gamma(x)\| \le C \qquad \text{for almost all } x \in \Omega,$$
$$|u_\gamma(x) \quad u(x)| \le C\gamma, \qquad x \in \Omega,$$
$$u_\gamma(x) \in C(\mathbb{R}^3; \mathbb{R}^3).$$

The estimate (3.73) now follows by the argument (3.43).

The microstructure represented by the deformations $u_\gamma(x)$ for $\gamma \to 0$ is austenite for $x \cdot m < 0$, and is finely twinned martensite for $x \cdot m > 0$ with volume fraction $1 - \lambda^*$ of the deformation gradient U_1 and volume fraction λ^* of the deformation gradient RU_2. The plane of the interface satisfies $x \cdot m = 0$. It is easily checked that any sequence of deformations $u_{\gamma_k}(x)$ with $\gamma_k \to 0$ has the Young measure

$$\mu_x = \begin{cases} \delta_Q & \text{if } x \cdot m < 0, \\ (1 - \lambda^*)\delta_{U_1} + \lambda^*\delta_{RU_2} & \text{if } x \cdot m > 0. \end{cases}$$

Note that $u(x)$ is not an energy-minimizing deformation, since by Lemma 9

$$\phi((1 - \lambda^*)U_1 + \lambda^*RU_2, \theta_T) > \phi(U_1, \theta_T) - \phi(RU_2, \theta_T).$$

The austenitic–martensitic phase transformation has been the subject of many numerical studies (Collins and Luskin 1989, Klouček and Luskin 1994*a*, Klouček and Luskin 1994*b*) since it is one of the primary mechanisms for the creation of microstructure. These numerical studies have been three-dimensional since the following lemma does not seem to allow for an adequate two-dimensional model. Two-dimensional models (Collins et al. 1993) usually represent the martensitic variants by $SO(2)U_i$ where the eigenvalues $\tilde{\nu}_1^2$, $\tilde{\nu}_2^2$ of $U_i^T U_i$ satisfy $0 < \tilde{\nu}_1^2 \le 1$ and $\tilde{\nu}_2^2 \ge 1$, so the following lemma shows that these variants have a rank-one connection to the matrices $SO(2)$, which represent the austenitic phase.

Lemma 10 If $U \in \mathbb{R}^{2 \times 2}$ and the eigenvalues $\tilde{\nu}_1^2$, $\tilde{\nu}_2^2$ of $U^T U$ satisfy $0 < \tilde{\nu}_1^2 \leq 1$ and $\tilde{\nu}_2^2 \geq 1$, then there exist a rotation $R \in SO(2)$ and vectors $a \in \mathbb{R}^2$ and $n \in \mathbb{R}^2$, $|n| = 1$, such that

$$RU = I + a \otimes n. \tag{3.74}$$

Proof. Since $U^T U \in \mathbb{R}^{2 \times 2}$ has eigenvalues $\tilde{\nu}_1^2$, $\tilde{\nu}_2^2$ such that $0 < \tilde{\nu}_1^2 \leq 1$ and $\tilde{\nu}_2^2 \geq 1$, there exists a $v \in \mathbb{R}^2$, $v \neq 0$, such that

$$|Uv| = |v|.$$

So, there exists $R \in SO(2)$ such that

$$RUv = v.$$

Hence, for $n \in \mathbb{R}^2$ satisfying $n \cdot v = 0$ and $|n| = 1$, we have that (3.74) holds with $a = RUn - n$. $\square$

3.10. Higher-order laminates

Higher-order laminates of layers within layers are common in martensitic materials. For example, the photomicrograph in Fig. 2 shows a second-order laminate that has been explained by Chu and James (1995) using the geometrically nonlinear continuum theory. More general treatments of higher-order laminates can be found in Kohn and Strang (1986), Kohn (1991) and Pedregal (1993).

Collins (1993a) has reported computational results for affine boundary conditions that have a second-order laminate as an optimal microstructure, but do not have a first-order laminate as an optimal laminate. He reported that his algorithm computed a first-order laminate until the mesh was sufficiently fine. He explained this by an argument that the energy associated with the lack of compatibility of the first-order laminate with the boundary conditions is less than the additional energy associated with the additional interfaces needed to represent the second-order laminate until the mesh is sufficiently fine.

We will construct a second-order laminate by layering two first-order laminates. To construct the first-order laminates, we assume that F_{00}, $F_{01} \in \mathcal{U}$ and F_{10}, $F_{11} \in \mathcal{U}$ are pairs of rank-one connected matrices, that is, we assume that there exist a_0, $n_0 \in \mathbb{R}^3$, $|n_0| = 1$, and a_1, $n_1 \in \mathbb{R}^3$, $|n_1| = 1$, such that

$$F_{01} = F_{00} + a_0 \otimes n_0,$$
$$F_{11} = F_{10} + a_1 \otimes n_1.$$

We can construct first-order laminates with layer thickness $\gamma_1 > 0$ and a mixture of F_{i0} with volume fraction $1 - \lambda_i$ and F_{i1} with volume fraction λ_i

following (3.36) by

$$w_{\gamma_1}^{[i]}(x) = \gamma_1 w^{[i]}\left(\frac{x}{\gamma_1}\right) \qquad \text{for all } x \in \mathbb{R}^3$$

where

$$w^{[i]}(x) = F_{i0}x + \left[\int_0^{x \cdot n_i} \chi_i(s)\, \mathrm{d}s\right] a_i \qquad \text{for all } x \in \mathbb{R}^3$$

for $i = 0,\, 1$ and where $\chi_i(s) : \mathbb{R} \to \mathbb{R}$ is the characteristic function with period 1 defined by

$$\chi_i(s) = \begin{cases} 0 & \text{for all} \quad 0 \leq s \leq 1 - \lambda_i, \\ 1 & \text{for all} \quad 1 - \lambda_i < s < 1. \end{cases}$$

We recall that by (3.37) we have that

$$\left|w_{\gamma_1}^{[i]}(x) - F_{i\lambda_i}x\right| = \gamma_1 \left|w^{[i]}\left(\frac{x}{\gamma_1}\right) - F_{i\lambda_i}\left(\frac{x}{\gamma_1}\right)\right| \leq \lambda_i(1-\lambda_i)|a_i|\gamma_1 \quad (3.75)$$

for all $x \in \mathbb{R}^3$ where

$$F_{i\lambda_i} = (1-\lambda_i)F_{i0} + \lambda_i F_{i1} = F_{i0} + \lambda_i a_i \otimes n_i$$

for $i = 0, 1$.

We can construct a second-order laminate from the first-order laminates $w_{\gamma_1}^{[i]}(x)$ if there exist $0 \leq \lambda_0,\, \lambda_1 \leq 1$ such that $F_{0\lambda_0} \in \mathbb{R}^{3\times3}$ and $F_{1\lambda_1} \subset \mathbb{R}^{3\times3}$ are rank-one connected, that is, there exist $a,\, n \in \mathbb{R}^3$, $|n| = 1$, such that

$$F_{1\lambda_1} = F_{0\lambda_0} + a \otimes n. \qquad (3.76)$$

If (3.76) holds, then for $2\gamma_1 < \min\{1-\lambda, \lambda\}$ we can construct a second-order laminate for any $0 < \lambda < 1$ by the periodic extension to $\mathbb{R}^3$ of the continuous deformation

$$w_{\gamma_1}(x) = \begin{cases} \psi_{\gamma_1}(x)w_{\gamma_1}^{[0]}(x) + (1-\psi_{\gamma_1}(x))F_{0\lambda_0}x & \text{for } 0 < x \cdot n < 1-\lambda, \\ \psi_{\gamma_1}(x)w_{\gamma_1}^{[1]}(x) + (1-\psi_{\gamma_1}(x))F_{1\lambda_1}x & \text{for } 1-\lambda < x \cdot n < 1, \end{cases}$$

where $w_{\gamma_1}^{[i]}(x)$ is the first-order laminate defined by (3.75), and where

$$\psi_{\gamma_1}(x) = \begin{cases} \gamma_1^{-1}x \cdot n & \text{if } 0 < x \cdot n < \gamma_1, \\ 1 & \text{if } \gamma_1 < x \cdot n < 1 - \lambda - \gamma_1, \\ \gamma_1^{-1}|x \cdot n - (1-\lambda)| & \text{if } |x \cdot n - (1-\lambda)| < \gamma_1, \\ 1 & \text{if } 1 - \lambda + \gamma_1 < x \cdot n < 1 - \gamma_1, \\ (\gamma_1)^{-1}|x \cdot n - 1| & \text{if } 1 - \gamma_1 < x \cdot n < 1. \end{cases}$$

We can scale the second-order laminate $w_{\gamma_1}(x)$ by $\gamma_2 > 0$ to obtain the second-order laminate $w_{\gamma_1\gamma_2}(x)$ defined by

$$w_{\gamma_1\gamma_2}(x) = \gamma_2 w_{\gamma_1}\left(\frac{x}{\gamma_2}\right) \qquad \text{for all } x \in \mathbb{R}^3.$$

As $\gamma_1 \to 0$, the second-order laminate $w_{\gamma_1\gamma_2}(x)$ converges to a mixture with layer thickness γ_2 of the first-order laminate $w_{\gamma_1}^{[0]}(x)$ with volume fraction $1-\lambda$ and of the first-order laminate $w_{\gamma_1}^{[1]}(x)$ with volume fraction λ. The analysis in Section 3.5 can then be used to prove that

$$\left| w_{\gamma_1\gamma_2}(x) - \hat{F}_\lambda x \right| \leq \max \left\{ \lambda_1(1-\lambda_1)|a_1|,\ \lambda_2(1-\lambda_2)|a_2| \right\} \gamma_1\gamma_2$$

for all $x \in \mathbb{R}^3$ where

$$\hat{F}_\lambda = (1-\lambda)F_{0\lambda_0} + \lambda F_{1\lambda_1}.$$

We can check that

$$\|\nabla w_{\gamma_1\gamma_2}(x)\| \leq C \qquad \text{for almost all } x \in \mathbb{R}^3, \tag{3.77}$$

and that

$$\nabla w_{\gamma_1\gamma_2}(x) \in \{ F_{00},\ F_{01},\ F_{10},\ F_{11} \} \subset \mathcal{U} \qquad \text{for all } x \in \mathbb{R}^3 \setminus \hat{\Omega}_{\gamma_2} \tag{3.78}$$

where

$$\hat{\Omega}_{\gamma_2} = \bigcup_{j \in \mathbb{Z}} \left\{ x \in \mathbb{R}^3 : |x \cdot n - j\gamma_2| \leq \gamma_1\gamma_2 \text{ or } |x \cdot n - (j+1-\lambda)\gamma_2| \leq \gamma_1\gamma_2 \right\}.$$

Since $\Omega \subset \mathbb{R}^3$ is a bounded domain,

$$\text{meas}\,(\Omega \cap \hat{\Omega}_{\gamma_2}) \leq C\gamma_1, \tag{3.79}$$

because $\Omega \cap \hat{\Omega}_{\gamma_2}$ is the union of $\mathcal{O}(\gamma_2^{-1})$ non-empty planar layers of thickness $\gamma_1\gamma_2$. (Note that only $\mathcal{O}(\gamma_2^{-1})$ of the sets in the definition of $\hat{\Omega}_{\gamma_2}$ have a non-empty intersection with Ω.) We thus have from (3.77)–(3.79) that for $\theta \leq \theta_T$

$$\int_\Omega \phi(\nabla w_{\gamma_1\gamma_2}(x), \theta)\,\mathrm{d}x$$
$$= \int_{\Omega \setminus \hat{\Omega}_{\gamma_2}} \phi(\nabla w_{\gamma_1\gamma_2}(x), \theta)\,\mathrm{d}x + \int_{\Omega \cap \hat{\Omega}_{\gamma_2}} \phi(\nabla w_{\gamma_1\gamma_2}(x), \theta)\,\mathrm{d}x$$
$$\leq \phi_{\min}(\theta)\,\text{meas}\,\Omega + C\gamma_1.$$

It can also be shown that for any pair of sequences such that $\gamma_{1k} \to 0$ and $\gamma_{2k} \to 0$ as $k \to \infty$ we have that the sequence of deformations $w_{\gamma_{1k}\gamma_{2k}}(x)$ has the Young measure

$$(1-\lambda)(1-\lambda_0)\delta_{F_{00}} + (1-\lambda)\lambda_0\delta_{F_{01}} + \lambda(1-\lambda_1)\delta_{F_{10}} + \lambda\lambda_1\delta_{F_{11}}.$$

Higher-order laminates than second-order can be constructed by iterating the above construction.

4. Finite element methods

We wish to compute an approximation to the microstructure defined by energy-minimizing sequences of deformations to the problem

$$\inf_{y \in \mathcal{A}} \int_{\Omega} \phi\left(\nabla y(x), \theta\right) \, \mathrm{d}x, \tag{4.1}$$

where $\mathcal{A}$ denotes a set of admissible deformations. The most accurate finite element method depends on the scale of the microstructure relative to the scale of the mesh and whether it is possible to align the mesh with the microstructure.

4.1. Conforming finite elements

The most commonly used finite element spaces in solid mechanics are conforming spaces that approximate the admissible set of deformations $\mathcal{A}$ by a finite-dimensional subset $\mathcal{A}_h \subset \mathcal{A}$ of continuous deformations which are piecewise polynomials with respect to a finite element mesh. We can compute approximations to energy-minimizing sequences of deformations for problem (4.1) by computing energy-minimizing deformations of the problem

$$\min_{y_h \in \mathcal{A}_h} \int_{\Omega} \phi\left(\nabla y_h(x), \theta\right) \, \mathrm{d}x. \tag{4.2}$$

We note that, since $\mathcal{A}_h$ is finite-dimensional and the energy

$$\mathcal{E}(y_h) = \int_{\Omega} \phi\left(\nabla y_h(x), \theta\right) \, \mathrm{d}x$$

is continuous, the *infimum* of the energy $\mathcal{E}(y_h)$ is attained for at least one finite element deformation $y_h \in \mathcal{A}_h$, since it follows from the growth property (2.1) that $\phi(F, \theta) \to \infty$ as $\|F\| \to \infty$. The lack of attainment of the *infimum* for the continuous problem (4.1) is the result of the development of arbitrarily fine oscillations by the gradient of energy-minimizing sequences of deformations. The restriction of the admissible deformations to a finite element space limits the possible fineness of the oscillations to the scale of the mesh; therefore, the *infimum* of the energy is attained among deformations which are constrained to lie in the finite element space.

Since deformations with microstructure are typically approximately piecewise linear, the use of piecewise linear or piecewise trilinear elements is a good choice of finite element space for the approximation of microstructure. Although these spaces of continuous finite elements effectively approximate microstructure with layers that are parallel to the planes across which the finite element deformation gradients can be discontinuous, they have difficulty approximating microstructure on the scale of the mesh when the layers are not oriented with respect to the mesh. Computational experiments with the continuous, piecewise linear element for a two-dimensional model have shown

that numerical solutions for microstructure given by conforming spaces have
a layer thickness that depends on the orientation of the microstructure with
respect to the mesh; see Fig. 3 (below) and Collins (1994).

However, we proved in Section 3 that the number of families of parallel
planes (the 'twin planes') across which the deformation gradients of energy-
minimizing deformations can be discontinuous is finite, so it is possible for
many problems to orient the mesh to the possible twin planes. (By Lemma 5
there are two families of twin planes for the orthorhombic to monoclinic transformation (2.13) and by Lemma 6 there are six families of twin planes for the
cubic to tetragonal transformation (2.15).)

Luskin (1996a, 1996b) has given the use of conforming methods a theoretical
validation by giving error estimates for the convergence of the conforming
finite element approximation of a laminated microstructure for the rotationally
invariant, double well problem ($\mathcal{U} = \mathcal{U}_1 \cup \mathcal{U}_2$), and Li and Luskin (1996) have
given error estimates for the finite element approximation of a laminated
microstructure for the cubic to tetragonal transformation (2.15). We will give
error estimates for this convergence in Section 6.

4.2. Optimization and local minima

It would be most correct to pose the problem of interest as the computa-
tion of local minima of the non-convex energy $\mathcal{E}(y) = \int_\Omega \phi(\nabla y(x), \theta) \, dx$
which represent physically observable equilibrium states. The continuous
problem (4.1) can be expected to have multiple local minima (Ball et al.
1991, Truskinovsky and Zanzotto 1995, Truskinovsky and Zanzotto 1996),
only some of which represent states that can be observed in the laboratory.
However, the restriction of our computational interest to global minima is
not appropriate, since martensitic crystals typically exhibit hysteresis and
meta-stability (Abeyaratne et al. 1994, Ball et al. 1995).

In addition to the local minima which the finite-dimensional problem (4.2)
inherits from the continuous problem, there are also local minima created
by the numerical discretization, which are the representation of the same
microstructure on different length scales and which give the same macroscopic
properties.

Gradient iterative methods, which reduce the energy at each iteration, can
be used to compute the local minimum corresponding to the energy well of
the initial state. Conjugate gradient and other accelerations can be used to
develop more efficient iterative methods (Collins and Luskin 1989, Collins
1993a, Collins et al. 1993). Since the iterates of gradient methods remain in
the energy well of the initial state, the addition of random perturbations to an
initial state can be used to explore new local minima (Collins 1993a, Collins
et al. 1993).

The addition of random perturbations to the initial states for gradient methods suggests the use of more systematic Monte Carlo techniques. Luskin and Ma (1993) used a variant of the simulated annealing algorithm to compute microstructures of fine domains in ferromagnetic crystals. They constructed a discrete set of magnetizations that were close to the set of local minima and then utilized a gradient method to compute the optimal solution within the energy well they had computed with the simulated annealing. The key to the generalization of this algorithm to the case of martensitic crystals is the construction of a discrete set of deformations that represent the energy wells of the martensitic crystal. Kartha et al. (1994) have used a Monte Carlo method to investigate the properties of a two-dimensional model of martensite, and Gremaud (1995) has developed a Monte Carlo method to compute global minima of two-dimensional variational problems with local minima.

To ensure that one computes physically observed states in a quasi-static or dynamical process, one should start with a physically observed state and then compute the change in the state as environmental conditions such as boundary conditions or temperature are varied. For quasi-static processes, continuation methods can be used. For example, Kinderlehrer and Ma (1994a, 1994b) have used a continuation method to compute hysteresis in the response of a ferromagnetic crystal to changes in the applied magnetic field. The techniques reported in Klouček and Luskin (1994a, 1994b) for the computation of the dynamics of martensitic crystals offer another possibility for exploring physically observed local minima and hysteretic phenomena by computing the physical dynamics of the response of the crystal to changes in its environment.

4.3. Rotation of the coordinate system

We discussed in Section 4.1 that it can be advantageous to orient the mesh with respect to the planes across which the gradients of energy-minimizing deformations are allowed to be discontinuous. This can often be achieved by rotating the coordinate system describing the reference domain with the mesh fixed in the coordinate system. It is also convenient to rotate the coordinate system with the mesh fixed in the coordinate system to test the effect of the orientation of the finite element mesh with respect to the microstructure.

If we rotate the coordinate system of the reference domain by the rotation $\hat{R}^T$ where $\hat{R} \subset SO(3)$, then the energy density for the crystal in the rotated coordinate system is given by

$$\hat{\phi}(F, \theta) = \phi(F\hat{R}, \theta).$$

For the transformed energy density $\hat{\phi}(F, \theta)$, it follows from (2.18) that for $\theta < \theta_T$ we have that $\hat{\phi}_{\min}(\theta) = \phi_{\min}(\theta)$ and that

$$\hat{\phi}(F, \theta) = \hat{\phi}_{\min}(\theta) \quad \text{if and only if} \quad F \in SO(3)\hat{U}_1 \cup \cdots \cup SO(3)\hat{U}_M$$

where

$$\hat{U}_i = \hat{R} U_i \hat{R}^T \qquad \text{for } i = 1, \ldots, M.$$

We also note that we have that

$$QU_i = U_j + a \otimes n \quad \text{if and only if} \quad \hat{Q}\hat{U}_i = \hat{U}_j + \hat{a} \otimes \hat{n}$$

for $Q \in \mathrm{SO}(3)$, $a \in \mathbb{R}^3$, and $n \in \mathbb{R}^3$, where

$$\hat{Q} = \hat{R} Q \hat{R}^T, \quad \hat{a} = \hat{R}a, \quad \hat{n} = \hat{R}n.$$

Hence, it follows that $\hat{n} = \hat{R}n$ is the normal to a plane across which the gradient of an energy-minimizing deformation for the energy density $\hat{\phi}(F, \theta)$ can be discontinuous if and only if n is the normal to a plane across which the gradient of an energy-minimizing deformation for the energy density $\phi(F, \theta)$ can be discontinuous.

4.4. Visualization techniques

The development of techniques to visualize the results of the computation of microstructure has been important to the study of microstructure. It is possible to visualize the deformation by displaying the transformation of the finite element mesh (Collins and Luskin 1989). However, it is generally easier to study microstructure by displaying the deformation gradient.

Several techniques have been developed to visualize the deformation gradient. Collins and Luskin (1989) developed the technique of colouring elements according to the closest energy well to the deformation gradient. They assigned the martensitic variant U_l to a given element K with right Cauchy–Green strain $C(x) = (\nabla y(x))^T \nabla y(x)$ if and only if

$$|C - C_l|_K = \min\{|C - C_1|_K, \ldots, |C - C_M|_K, |C - I|_K, \tau\},$$

where $C_i = U_i^T U_i$, where $\tau > 0$ is a user-supplied sensitivity, and where the matrix norm $|C|_K$ is defined by

$$|C|_K = \left[\frac{1}{\mathrm{meas}\, K} \int_K \|C(x)\|^2 \, \mathrm{d}x\right]^{1/2}.$$

They assigned the austenitic phase I to the element K if and only if

$$|C - I|_K = \min\{|C - C_1|_K, \ldots, |C - C_M|_K, |C - I|_K, \tau\}.$$

Finally, they assigned the 'unidentified phase' to the element K if it is not assigned to the austenitic or martensitic phases by the above formulae. The different variants of martensite and austenite are then represented by distinct colours or shades of grey. Collins and Luskin (1989) visualized the gradients of three-dimensional deformations by displaying the gradients on a series of parallel cross-sections.

We know from Ball and James (1992), Luskin (1996a) and Li and Luskin

(1996) or Theorem 7 that the microstructure which minimizes the energy among deformations constrained on the boundary by the condition

$$y(x) = [(1 - \lambda)F_0 + \lambda F_1] x \qquad \text{for all } x \in \partial\Omega$$

is a mixture only of the deformation gradients F_0 and F_1 for the orthorhombic to monoclinic transformation (2.13) and for the cubic to tetragonal transformation (2.15) when F_0 and F_1 are rank-one connected, $F_0, F_1 \in \mathcal{U}$, and $\theta < \theta_T$. Thus, for this problem Collins, Luskin and Riordan (1991b) and Collins et al. (1993) displayed the interpolant of the function

$$\psi(F)_K = \frac{|F^T F - F_0^T F_0|_K}{|F^T F - F_0^T F_0|_K + |F^T F - F_1^T F_1|_K}$$

defined at the centre of gravity of the elements K to display the proximity of the deformation gradient to the energy wells corresponding to F_0 (where $\psi = 0$) and to F_1 (where $\psi = 1$). They represented the function ψ by a map of $(0, 1)$ into colour space or into a grey scale. Other useful variants of the function ψ are given by

$$\tilde{\psi}(F)_K = \frac{|F^T F - F_0^T F_0|_K^2}{|F^T F - F_0^T F_0|_K^2 + |F^T F - F_1^T F_1|_K^2}, \qquad (4.3)$$

which increases the range of deformations that are represented to be nearly in the energy wells of F_0 and F_1, and

$$\hat{\psi}(F)_K = \frac{|F - F_0|_K}{|F - F_0|_K + |F - F_1|_K}, \qquad (4.4)$$

which measures the proximity of the deformation gradient to F_0 and F_1 rather than to their respective energy wells.

The use of isosurfaces of the energy density and surface energy density was developed and used in Klouček and Luskin (1994a, 1994b) to identify the austenitic–martensitic interface.

4.5 Numerical experiments for the continuous, piecewise linear approximation of a two-dimensional model

We can investigate the computation of a simple laminated microstructure by a two-dimensional model (Collins and Luskin 1990, Collins et al. 1991b, Collins 1993a, Collins et al. 1993, Collins 1994). For the two-dimensional model, we have that the reference configuration $\Omega \subset \mathbb{R}^2$, the deformation $y(x)$: $\mathbb{R}^2 \to \Omega$, and the energy density $\phi(F) : \mathbb{R}^{2\times 2} \to \mathbb{R}$ (where we suppress the dependence of the energy density on temperature). We present in Figs 3 and 4 the results of two-dimensional computations by C. Collins using the continuous, piecewise linear finite element for the problem that will next be described.

The three-dimensional orthorhombic to monoclinic problem (2.13) can be modelled in two dimensions by the energy density

$$\phi(F) = \kappa_1 \left(C_{11} - (1 + \eta^2) \right)^2 + \kappa_2 (C_{22} - 1)^2 + \kappa_3 (C_{12}^2 - \eta^2)^2, \qquad (4.5)$$

where $C = F^T F$ is the right Cauchy–Green strain and where η, κ_1, κ_2, κ_3 are positive constants. It can then be checked that

$$\phi(F) \geq 0 \qquad \text{for all } F \in \mathbb{R}^{2\times 2},$$

and

$$\phi(F) = 0 \quad \text{if and only if} \quad F \in \mathrm{SO}(2)U_1 \cup \mathrm{SO}(2)U_2 \qquad (4.6)$$

where

$$U_1 = I - \eta e_2 \otimes e_1 \qquad \text{and} \qquad U_2 = I + \eta e_2 \otimes e_1$$

for $e_1 \in \mathbb{R}^2$ and $e_2 \in \mathbb{R}^2$ given by the canonical basis

$$e_1 = (1, 0) \qquad \text{and} \qquad e_2 = (0, 1).$$

The proof of Lemma 5 can be used to show that there exists a continuous deformation with a linear interface with normal n separating two regions with constant deformation gradients $F_0 \in \mathrm{SO}(2)U_1$ and $F_1 \in \mathrm{SO}(2)U_2$ if and only if $n = e_1$ or $n = e_2$. It can be checked that the energy density (4.5) does not have a local minimum at deformations $F \in \mathrm{SO}(2)$ representing the austenitic phase. This is a desired property for a two-dimensional model, since otherwise, by Lemma 10, there would be rank-one connections between stress-free deformation gradients representing the martensitic and austenitic phases.

To allow for interfaces with arbitrary orientation with respect to a fixed mesh or coordinate system (see Section 4.3), we define for the rotation $\hat{R} \in \mathrm{SO}(2)$ the energy density

$$\hat{\phi}(F) = \phi(F\hat{R}) \qquad \text{for all } F \in \mathbb{R}^{2\times 2}. \qquad (4.7)$$

For this energy density, it follows from (4.6) that

$$\hat{\phi}(F) = 0 \quad \text{if and only if} \quad F \in \mathrm{SO}(2)\hat{U}_1 \cup \mathrm{SO}(2)\hat{U}_2$$

where

$$\hat{U}_1 = I - \eta \hat{e}_2 \otimes \hat{e}_1 \qquad \text{and} \qquad \hat{U}_2 = I + \eta \hat{e}_2 \otimes \hat{e}_1$$

for

$$\hat{e}_1 = \hat{R}e_1 \qquad \text{and} \qquad \hat{e}_2 = \hat{R}e_2.$$

It follows by the above that there exists a continuous deformation with a linear interface with normal $\hat{n}$ separating two regions with constant deformation gradients $F_0 \in \mathrm{SO}(2)\hat{U}_1$ and $F_1 \in \mathrm{SO}(2)\hat{U}_2$ if and only if $\hat{n} = \hat{e}_1 = \hat{R}e_1$ or $\hat{n} = \hat{e}_2 = \hat{R}e_2$.

We now give computational results for the approximations to the energy-minimizing microstructure for the energy

$$\int_\Omega \hat\phi(\nabla y(x))\,\mathrm{d}x \qquad (4.8)$$

for the reference configuration $\Omega = (0,\,1)\times(0,\,1)$ where the deformation $y(x)$ is constrained on the boundary by

$$y(x) = \left[\frac{1}{2}\hat U_1 + \frac{1}{2}\hat U_2\right]x, \qquad x\in\partial\Omega. \qquad (4.9)$$

All of the results in Section 6 hold for the two-dimensional problem (4.7)–(4.9), so we can conclude that the gradients of energy-minimizing sequences of deformations to the two-dimensional problem (4.7)–(4.9) computed using the continuous, piecewise linear finite element approximation on a uniform mesh converge to the Young measure

$$\nu_x = \frac{1}{2}\delta_{\hat U_1} + \frac{1}{2}\delta_{\hat U_2}.$$

In Fig. 3, we present Collins' numerical results for the approximation of an energy-minimizing microstructure to the problem (4.7)–(4.9) with $\hat R = R(45°)$ (where $R(\theta)$ denotes the rotation matrix of θ degrees) by the piecewise linear finite element approximation on a uniform mesh of size $h = 1/N$ where $N = 16,\ 32,\ 64$. Thus, we have that the lines that can separate regions with constant deformation gradients $\hat U_1$ and $\hat U_2$ have normal

$$\hat n = \hat e_1 = \frac{1}{\sqrt 2}\,(e_1 + e_2),$$

and are parallel to lines along which the gradients of deformations in the finite element space are allowed to be discontinuous.

The optimization problem was solved by the Polak–Ribière conjugate gradient method (Polak 1971, Glowinski 1984) with initial data

$$y_{init}(x) = \left[\frac{1}{2}\hat U_1 + \frac{1}{2}\hat U_2\right]x + \frac{1}{2}\eta hr(x) \qquad \text{for all } x\in\Omega, \qquad (4.10)$$

where h is the mesh size and where $r(x) = (r_1(x),\,r_2(x))$ was obtained by getting values for $r_i(x)$ on the interior vertices from a random number generator for the interval $(-1,\,1)$ and then extending $r_i(x)$ to all of Ω by interpolation. We note that $\|\nabla\lfloor\eta hr(x)\rfloor\| = \mathcal{O}(1)$, so the deformation gradients of the initial state need not be close to the energy wells.

To visualize the results of the computations of microstructure, we use the function $\hat\psi$ defined by (4.4) with $F_0 = U_1$ and $F_1 = U_2$ and enhanced by the continuous function

$$g(\varsigma) = \begin{cases} \frac{1}{2}(2\varsigma)^2 & \text{for } 0\le\varsigma<\frac{1}{2}, \\ 1 - \frac{1}{2}(2(1-\varsigma))^2 & \text{for } \frac{1}{2}\le\varsigma\le 1. \end{cases}$$

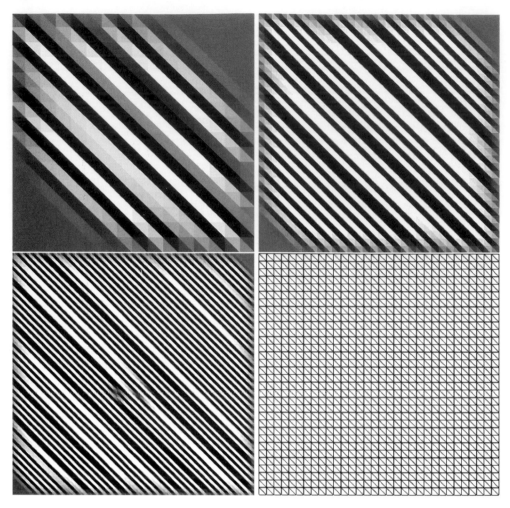

Fig. 3. Deformation gradients for the problem (4.7)–(4.9) with $\eta = .1$ and $\hat{R} = R(45°)$ computed by continuous, piecewise linear finite elements for a uniform finite element mesh of size $h = 1/N$ with $N = 16, 32, 64$. The finite element mesh for $N = 32$ is shown. Courtesy of C. Collins.

We display a map from $g(\hat{\psi}(F))$ into a grey scale so that elements are coloured white if $g(\hat{\psi}(F)) = 0$ (corresponding to $F = U_1$) and elements are coloured black if $g(\hat{\psi}(F)) = 1$ (corresponding to $F = U_2$).

We see in Fig. 3 that microstructure has been obtained on the scale of each successively refined mesh. Since the computed microstructure shown in Fig. 3 is not completely regular, a local minimum of the finite element optimization problem has been computed and not a global minimum. However, the energy of the computed local minimum is close enough to that of a global minimum to give the microstructure and the macroscopic properties of a global minimum.

Fig. 4. Deformation gradients for the problem (4.7)–(4.9) with $\eta = .5$ and the mesh $N = 64$ for the orientation defined by $\hat{R} = R(\theta)$ with $\theta = 25°, 0°, -25°, -45°$. Courtesy of C. Collins.

The results in Fig. 4 illustrate the effect of mesh orientation with respect to the lines of discontinuity of the deformation gradient. We see that the layers are several mesh widths thick when they are not oriented with respect to the mesh.

4.6. Nonconforming finite elements

An alternative approach is that given by the use of non-conforming finite elements (Ciarlet 1978, Quarteroni and Valli 1994), that is, $\mathcal{A}_h \not\subset \mathcal{A}$. The use of non-conforming finite elements is intuitively appealing for problems with microstructure because the admissible finite element deformations should

all of the results in this section give related results for the convergence of energy-minimizing sequences. In Section 6 we will give an estimate for

$$\inf_{v_h \in \mathcal{A}_{h,F_\lambda}} \mathcal{E}(v_h),$$

where $\mathcal{A}_{h,F_\lambda}$ is a conforming finite element approximation to $W_{F_\lambda}^\phi$, which is then used to give estimates for the finite element approximation of microstructure.

We shall also assume that ϕ grows quadratically away from the energy wells, that is, we shall assume that there exists $\kappa > 0$ such that

$$\phi(F) \geq \kappa \|F - \pi(F)\|^2 \qquad \text{for all } F \in \mathbb{R}^{3\times 3} \tag{5.6}$$

where $\pi : \mathbb{R}^{3\times 3} \to \mathcal{U}$ is a Borel measurable projection defined by

$$\|F - \pi(F)\| = \min_{U \in \mathcal{U}} \|F - U\|.$$

The projection π exists since $\mathcal{U}$ is compact, although the projection is not uniquely defined at $F \in \mathbb{R}^{3\times 3}$ where the minimum above is attained at more than one $U \in \mathcal{U}$. We also define the Borel measurable projection $\pi_{1,2} : \mathbb{R}^{3\times 3} \to \mathcal{U}_1 \cup \mathcal{U}_2$ by

$$\|F - \pi_{1,2}(F)\| = \min_{U \in \mathcal{U}_1 \cup \mathcal{U}_2} \|F - U\|. \tag{5.7}$$

We note that $\pi = \pi_{1,2}$ in the double well case (5.3), but that $\pi \neq \pi_{1,2}$ in the cubic to tetragonal case (5.4) since $\mathcal{U} \neq \mathcal{U}_1 \cup \mathcal{U}_2$. We shall also find it useful to utilize the operators $R(F) : \mathbb{R}^{3\times 3} \to \mathrm{SO}(3)$ and $\Pi : \mathbb{R}^{3\times 3} \to \{F_0, F_1\}$, which are defined by the relation

$$\pi_{1,2}(F) = R_{1,2}(F)\Pi_{1,2}(F) \qquad \text{for all } F \in \mathbb{R}^{3\times 3}. \tag{5.8}$$

The following theorem demonstrates that the directional derivatives orthogonal to n (where $F_1 = F_0 + a \otimes n$) of sequences of energy-minimizing deformations converge strongly in L^2. It is crucial to the proof of all of the other results.

Theorem 3 If $w \in \mathbb{R}^3$ satisfies $w \cdot n = 0$, then there exists a positive constant C such that

$$\int_\Omega |(\nabla v(x) - F_\lambda)w|^2 \, \mathrm{d}x \leq C\mathcal{E}(v)^{1/2} + C\mathcal{E}(v) \qquad \text{for all } v \in W_{F_\lambda}^\phi. \tag{5.9}$$

Proof. See Luskin (1996a) for the case of two rotationally invariant energy wells (5.3) and Li and Luskin (1996) for the case of three rotationally invariant energy wells given by the cubic to tetragonal transformation (5.4). $\square$

It follows from the convergence of the directional derivatives orthogonal to n of energy-minimizing sequences of deformations and the Poincaré inequality (Wloka 1987) that energy-minimizing sequences of deformations converge in L^2.

Corollary 1 There exists a positive constant C such that

$$\int_\Omega |v(x) - F_\lambda x|^2 \, dx \leq C\mathcal{E}(v)^{1/2} + C\mathcal{E}(v) \qquad \text{for all } v \in W_{F_\lambda}^\phi. \qquad (5.10)$$

For the double well case (5.3), it follows trivially from the quadratic growth of the energy density away from the energy wells (5.6) that the deformation gradients of energy-minimizing sequences converge to the union of the energy wells $\mathcal{U} = \mathcal{U}_1 \cup \mathcal{U}_2$. However, for the cubic to tetragonal case (5.4) the proof of this result relies on the bound for the directional derivatives orthogonal to n given by Theorem 3. We state this result in the following Theorem.

Theorem 4 For the double well case (5.3) we have the estimate

$$\int_\Omega \|\nabla v(x) - \pi_{1,2}(\nabla v(x))\|^2 \, dx \leq \kappa^{-1}\mathcal{E}(v) \qquad \text{for all } v \in W_{F_\lambda}^\phi.$$

For the cubic to tetragonal transformation (5.4), there exists a positive constant C such that

$$\int_\Omega \|\nabla v(x) - \pi_{1,2}(\nabla v(x))\|^2 \, dx \leq C\mathcal{E}^{1/2}(v) + C\mathcal{E}(v) \qquad \text{for all } v \in W_{F_\lambda}^\phi.$$

Proof. The proof for the double well case (5.3) follows trivially from the quadratic growth of the energy density away from the energy wells (5.6). See Li and Luskin (1996) for the proof in the cubic to tetragonal case (5.4). □

The next theorem shows that the gradients of energy-minimizing sequences of deformations converge weakly to F_λ. It is a consequence of the convergence of the deformations in L^2.

Theorem 5 If $\omega \subset \Omega$ is a smooth domain, then there exists a positive constant C such that

$$\left\|\int_\omega (\nabla v(x) - F_\lambda) \, dx\right\| \leq C\mathcal{E}(v)^{1/8} + C\mathcal{E}(v)^{1/2} \qquad \text{for all } v \in W_{F_\lambda}^\phi.$$

Proof. The proof for the double well case (5.3) is given in Luskin (1996a), and the proof for the cubic to tetragonal transformation (5.4) is given in Li and Luskin (1996). □

The following theorem shows that the gradients of energy-minimizing sequences converge to the set $\{F_0, F_1\}$. The proof relies on the bound for the directional derivatives orthogonal to n given in Theorem 3.

Theorem 6 We have the estimate

$$\int_\Omega \|\nabla v(x) - \Pi_{1,2}(\nabla v(x))\|^2 \, dx \leq C\mathcal{E}(v)^{1/2} + C\mathcal{E}(v) \qquad \text{for all } v \in W_{F_\lambda}^\phi.$$

Proof. See Luskin (1996a) for the case of two rotationally invariant energy wells (5.3) and Li and Luskin (1996) for the case of three rotationally invariant energy wells given by the cubic to tetragonal transformation (5.4). □

The next theorem states that in any smooth domain $\omega \subset \Omega$ and for any energy-minimizing sequence the volume fraction that $\nabla v(x)$ is near F_0 converges to $1 - \lambda$ and the volume fraction that $\nabla v(x)$ is near F_1 converges to λ. This result follows from the weak convergence of the deformation gradients (see Theorem 5) and the convergence of the deformation gradients to the set $\{F_0, F_1\}$ (see Theorem 6). We recall from Theorem 2 that the result of the following theorem implies that there does not exist an energy-minimizing deformation $y \in W_{F_\lambda}^\phi$ to the problem (5.1).

To make the result of the following theorem precise, we define for any smooth domain $\omega \subset \Omega$, $\rho > 0$, and $v \in W_{F_\lambda}^\phi$, the sets

$$\omega_\rho^0 = \omega_\rho^0(v) = \{ x \in \omega : \Pi_{1,2}(\nabla v(x)) = F_0 \text{ and } \|F_0 - \nabla v(x)\| < \rho \},$$
$$\omega_\rho^1 = \omega_\rho^1(v) = \{ x \in \omega : \Pi_{1,2}(\nabla v(x)) = F_1 \text{ and } \|F_1 - \nabla v(x)\| < \rho \}.$$

We can then use Theorem 5 and Theorem 6 to prove the following theorem which describes the convergence of the microstructure (or Young measure) of the deformation gradients of energy minimizing sequences.

Theorem 7   For any smooth domain $\omega \subset \Omega$ and any $\rho > 0$ we have that

$$\left| \frac{\text{meas } \omega_\rho^0(v)}{\text{meas } \omega} - (1 - \lambda) \right| \le C\mathcal{E}(v)^{1/8} + C\mathcal{E}(v)^{1/2}, \tag{5.11}$$

$$\left| \frac{\text{mcas } \omega_\rho^1(v)}{\text{meas } \omega} - \lambda \right| \le C\mathcal{E}(v)^{1/8} + C\mathcal{E}(v)^{1/2} \tag{5.12}$$

for all $v \in W_{F_\lambda}^\phi$. The constants C in the estimates (5.11) and (5.12) are independent of $v \in W_{F_\lambda}^\phi$, but they depend on ω and ρ.

Proof.   The proof for the double well case (5.3) is given in Luskin (1996a), and the proof for the cubic to tetragonal transformation (5.4) is given in Li and Luskin (1996). $\square$

We have by the compactness of $\mathrm{SO}(3)U_1$ and $\mathrm{SO}(3)U_2$ that there exists a positive constant ρ_0 such that

$$\mathrm{dist}\,(\mathrm{SO}(3)U_1, \mathrm{SO}(3)U_2) = \rho_0 > 0 \tag{5.13}$$

where

$$\mathrm{dist}\,(\mathrm{SO}(3)U_1, \mathrm{SO}(3)U_2) = \min \{\|V_1 - V_2\| : V_1 \in \mathrm{SO}(3)U_1, V_2 \in \mathrm{SO}(3)U_2\}.$$

By the definition of $\pi_{1,2}$ (see (5.7)) and the definition of $\Pi_{1,2}$ (see (5.8)), we have for $0 < \rho < \rho_0/2$ that

$$\|F_i - F\| < \rho \qquad \text{implies that} \qquad \Pi_{1,2}(F) = F_i$$

for all $F \in \mathbb{R}^{3\times3}$ and $i \in \{0, 1\}$. Thus, for any $0 < \rho < \rho_0/2$, any smooth domain $\omega \subset \Omega$, any $v \in W^\phi$, and any $i \in \{0, 1\}$ we have that

$$\omega_\rho^i(v) = \{ x \in \omega : \nabla v(x) \in \mathcal{B}_\rho(F_i) \} ,$$

where the set $\mathcal{B}_\rho(F)$ for $\rho > 0$ and $F \in \mathbb{R}^{3\times3}$ is defined by

$$\mathcal{B}_\rho(F) = \left\{ G \in \mathbb{R}^{3\times3} : \|G - F\| < \rho \right\} .$$

Hence, it follows from the definition of the probability measure $\mu_{x,R,\nabla v}(\Upsilon)$ given by (3.53) that we have for $x \in \Omega$, $R > 0$, $v \in W^\phi$, and $0 < \rho < \rho_0/2$ that

$$\mu_{x,R,\nabla v}\left(\mathcal{B}_\rho(F_i)\right) = \frac{\operatorname{meas} \omega_\rho^i(v)}{\operatorname{meas} \omega} \tag{5.14}$$

for $\omega = B_R(x)$.

The following corollary is a direct consequence of Theorem 7 and the identity (5.14) and implies the result on the uniqueness of the Young measure for energy-minimizing sequences of the problem (5.1) that was given by Ball and James (1992).

Corollary 2   If $x \in \Omega$, $R > 0$, and $\rho < \rho_0/2$, where ρ_0 is given by (5.13), then there exists a positive constant C such that

$$|\mu_{x,R,\nabla v}\left(\mathcal{B}_\rho(F_0)\right) - (1 - \lambda)| \leq C\mathcal{E}(v)^{1/8} + C\mathcal{E}(v)^{1/2},$$
$$|\mu_{x,R,\nabla v}\left(\mathcal{B}_\rho(F_1)\right) - \lambda| \leq C\mathcal{E}(v)^{1/8} + C\mathcal{E}(v)^{1/2}$$

for all $v \in W_{F_\lambda}^\phi$.

Next, we show that the estimates for the weak convergence of the deformation gradients (see Theorem 5) and the convergence of the deformation gradients to the set $\{F_0, F_1\}$ (see Theorem 6) can be used to give estimates for the nonlinear integrals of $\nabla v(x)$ that approximate macroscopic densities.

For linear transformations $\mathcal{L} : \mathbb{R}^{3\times3} \to \mathbb{R}$ we define the operator norm

$$|\mathcal{L}| = \max_{\|F\|=1} |\mathcal{L}(F)|,$$

and for uniformly Lipschitz functions $g(F) : \mathbb{R}^{3\times3} \to \mathbb{R}$ we define the function norm

$$\left\| \frac{\partial g}{\partial F} \right\|_{L^\infty} = \operatorname{ess\,sup}_{B\in\mathbb{R}^{3\times3}} \left\| \frac{\partial g}{\partial F}(B) \right\|.$$

We will give estimates of nonlinear integrals of $\nabla v(x)$ for the Sobolev space $\mathcal{V}$ of measurable functions $f(x, F) : \Omega \times \mathbb{R}^{3\times3} \to \mathbb{R}$ such that

$$\int_\Omega \left[\left\| \frac{\partial f}{\partial F}(x, \cdot) \right\|_{L^\infty}^2 + |\nabla G(x) \cdot n|^2 + G(x)^2 \right] dx < \infty$$

where

$$G(x) = f(x, F_1) - f(x, F_0).$$

We note that if $f(x, F) \in \mathcal{V}$, then $f(x, F)$ is Lipschitz continuous as a function of $F \in \mathbb{R}^{3 \times 3}$ for almost all $x \in \Omega$.

Theorem 8 We have for all $v \in W_{F_\lambda}^\phi$ and all functions $f(x, F) \in \mathcal{V}$ that

$$\left| \int_\Omega f(x, \nabla v(x)) - [(1 - \lambda) f(x, F_0) + \lambda f(x, F_1)] \, dx \right| \leq$$

$$C \left\{ \int_\Omega \left\| \frac{\partial f}{\partial F}(x, \cdot) \right\|_{L^\infty}^2 + |\nabla G(x) \cdot n|^2 + G(x)^2 \, dx \right\}^{1/2} \left(\mathcal{E}(v)^{1/4} + \mathcal{E}(v)^{1/2} \right)$$

where

$$G(x) = f(x, F_1) - f(x, F_0).$$

Proof. See Luskin (1996a) for the case of two rotationally invariant energy wells (5.3) and Li and Luskin (1996) for the case of three rotationally invariant energy wells given by the cubic to tetragonal transformation (5.4). $\square$

6. Numerical analysis of microstructure

We shall give in this section error estimates for the finite element approximation of a laminated microstructure for rotationally invariant, double well energy densities (Luskin 1996a, 1996b) and for energy densities for the cubic to tetragonal transformation (Li and Luskin 1996). These error estimates follow directly from the approximation theory given in Section 5 and the theorem proved in this section for the *infimum* of the energy

$$\inf_{v_h \in \mathcal{A}_{h, F_\lambda}} \mathcal{E}(v_h)$$

where $\mathcal{A}_{h, F_\lambda}$ is a conforming finite element space. We shall assume that all of the assumptions described in Section 5 hold.

6.1. Properties of the conforming finite element approximation

We now define the properties of conforming finite element spaces required for our analysis of microstructure in Section 6. We assume that τ_h for $0 < h < h_0$ is a family of decompositions of Ω into polyhedra $\{K\}$ such that (Quarteroni and Valli 1994):

1   $\bar{\Omega} = \cup_{K \in \tau_h} K$;
2   interior $K_1 \cap$ interior $K_2 = \emptyset$ if $K_1 \neq K_2$ for $K_1, K_2 \in \tau_h$;
3   if $S = K_1 \cap K_2 \neq \emptyset$ for $K_1 \neq K_2$, $K_1, K_2 \in \tau_h$, then S is a common face, edge, or vertex of K_1 and K_2;
4   diam $K \leq h$ for all $K \in \tau_h$.

The admissible deformations have finite energy and are constrained on the part of the boundary where the deformation of the crystal is given. Hence, we have by (2.4) that our family of conforming finite element spaces, $\mathcal{A}_h$, defined for mesh diameters in the range $0 < h < h_0$, satisfies

$$\mathcal{A}_h \subset \mathcal{A} \subset W^\phi \subset W^{1,p}(\Omega; \mathbb{R}^3) \subset C(\bar{\Omega}; \mathbb{R}^3)$$

for $0 < h < h_0$.

We assume that there exists an interpolation operator $\mathcal{I}_h : W^{1,\infty}(\Omega; \mathbb{R}^3) \to \mathcal{A}_h$ such that

$$\operatorname{ess\,sup}_{x\in\Omega}\|\nabla\mathcal{I}_h v(x)\| \le C \operatorname{ess\,sup}_{x\in\Omega}\|\nabla v(x)\| \tag{6.1}$$

for all $v \in W^{1,\infty}(\Omega; \mathbb{R}^3)$, where the constant C in (6.1) and below will always denote a generic positive constant independent of h. We also assume for $v \in W^{1,\infty}(\Omega; \mathbb{R}^3)$ that

$$\mathcal{I}_h v(x)|_K = v(x)|_K \text{ for any } K \in \tau_h \text{ such that } v(x)|_K \in \left\{P^1(K)\right\}^3 \tag{6.2}$$

where $\left\{P^1(K)\right\}^3 \equiv P^1(K) \times P^1(K) \times P^1(K)$ and $P^1(K)$ denotes the space of linear polynomials defined on K.

We denote the finite element space of admissible functions satisfying the boundary condition

$$v_h(x) = Fx \qquad \text{for all } x \in \partial\Omega$$

for $F \in \mathbb{R}^{3\times3}$ by

$$\mathcal{A}_{h,F} = \mathcal{A}_h \cap W_F^\phi = \{ v_h \in \mathcal{A}_h : v_h(x) = Fx \text{ for } x \in \partial\Omega \} ,$$

and we further assume that the interpolation operator $\mathcal{I}_h$ satisfies the property that

$$\mathcal{I}_h v \in \mathcal{A}_{h,F} \qquad \text{if} \qquad v \in W_F^\phi. \tag{6.3}$$

The most widely used conforming finite element methods based on continuous, piecewise polynomial spaces have interpolation operators $\mathcal{I}_h$ satisfying (6.1) (for quasi-regular meshes), (6.2), and (6.3) (see Ciarlet 1978, Quarteroni and Valli 1994). In particular, (6.1)–(6.3) are valid for trilinear elements defined on rectangular parallelepipeds as well as for linear elements defined on tetrahedra.

6.2. Approximation of the infimum of the energy

Our analysis of the approximation of microstructure begins with an estimate on the minimization of the energy over deformations $v_h \in \mathcal{A}_h$ that are constrained to satisfy the boundary condition

$$v_h(x) = [(1-\lambda)F_0 + \lambda F_1]\, x = F_\lambda x \qquad \text{for all } x \in \partial\Omega \tag{6.4}$$

for $F_0 \in \mathcal{U}$ and $F_1 \in \mathcal{U}$ rank-one connected as in (3.3) and $\theta < \theta_T$. We recall by (2.18) and (5.5) that

$$\phi_{\min}(\theta) = \phi(F_0, \theta) = \phi(F_1, \theta) = 0 \qquad (6.5)$$

if F_0, $F_1 \in \mathcal{U}$ and $\theta < \theta_T$. The following estimate is an extension of similar results in Gremaud (1994), Chipot et al. (1995) and Luskin (1996a). We note that improved estimates for all of the results in this section can be obtained for finite element meshes that are aligned with the microstructure.

Theorem 9 If $F_0 \in \mathcal{U}$ and $F_1 \in \mathcal{U}$ are rank-one connected as in (3.3) and $\theta < \theta_T$, then

$$\inf_{v_h \in \mathcal{A}_{h, F_\lambda}} \mathcal{E}(v_h) \leq Ch^{1/2} \qquad \text{for all } 0 < h < h_0 \qquad (6.6)$$

Proof. By (6.3), we can define the deformation $v_h(x) \in \mathcal{A}_{h, F_\lambda}$ by

$$v_h(x) = \mathcal{I}_h\left(\hat{w}_\gamma(x)\right)$$

for $\gamma = h^{1/2}$ where $\hat{w}_\gamma(x) \in W_{F_\lambda}^\phi$ is defined by (3.41) in Theorem 1. It follows from property (6.2) of the interpolation operator $\mathcal{I}_h$ that

$$v_h(x) = \hat{w}_\gamma(x) = w_\gamma(x) \qquad \text{for all } x \in \Omega_h \qquad (6.7)$$

for (recalling that $|n|=1$)

$$\Omega_h = \Omega_h^2 \setminus \Lambda_h$$

where

$$\Omega_h^2 = \left\{ x \in \Omega : \text{dist}(x, \partial\Omega) > vh^{1/2} + h \right\},$$

$$\Lambda_h = \bigcup_{j \in \mathbb{Z}} \left\{ x \in \Omega_h^2 : |x \cdot n - jh^{1/2}| \leq h \text{ or } |x \cdot n - (j+1-\lambda)h^{1/2}| \leq h \right\}.$$

Now $\text{meas}(\Omega \setminus \Omega_h^2) \leq Ch^{1/2}$, since $\Omega \setminus \Omega_h^2$ is a layer of width $vh^{1/2} + h$ around the boundary of Ω, and $\text{meas}(\Lambda_h) \leq Ch^{1/2}$, since Λ_h is the union of $\mathcal{O}(h^{-1/2})$ planar layers of thickness h. (Note that only $\mathcal{O}(h^{-1/2})$ of the sets in the definition of Λ_h are non-empty.) So, since $\Omega \setminus \Omega_h = \{\Omega \setminus \Omega_h^2\} \cup \Lambda_h$, we have that

$$\text{meas}(\Omega \setminus \Omega_h) \leq Ch^{1/2}, \qquad (6.8)$$

and we have by (6.1), (3.38), and (6.7) that

$$\nabla v_h(x) \in \{F_0, F_1\} \subset \mathcal{U}, \quad \text{for almost all } x \in \Omega_h,$$
$$\|\nabla v_h(x)\| \leq C, \qquad \text{for almost all } x \in \Omega. \qquad (6.9)$$

Since ϕ is continuous, it is bounded on bounded sets in $\mathbb{R}^{3 \times 3}$. Thus, it follows from (6.5), (6.8) and (6.9) that

$$\int_\Omega \phi(\nabla v_h(x)) = \int_{\Omega \setminus \Omega_h} \phi(\nabla v_h(x)) \, dx \leq C \operatorname{meas}(\Omega \setminus \Omega_h) \leq Ch^{1/2}.$$

$\square$

We have seen in Section 4.2 that we generally expect to compute local minima of the problem

$$\inf_{v_h \in \mathcal{A}_{h,F_\lambda}} \mathcal{E}(v_h)$$

rather than global minima. The local minima that we compute often represent the energy-minimizing microstructure on a length scale $2h$ rather than h. So, it is reasonable to give error estimates for finite element approximations $u_h \in \mathcal{A}_{h,F_\lambda}$ satisfying the quasi-optimality condition

$$\mathcal{E}(u_h) \leq C \inf_{v_h \in \mathcal{A}_{h,F_\lambda}} \mathcal{E}(v_h) \tag{6.10}$$

for some constant $C > 1$ independent of h. For instance, if we compute a local minimum that oscillates on a scale of $2h$, then it is reasonable from Theorem 9 to take $C = \sqrt{2}$.

The following corollaries are direct consequences of the estimate given in Theorem 9 and the bounds given in Section 5. We note that the results in this section hold for both the case of a double well energy density (5.3) and the case of an energy density for the cubic to tetragonal transformation (5.4).

We recall that these estimates hold for general finite element meshes satisfying only the conditions given at the beginning of this section. Improved estimates are possible for meshes which are aligned with the microstructure.

Corollary 3 If u_h satisfies the quasi-optimality condition (6.10) and $\omega \subset \Omega$ is a smooth domain, then there exists a positive constant C such that

$$\int_\Omega |(\nabla u_h(x) - F_\lambda)w|^2 \, dx \leq Ch^{1/4}.$$

Corollary 4 If u_h satisfies the quasi-optimality condition (6.10), then there exists a positive constant C such that

$$\int_\Omega |u_h(x) - F_\lambda x|^2 \, dx \leq Ch^{1/4}.$$

Corollary 5 If u_h satisfies the quasi-optimality condition (6.10) and $\omega \subset \Omega$ is a smooth domain, then there exists a positive constant C such that

$$\left\| \int_\omega (\nabla v(x) - F_\lambda) \, dx \right\| \leq Ch^{1/16} \qquad \text{for all } v \in W_{F_\lambda}^\phi.$$

Corollary 6 If u_h satisfies the quasi-optimality condition (6.10) and $\omega \subset \Omega$ is a smooth domain, then there exists a positive constant C such that

$$\left| \frac{\text{meas}\left(\omega_\rho^0(u_h)\right)}{\text{meas}(\omega)} - (1-\lambda) \right| \leq Ch^{\frac{1}{16}}, \qquad \left| \frac{\text{meas}\left(\omega_\rho^1(u_h)\right)}{\text{meas}(\omega)} - \lambda \right| \leq Ch^{\frac{1}{16}}.$$

Corollary 7 If u_h satisfies the quasi-optimality condition (6.10), then there exists a positive constant C such that

$$\left| \int_\Omega f\left(x, \nabla u_h(x)\right) - \left[(1-\lambda)f(x, F_0) + \lambda f(x, F_1)\right] \, dx \right|$$

$$\leq C \left\{ \int_\Omega \left[\left\| \frac{\partial f}{\partial F}(x, \cdot) \right\|^2_{L^\infty} + |\nabla G(x) \cdot n|^2 + G(x)^2 \right] dx \right\}^{1/2} h^{1/8}$$

for all $f(x, F) \in \mathcal{V}$ where

$$G(x) = f(x, F_1) - f(x, F_0).$$

7. Relaxation

We have seen that the deformation gradients of energy-minimizing sequences of the non-convex energy $\mathcal{E}(y)$ develop oscillations that allow the energy to converge to the lowest possible value. The minimum energy attainable by a microstructure that is constrained by the boundary condition $y(x) = Fx$ for $x \in \partial\omega$, where $\omega \subset \mathbb{R}^3$ is a bounded domain, is given by the relaxed energy density $Q\phi(F)$, which can be defined by

$$Q\phi(F) = \inf \left\{ \frac{1}{\text{meas}\,\omega} \int_\omega \phi(\nabla v(x)) \, dx : \right.$$

$$\left. v \in W^{1,\infty}(\omega; \mathbb{R}^3) \text{ and } v(x) = Fx \text{ for } x \in \partial\omega \right\}. \tag{7.1}$$

The definition of $Q\phi(F)$ can be shown to be independent of ω (Dacorogna 1989).

An energy density $\psi(F)$ is defined to be *quasi-convex* if $Q\psi(F) = \psi(F)$ for all $F \in \mathbb{R}^{3 \times 3}$. It can be shown that $Q\phi(F)$ is quasi-convex and that $Q\phi(F)$ is the quasi-convex envelope of $\phi(F)$ since

$$Q\phi = \sup \left\{ \psi \leq \phi : \psi \text{ quasi-convex} \right\}.$$

We note that in general the relaxed energy density $Q\psi(F)$ is not convex (Kohn 1991).

To make the following discussion simple, we will assume that the energy density satisfies the growth condition that for positive constants C_0, C_1, C_2, C_3 and $p > 3$ we have

$$C_1 \|F\|^p - C_0 \leq \phi(F, \theta) \leq C_2 \|F\|^p + C_3 \qquad \text{for all } F \in \mathbb{R}^{3 \times 3}. \tag{7.2}$$

Hence, we have that

$$W^\phi = W^{1,p}(\Omega; \mathbb{R}^3).$$

We shall also assume that the admissible deformations belong to the set

$$\mathcal{A} = \left\{ y \in W^\phi : y(x) = y_0(x) \text{ for } x \in \partial\Omega \right\}$$

for $y_0(x) \in W^\phi$.

It can then be shown under appropriate conditions on the energy density $\phi(F)$ that

$$\inf_{y \in \mathcal{A}} \int_\Omega Q\phi(\nabla y(x)) \, dx = \inf_{y \in \mathcal{A}} \int_\Omega \phi(\nabla y(x)) \, dx \qquad (7.3)$$

and that there exists an energy-minimizing deformation $\bar{y}(x) \in \mathcal{A}$ for the relaxed energy density $Q\phi(F)$ such that

$$\int_\Omega Q\phi(\nabla \bar{y}(x)) \, dx = \inf_{y \in \mathcal{A}} \int_\Omega Q\phi(\nabla y(x)) \, dx. \qquad (7.4)$$

Further, it can be shown that there exists an energy-minimizing sequence $\{y_k\} \subset \mathcal{A}$ for the energy density ϕ such that

$$\lim_{k \to \infty} \int_\Omega \phi(\nabla y_k(x)) \, dx = \int_\Omega Q\phi(\nabla \bar{y}(x)) \, dx$$

and that

$$y_k(x) \rightharpoonup \bar{y}(x) \qquad \text{weakly in } W^{1,p}(\Omega; \mathbb{R}^3)$$

as $k \to \infty$ (Dacorogna 1989).

It is natural to consider the computation of the numerical solution of (7.4) for the deformation $\bar{y}(x)$, that is, the macroscopic deformation for the energy-minimizing microstructure defined by the sequence $\{y_n\}$. We can also consider the computation of a microstructure at each $\bar{x} \in \Omega$ by computing the energy-minimizing microstructure for the problem (7.1), which defines the relaxed energy density $Q\phi(\nabla \bar{y}(\bar{x}))$. However, explicit formulae or effective algorithms to compute the relaxed energy density (7.1) for the energy densities used to model martensitic crystals have not been found. (See Kohn 1991, though, for an explicit solution to (7.1) for a double well energy density with a special 'Hooke's law'.)

We can approximate (7.1) by considering as test functions the first-order laminates $v(x) = \hat{w}_\gamma(x)$ defined by (3.41) with boundary values $\hat{w}_\gamma(x) = Fx$ for $x \in \partial\omega$. To construct the class of all first-order laminates $v(x) = \hat{w}_\gamma(x)$ with boundary values $\hat{w}_\gamma(x) = Fx$ for $x \in \partial\omega$ we consider all $F_0, F_1 \in \mathbb{R}^{3 \times 3}$ and all $0 \leq \lambda \leq 1$ such that

$$F = (1 - \lambda)F_0 + \lambda F_1, \qquad (7.5)$$

where

$$F_1 = F_0 + a \otimes n \qquad (7.6)$$

for $a, n \in \mathbb{R}^3$, $|n| = 1$. We note that it follows from (7.5) and (7.6) that

$$F_0 = F - \lambda a \otimes n \qquad \text{and} \qquad F_1 = F + (1 - \lambda)a \otimes n.$$

The volume fraction that $\hat{w}_\gamma(x)$ has deformation gradient F_0 converges to

$1 - \lambda$ as $\gamma \to 0$, and the volume fraction that $\hat{w}_\gamma(x)$ has deformation gradient F_1 converges to λ as $\gamma \to 0$. Thus it follows from the proof of Theorem 1 that

$$\lim_{\gamma \to 0} \frac{1}{\text{meas}\,\omega} \int_\omega \phi(\nabla \hat{w}_\gamma(x))\,\mathrm{d}x = (1 - \lambda)\phi(F_0) + \lambda\phi(F_1)$$
$$= (1 - \lambda)\phi(F - \lambda a \otimes n) + \lambda\phi(F + (1 - \lambda)a \otimes n).$$

If we optimize (7.1) by restricting $v \in W^{1,\infty}(\omega; \mathbb{R}^3)$ to the first-order laminates of the form $\hat{w}_\gamma(x)$ discussed in the preceding paragraph, then we obtain the energy density $R_1\phi(F)$ defined by

$$R_1\phi(F) =$$
$$\inf\{(1 - \lambda)\phi(F - \lambda a \otimes n) + \lambda\phi(F + (1 - \lambda)a \otimes n) :$$
$$0 \le \lambda \le 1,\, a,\, n \in \mathbb{R}^3,\, |n| = 1\}$$

for all $F \in \mathbb{R}^{3 \times 3}$. We can more generally optimize (7.1) over the laminates of order k discussed in Section 3.10 and obtain the energy density $R_k\phi(F)$, which can be defined by $R_0\phi(F) = \phi(F)$ and inductively for $k = 1, \ldots$ by

$$R_k\phi(F) =$$
$$\inf\{(1 - \lambda)R_{k-1}\phi(F - \lambda a \otimes n) + \lambda R_{k-1}\phi(F + (1 - \lambda)a \otimes n) :$$
$$0 \le \lambda \le 1,\, a,\, n \in \mathbb{R}^3,\, |n| = 1\}$$

for all $F \in \mathbb{R}^{3 \times 3}$ (Kohn and Strang 1986).

It can be seen that

$$Q\phi(F) \le R_k\phi(F) \le \ldots \le R_1\phi(F) \le \phi(F) \qquad \text{for all } F \in \mathbb{R}^{3 \times 3},$$

so we can conclude from (7.3) that

$$\inf_{y \in \mathcal{A}} \int_\Omega Q\phi(\nabla y(x))\,\mathrm{d}x = \inf_{y \in \mathcal{A}} \int_\Omega R_k\phi(\nabla y(x))\,\mathrm{d}x = \inf_{y \in \mathcal{A}} \int_\Omega \phi(\nabla y(x))\,\mathrm{d}x.$$
$$(7.7)$$

An energy density $\psi(F) : \mathbb{R}^{3 \times 3} \to R$ is *rank-one convex* if

$$\psi((1 - \lambda)F_0 + \lambda F_1) \le (1 - \lambda)\psi(F_0) + \lambda\psi(F_1)$$

for all $0 \le \lambda \le 1$ and all $F_1,\, F_0 \in \mathbb{R}^{3 \times 3}$ such that $\text{rank}\,(F_1 - F_0) \le 1$. The rank-one convex envelope $R\phi(F)$ is then defined by

$$R\phi = \sup\{\psi \le \phi : \psi \text{ rank-one convex}\}.$$

We note that Kohn and Strang (1986) have shown that

$$R\phi(F) = \lim_{k \to \infty} R_k\phi(F) \qquad \text{for all } F \in \mathbb{R}^{3 \times 3},$$

and that Šverák (1992) has shown that in general $Q\phi(F) \ne R\phi(F)$.

The approximation

$$\inf_{y \in \mathcal{A}_h} \int_\Omega R_k\phi(\nabla y(x))\,\mathrm{d}x$$

for finite element spaces $\mathcal{A}_h \subset \mathcal{A}$ has been considered in Nicolaides and Walkington (1993), Roubíček (1994), Carstensen and Plecháč (1995), Roubíček (1996 a), Pedregal (1996), Pedregal (1995), Kružík (1995).

An energy density $\psi(F) : \mathbb{R}^{3\times 3} \to R$ is *polyconvex* if it is a convex function of the minors of $F \in \mathbb{R}^{3\times 3}$ (Ball 1977, Dacorogna 1989). The polyconvex envelope $P\phi(F)$ is then defined by

$$P\phi = \sup \{ \, \psi \leq \phi : \psi \text{ polyconvex} \, \} \, .$$

Since a polyconvex energy density is always quasi-convex by Jensen's inequality, we have that $P\phi(F) \leq Q\phi(F)$ for all $F \in \mathbb{R}^{3\times 3}$. It can be shown that in general $P\phi(F) \neq Q\phi(F)$. Representations of the polyconvex envelope $P\phi(F)$, especially that due to Dacorogna (1989), have been used to develop numerical approximations of the lower bound for the energy given by

$$\inf_{y \in \mathcal{A}} \int_\Omega P\phi\left(\nabla y(x)\right) \, \mathrm{d}x$$

(Roubíček 1996 a, Pedregal 1996, Pedregal 1995, Kružík 1995).

8. Acknowledgments

The author would like to thank Pavel Belik, Tim Brule, Antonio DeSimone, Gero Friesecke, Richard James, Robert Kohn, Bo Li, Julia Liakhova, Pablo Pedregal, Vladimir Šverak, and Giovanni Zanzotto for their many helpful comments and suggestions. He would also like to thank Chunhwa Chu and Richard James for providing the photographs of the microstructure in Figs 1 and 2 and Charles Collins for providing the computational results displayed in Figs 3 and 4.

REFERENCES

R. Abeyaratne and J. Knowles (1994), 'Dynamics of propagating phase boundaries: thermoelastic solids with heat conduction', *Arch. Rat. Mech. Anal.* **126**, 203–230.

R. Abeyaratne, C. Chu and R. James (1994), Kinetics and hysteresis in martensitic single crystals, in *Proc. Symposium on the Mechanics of Phase Transformations and Shape Memory Alloys*, ASME.

R. Adams (1975), *Sobolev Spaces*, Academic Press, New York.

H.-W. Alt, K.-H. Hoffmann, M. Niezgódka and J. Sprekels (1985), A numerical study of structural phase transitions in shape memory alloys, Technical Report 90, Institut für Mathematik, Augsburg University.

G. Arlt (1990), 'Twinning in ferroelectric and ferroelastic ceramics: stress relief', *J. Mat. Sci.* **22**, 2655–2666.

J. Ball (1977), 'Convexity conditions and existence theorems in nonlinear elasticity', *Arch. Ration. Mech. Anal.* **63**, 337–403.

J. Ball (1989), A version of the fundamental theorem for Young measures, in *PDEs and Continuum Models of Phase Transition* (M. Rascle, D. Serre and M. Slemrod, eds), Springer, New York, pp. 207–215. Lecture Notes in Physics, vol. 344.

J. Ball and R. James (1987), 'Fine phase mixtures as minimizers of energy', *Arch. Rational Mech. Anal.* **100**, 13–52.

J. Ball and R. James (1992), 'Proposed experimental tests of a theory of fine microstructure and the two-well problem', *Phil. Trans. R. Soc. Lond. A* **338**, 389–450.

J. Ball and R. James (1993), Theory for the microstructure of martensite and applications, in *Proceedings of the International Conference on Martensitic Transformations* (Perkins and Wayman, eds), Monterey Institute of Advanced Studies, Carmel, California, pp. 65–76.

J. Ball, C. Chu and R. James (1994), Metastability of martensite. Manuscript.

J. Ball, C. Chu and R. James (1995), Hysteresis during stress-induced variant rearrangement, in *Proceedings of the International Conference on Martensitic Transformations*.

J. Ball, P. Holmes, R. James, R. Pego and P. Swart (1991), 'On the dynamics of fine structure', *J. Nonlinear Sci.* **1**, 17–70.

G. Barsch, B. Horovitz and J. Krumhansl (1987), 'Dynamics of twin boundaries in martensites', *Phys. Rev. Lett.* **59**, 1251–1254.

Z. S. Basinski and J. W. Christian (1954), 'Experiments on the martensitic transformation in single crystals of indium-thallium alloys', *Acta Metall.* **2**, 148–166.

K. Bhattacharya (1991), 'Wedge-like microstructure in martensite', *Acta Metall. Mater.* **39**, 2431–2444.

K. Bhattacharya (1992), 'Self accommodation in martensite', *Arch. Rat. Mech. Anal.* **120**, 201–244.

K. Bhattacharya (1993), 'Comparison of the geometrically nonlinear and linear theories of martensitic transformation', *Continuum Mechanics and Thermodynamics* **5**, 205–242.

K. Bhattacharya and R. Kohn (1995), Elastic energy minimization and the recoverable strains of polycrystalline shape-memory materials, Technical Report 1366, IMA.

K. Bhattacharya and R. Kohn (1996), 'Symmetry, texture, and the recoverable strain of shape memory polycrystals', *Acta Metall. Mater.*

K. Bhattacharya, N. Firoozye, R. James and R. Kohn (1994), 'Restrictions on microstructure', *Proc. Roy. Soc. Edinburgh A* **124A**, 843–878.

K. Bhattacharya, R. James and P. Swart (1993), A nonlinear dynamic model for twin relaxation with applications to Au 47.5at.%Cd and other shape-memory materials, in *Twinning in Advanced Materials* (M. Yoo and M. Wuttig, eds), Theoretical Materials Science.

D. Brandon, T. Lin and R. Rogers (1995), 'Phase transitions and hysteresis in nonlocal and order-parameter models', *Meccanica* **30**, 541–565.

B. Brighi and M. Chipot (1994), 'Approximated convex envelope of a function', *SIAM J. Numer. Anal.* **31**, 128–148.

O. Bruno (1995), 'Quasistatic dynamics and pseudoelasticity in polycrystalline shape memory wires', *Smart Mater. Struct.* **4**, 7–13.

O. P. Bruno, P. H. Leo and F. Reitich (1995a), 'Free boundary conditions at austenite–martensite interfaces', *Physics Review Letters* **74**, 746–749.

O. P. Bruno, F. Reitich and P. H. Leo (1995b), The overall elastic energy of poly-crystalline martensitic solids. Manuscript.

M. W. Burkart and T. A. Read (1953), 'Diffusionless phase changes in the indium-thallium system', *Trans. AIME J. Metals* **197**, 1516–1524.

C. Carstensen and P. Plecháč (1995), Numerical solution of the scalar double-well problem allowing microstructure, Technical Report 1752, Technische Hochschule Darmstadt.

M. Chipot (1991), 'Numerical analysis of oscillations in nonconvex problems', *Numer. Math.* **59**, 747–767.

M. Chipot and C. Collins (1992), 'Numerical approximations in variational problems with potential wells', *SIAM J. Numer. Anal.* **29**, 1002–1019.

M. Chipot and D. Kinderlehrer (1988), 'Equilibrium configurations of crystals', *Arch. Rat. Mech. Anal.* **103**, 237–277.

M. Chipot, C. Collins and D. Kinderlehrer (1995), 'Numerical analysis of oscillations in multiple well problems', *Numer. Math.* **70**, 259–282.

C. Chu and R. James (1995), Analysis of microstructures in Cu-14% Al-3.9% Ni by energy minimization, in *Proceedings of the International Conference on Martensitic Transformations*.

P. Ciarlet (1978), *The Finite Element Method for Elliptic Problems*, North-Holland, Amsterdam.

P. Colli (1995), 'Global existence for the three-dimensional Frémond model of shape memory alloys', *Nonlinear analysis, theory, methods, and applications* **24**, 1565–1579.

P. Colli, M. Frémond and A. Visintin (1990), 'Thermo-mechanical evolution of shape memory alloys', *Quart. Appl. Math.* **48**, 31–47.

C. Collins (1993a), Computation of twinning, in *Microstructure and Phase Transitions* (J. Ericksen, R. James, D. Kinderlehrer and M. Luskin, eds), Springer, New York, pp. 39–50. IMA Volumes in Mathematics and Its Applications, vol. 54.

C. Collins (1993b), Computations of twinning in shape-memory materials, in *Smart Structures and Materials 1993: Mathematics in Smart Structures* (H. T. Banks, ed.), Proc. SPIE 1919, pp. 30–37.

C. Collins (1994), Comparison of computational results for twinning in the two-well problem, in *Proceedings of the 2nd International Conference on Intelligent Materials* (C. Rogers and G. Wallace, eds), Technomic, Lancaster, PA, pp. 391–401.

C. Collins and M. Luskin (1989), The computation of the austenitic–martensitic phase transition, in *Partial Differential Equations and Continuum Models of Phase Transitions* (M. Rascle, D. Serre and M. Slemrod, eds), Springer, New York, pp. 34–50. Lecture Notes in Physics, vol. 344.

C. Collins and M. Luskin (1990), Numerical modeling of the microstructure of crystals with symmetry-related variants, in *Proceedings of the US–Japan Workshop on Smart/Intelligent Materials and Systems* (I. Ahmad, , M. Aizawa, A. Crowson and C. Rogers, eds), Technomic, Lancaster, PA, pp. 309–318.

C. Collins and M. Luskin (1991a), Numerical analysis of microstructure for crystals with a nonconvex energy density, in *Progress in Partial Differential Equations: the Metz Surveys* (M. Chipot and J. S. J. Paulin, eds), Longman, Harlow, UK, pp. 156–165. Pitman Research Notes in Mathematics Series, vol. 249.

C. Collins and M. Luskin (1991b), 'Optimal order estimates for the finite element approximation of the solution of a nonconvex variational problem', *Math. Comp.* **57**, 621–637.

C. Collins, D. Kinderlehrer and M. Luskin (1991a), 'Numerical approximation of the solution of a variational problem with a double well potential', *SIAM J. Numer. Anal.* **28**, 321–332.

C. Collins, M. Luskin and J. Riordan (1991b), Computational images of crystalline microstructure, in *Computing Optimal Geometries* (J. Taylor, ed.), Amer. Math. Soc., Providence, pp. 16–18. AMS Special Lectures in Mathematics and AMS Videotape Library.

C. Collins, M. Luskin and J. Riordan (1993), Computational results for a two-dimensional model of crystalline microstructure, in *Microstructure and Phase Transitions* (J. Ericksen, R. James, D. Kinderlehrer and M. Luskin, eds), Springer, New York, pp. 51–56. IMA Volumes in Mathematics and Its Applications, vol. 54.

B. Dacorogna (1989), *Direct methods in the calculus of variations*, Springer, Berlin.

A. DeSimone (1993), 'Energy minimizers for large ferromagnetic bodies', *Arch. Rat. Mech. Anal.* **125**, 99–143.

G. Dolzmann and S. Müller (1995), 'Microstructures with finite surface energy: the two-well problem', *Arch. Rat. Mech. Anal.* **132**, 101–141.

I. Ekeland and R. Temam (1974), *Analyse convexe et problèmes variationnels*, Dunod, Paris.

J. Ericksen (1980), 'Some phase transitions in crystals', *Arch. Rat. Mech. Anal.* **73**, 99–124.

J. Ericksen (1986), 'Constitutive theory for some constrained elastic crystals', *J. Solids and Structures* **22**, 951–964.

J. Ericksen (1987a), Some constrained elastic crystals, in *Material Instabilities in Continuum Mechanics and Related Problems* (J. Ball, ed.), Oxford University Press, pp. 119–137.

J. Ericksen (1987b), Twinning of crystals I, in *Metastability and Incompletely Posed Problems* (S. Antman, J. Ericksen, D. Kinderlehrer and I. Müller, eds), Springer, New York, pp. 77–96. IMA Volumes in Mathematics and Its Applications, vol. 3.

J. D. Eshelby (1961), Elastic inclusions and inhomogeneities, in *Progress in Solid Mechanics, vol. 2* (I. N. Sneddon and R. Hill, eds), pp. 87–140.

N. Firoozye (1993), 'Geometric parameters and the relaxation of multiwell energies', pp. 85–110. IMA Volumes in Mathematics and Its Applications, vol. 54.

I. Fonseca (1987), 'Variational methods for elastic crystals', *Arch. Rational Mech. Anal.* **97**, 189–220.

M. Frémond (1990), Shape memory alloys. a thermomechanical model, in *Free bounday problems: Theory and applications, vol. I* (K.-H. Hoffman and J. Sprekels, eds), Longman, Harlow, UK, pp. 295–306.

D. French (1991), 'On the convergence of finite element approximations of a relaxed variational problem', *SIAM J. Numer. Anal.* **28**, 419–436.

D. French and S. Jensen (1991), 'Behavior in the large of numerical solutions to one-dimensional nonlinear viscoelasticity by continuous time Galerkin methods', *Comp. Meth. Appl. Mech. Eng.* **86**, 105–124.

D. French and L. Walhbin (1993), 'On the numerical approximation of an evolution problem in nonlinear viscoelasticity', *Computer Methods in Applied Mechanics and Engineering* **107**, 101–116.

G. Friesecke (1994), 'A necessary and sufficient condition for nonattainment and formation of microstructure almost everywhere in scalar variational problems', *Proc. Roy. Soc. Edinb.* **124A**, 437–471.

G. Friesecke and G. Dolzmann (1996), 'Time discretization and global existence for a quasi-linear evolution equation with nonconvex energy', *SIAM J. Math. Anal.*

G. Friesecke and J. B. McLeod (submitted), 'Dynamic stability of nonminimizing phase mixtures', *Proc. Roy. Soc. London.*

G. Friesecke and J. B. McLeod (To appear), 'Dynamics as a mechanism preventing the formation of finer and finer microstructure', *Arch. Rat. Mech. Anal.*

R. Glowinski (1984), *Numerical Methods for Nonlinear Variational Problems*, Springer, New York.

J. Goodman, R. Kohn and L. Reyna (1986), 'Numerical study of a relaxed variational problem from optimal design', *Comp. Meth. in Appl. Mech. and Eng.* **57**, 107–127.

P. Gremaud (1993), 'On an elliptic-parabolic problem related to phase transitions in shape memory alloys', *Numer. Funct. Anal. and Optim.* **14**, 355–370.

P. Gremaud (1994), 'Numerical analysis of a nonconvex variational problem related to solid–solid phase transitions', *SIAM J. Numer. Anal.* **31**, 111–127.

P. Gremaud (1995), 'Numerical optimization and quasiconvexity', *Euro. J. of Applied Mathematics* **6**, 69–82.

M. Gurtin (1981), *Topics in Finite Elasticity*, SIAM, Philadelphia.

I. N. Herstein (1975), *Topics in Algebra*, 2nd edn, Wiley, New York.

K.-H. Hoffmann and M. Niezgódka (1990), 'Existence and uniqueness of global solutions to an extended model of the dynamical developments in shape memory alloys', *Nonlinear Analysis* **15**, 977–990.

K.-H. Hoffmann and S. Zheng (1988), 'Uniqueness for structured phase transitions in shape memory alloys', *Math. Methods Appl. Sciences* **10**, 145–151.

K.-H. Hoffmann and J. Zou (1995), 'Finite element approximations of Landau–Ginzburg's equation model for structural phase transitions in shape memory alloys', *Mathematical Modelling and Numerical Analysis* **29**, 629–655.

B. Horovitz, G. Barsch and J. Krumhansl (1991), 'Twin bands in martensites: statics and dynamics', *Phys. Rev. B* **43**, 1021–1033.

R. James (1986), 'Displacive phase transformations in solids', *Journal of the Mechanics and Physics of Solids* **34**, 359–394.

R. James (1987a), Microstructure and weak convergence, in *Material Instabilities in Continuum Mechanics and Related Mathematical Problems* (J. Ball, ed.), Oxford University Press, pp. 175–196.

R. James (1987b), The stability and metastability of quartz, in *Metastability and Incompletely Posed Problems* (S. Antman and J. Ericksen, eds), Springer, New York, pp. 147–176.

R. James (1989), Minimizing sequences and the microstructure of crystals, in *Proceedings of the Society of Metals Conference on Phase Transformations*, Cambridge University Press.

R. James and D. Kinderlehrer (1989), Theory of diffusionless phase transitions, in *PDE's and continuum models of phase transitions* (M. Rascle, D. Serre and M. Slemrod, eds), Springer, pp. 51–84. Lecture Notes in Physics, vol. 344.

S. Kartha, T. Castán, J. Krumhansl and J. Sethna (1994), 'The spin-glass nature of tweed precursors in martensitic transformations', *Phys. Rev. Lett.* **67**, 3630.

S. Kartha, J. A. Krumhansl, J. P. Sethna and L. K. Wickham (1995), 'Disorder-driven pretransitional tweed in martensitic transformations', *Phys. Rev. B* **52**, 803.

A. G. Khachaturyan (1967), 'Some questions concerning the theory of phase transformations in solids', *Soviet Phys. Solid State* **8**, 2163–2168.

A. G. Khachaturyan (1983), *Theory of structural transformations in solids*, Wiley, New York.

A. G. Khachaturyan and G. Shatalov (1969), 'Theory of macroscopic periodicity for a phase transition in the solid state', *Soviet Phys. JETP* **29**, 557–561.

D. Kinderlehrer (1987), 'Twinning in crystals II', pp. 185–212. IMA Volumes in Mathematics and Its Applications, vol. 3.

D. Kinderlehrer and L. Ma (1994a), 'Computational hysteresis in modeling magnetic systems', *IEEE. Trans. Magn.* **30.6**, 4380–4382.

D. Kinderlehrer and L. Ma (1994b), The simulation of hysteresis in nonlinear systems, in *Mathematics in Smart Structures and Materials* (H. T. Banks, ed.), SPIE, pp. 78–87. Vol. 2192.

D. Kinderlehrer and P. Pedregal (1991), 'Characterizations of gradient Young measures', *Arch. Rat. Mech. Anal.* **115**, 329–365.

S. Kirkpatrick, C. D. Gelatt, Jr., and M. P. Vecchi (1983), 'Optimization by simulated annealing', *Science* **220**, 671–680.

P. Klouček and M. Luskin (1994a), 'The computation of the dynamics of martensitic microstructure', *Continuum Mech. Thermodyn.* **6**, 209–240.

P. Klouček and M. Luskin (1994b), 'Computational modeling of the martensitic transformation with surface energy', *Mathematical and Computer Modelling* **20**, 101–121.

P. Klouček, B. Li and M. Luskin (1996), 'Analysis of a class of nonconforming finite elements for crystalline microstructures', *Math. Comput.* To appear.

R. Kohn (1991), 'Relaxation of a double-well energy', *Continuum Mechanics and Thermodynamics* **3**, 193–236.

R. Kohn and S. Müller (1992a), 'Branching of twins near an austenite/twinned-martensite interface', *Philosophical Magazine* **66A**, 697–715.

R. Kohn and S. Müller (1992b), 'Relaxation and regularization of nonconvex variational problems', *Rend. Sem. Mat. Fis. Univ. Milano* **62**, 89–113.

R. Kohn and S. Müller (1994), 'Surface energy and microstructure in coherent phase transitions', *Comm. Pure Appl. Math.* **47**, 405–435.

R. Kohn and G. Strang (1983), 'Explicit relaxation of a variational problem in optimal design', *Bull. A.M.S.* **9**, 211–214.

R. Kohn and G. Strang (1986), 'Optimal design and relaxation of variational problems I, II, and III', *Commun. Pure Appl. Math.* **39**, 113–137, 139–182, and 353–377.

M. Kružík (1995), Numerical approach to double well problems. Manuscript.

P. H. Leo, T. W. Shield and O. P. Bruno (1993), 'Transient heat transfer effects on the pseudoelastic hysteresis of shape memory wires', *Acta metallurgica* **41**, 2477–2485.

B. Li and M. Luskin (1996), Finite element analysis of microstructure for the cubic to tetragonal transformation, Technical Report 1373, IMA.

M. Luskin (1991), Numerical analysis of microstructure for crystals with a nonconvex energy density, in *Progress in Partial differential equations: the Metz Surveys* (M. Chipot and J. S. J. Paulin, eds), Longman, Harlow, UK, pp. 156–165.

M. Luskin (1996*a*), Approximation of a laminated microstructure for a rotationally invariant, double well energy density, Technical report. To appear.

M. Luskin (1996*b*), 'Numerical analysis of a microstructure for a rotationally invariant, double well energy', *Zeitschrift für Angewandte Mathematik und Mechanik*.

M. Luskin and L. Ma (1992), 'Analysis of the finite element approximation of microstructure in micromagnetics', *SIAM J. Numer. Anal.* **29**, 320–331.

M. Luskin and L. Ma (1993), Numerical optimization of the micromagnetics energy, in *Mathematics in Smart Materials*, SPIE, pp. 19–29.

L. Ma (1993), Computation of magnetostrictive materials, in *Mathematics in Smart Materials*, SPIE, pp. 47–54.

L. Ma and N. Walkington (1995), 'On algorithms for non-convex optimization', *SIAM J. Numer. Anal.* **32**, 900–923.

S. Müller (1993), 'Singular perturbations as a selection criterion for periodic minimizing sequences', *Calc. Var.* **1**, 169–204.

S. Müller and V. Šverák (1995), Attainment results for the two-well problem by convex integration, Technical Report SFB 256, Universität Bonn.

R. A. Nicolaides and N. Walkington (1993), 'Computation of microstructure utilizing Young measure representations', *J. Intelligent Material Systems and Structures* **4**, 457–462.

M. Niezgódka and J. Sprekels (1991), 'Convergent numerical approximations of the thermomechanical phase transitions in shape memory alloys', *Numer. Math.* **58**, 759–778.

M. Ortiz and G. Gioia (1994), 'The morphology and folding patterns of buckling-driven thin-film blisters', *J. Math. Phys. Solids* **42**, 531–559.

P. Pedregal (1993), 'Laminates and microstructure', *Europ. J. Appl. Math.* **4**, 121–149.

P. Pedregal (1995), On the numerical analysis of non-convex variational problems. Manuscript.

P. Pedregal (1996), 'Numerical approximation of parametrized measures', *Num. Funct. Anal. Opt.* **16**, 1049–1066.

R. Pego (1987), 'Phase transitions in one-dimensional nonlinear viscoelasticity', *Arch. Rat. Mech. Anal.* **97**, 353–394.

M. Pitteri (1984), 'Reconciliation of local and global symmetries of crystals', *J. Elasticity* **14**, 175–190.

M. Pitteri and G. Zanzotto (1996a), *Continuum models for twinning and phase transitions in crystals*, Chapman and Hall, London.

M. Pitteri and G. Zanzotto (1996b), Twinning in symmetry-breaking phase transitions. Manuscript.

E. Polak (1971), *Computational Methods in Optimization*, Academic Press.

A. Quarteroni and A. Valli (1994), *Numerical Approximation of Partial Differential Equations*, Springer, Berlin.

R. Rannacher and S. Turek (1992), 'Simple nonconforming quadrilaterial Stokes element', *Numer. Meth. for PDEs* **8**, 97–111.

A. Roitburd (1969), 'The domain structure of crystals formed in the solid phase', *Soviet Phys. Solid State* **10**, 2870–2876.

A. Roitburd (1978), 'Martensitic transformation as a typical phase transition in solids', *Solid State Physics* **34**, 317–390.

T. Roubíček (1994), 'Finite element approximation of a microstructure evolution', *Math. Methods in the Applied Sciences* **17**, 377–393.

T. Roubíček (1996a), 'Numerical approximation of relaxed variational problems', *J. Convex Analysis*.

T. Roubíček (1996b), *Relaxation in Optimization Theory and Variational Calculus*, Walter de Gruyter, Berlin.

W. Rudin (1987), *Real and Complex Analysis*, 3rd edn, McGraw-Hill, New York.

P. Rybka (1992), 'Dynamical modeling of phase transitions by means of viscoelasticity in many dimensions', *Proc. Royal Soc. Edinburgh* **120A**, 101–138.

P. Rybka (1995), Viscous damping prevents propagation of singularities in the system of viscoelasticity, Technical Report 184, Departament of Mathematics, Centro de Investigación y Estudios Avanzados del IPN, Mexico.

D. Schryvers (1993), 'Microtwin sequences in thermoelastic Ni_xAl_{100-x} martensite studied by conventional and high-resolution transmission electron microscopy', *Phil. Mag.* **A68**, 1017–1032.

J. P. Sethna, S. Kartha, T. Castan and J. A. Krumhansl (1992), 'Tweed in martensites: A potential new spin glass', *Physica Scripta* **T42**, 214.

M. Shearer (1986), 'Nonuniqueness of admissible solutions of Riemann initial value problems', *Arch. Rat. Mech. Anal.* **93**, 45–59.

S. Silling (1989), 'Phase changes induced by deformation in isothermal elastic crystals', *J. of the Mech. and Phys. of Solids* **37**, 293–316.

S. Silling (1992), 'Dynamic growth of martensitic plates in an elastic material', *Journal of Elasticity* **28**, 143–164.

M. Slemrod (1983), 'Admissibility criteria for propagating phase boundaries in a van der Waals fluid', *Arch. Rat. Mech. Anal.* **81**, 301–315.

P. Swart and P. Holmes (1992), 'Energy minimization and the formation of microstructure in dynamic anti-plane shear', *Arch. Rational Mech. Anal.* **121**, 37–85.

L. Tartar (1984), étude des oscillations dans les équations aux dérivées partielles nonlinéaires, in *Lecture Notes in Physics vol. 195*, Springer, pp. 384–412.

L. Tartar (1990), 'H-measures, a new approach for studying homogenization, oscillations, and concentration effects in partial differential equations', *Proc. Roy. Soc. Edinburgh* **115A**, 193–230.

L. Truskinovsky (1985), 'The structure of isothermal phase shock', *(Soviet Physics Doklady) Dokl. Acad. Nauk SSSR* **285** **(2)**, 309–315.

L. Truskinovsky (1987), 'Dynamics of non-equilibrium phase boundaries in the heat-conductive nonlinear elastic medium', *J. Appl. Math. & Mech. (PMM)* **51** **(6)**, 777–784.

L. Truskinovsky (1994), 'Transition to 'detonation' in dynamic phase changes', *Arch. for Rational Mech. Anal.* **125**, 375–397.

L. Truskinovsky and G. Zanzotto (1995), 'Finite-scale microstructures and metastability in one-dimensional elasticity', *Meccanica* **30**, 557–589.

L. Truskinovsky and G. Zanzotto (1996), 'Ericksen's bar revisited', *Mech. Phys. Solids.*

V. Šverák (1992), 'Rank-one convexity does not imply quasiconvexity', *Proc. Royal Soc. Edinburgh* **120A**, 185–189.

Y. Wang, L.-Q. Chen and A. G. Khachaturyan (1994), Computer simulation of microstructure evolution in coherent solids, in *Solid–Solid Phase Transformations* (W. C. Johnson, J. M. Howe, D. E. Laughlin and W. A. Soffa, eds), The Minerals, Metals & Materials Society, pp. 245–265.

S. Wen, A. G. Khachaturyan and J. W. Morris Jr. (1981), 'Computer simulation of a 'Tweed-Transformation' in an idealized elastic crystal', *Metallurgical Trans. A* **12A**, 581–587.

J. Wloka (1987), *Partial Differential Equations*, Cambridge University Press.

G. Zanzotto (1996), Weak phase transitions in simple lattices. Manuscript.

Acta Numerica (1996), pp. 259–307

Iterative solution of systems of linear differential equations

Ulla Miekkala and Olavi Nevanlinna

Helsinki University of Technology

Institute of Mathematics

Otakaari 1, 02150 Espoo, Finland

E-mail: Ulla.Miekkala@hut.fi, Olavi.Nevanlinna@hut.fi

CONTENTS

1	Introduction	259
2	Finite windows	260
3	Infinite windows	272
4	Acceleration techniques	277
5	Discretized iterations	287
6	Periodic problems	295
7	A case study: linear RC-circuits	296
	References	305

1. Introduction

Parallel processing has made iterative methods an attractive alternative for solving large systems of initial value problems. Iterative methods for initial value problems have a history of more than a century, and in the works of Picard (1893) and Lindelöf (1894) they were given a firm theoretical basis. In particular, the *superlinear* convergence on finite intervals is included in Lindelöf (1894).

In the early 1980s *waveform relaxation (WR)* was introduced for the simulation of electrical networks, by Lelarasmee, Ruehli and Sangiovanni-Vincentelli (1982). The methodology has been used in several application areas and has been extended to time-dependent PDEs. There are even books available: White and Sangiovanni-Vincentelli (1987) and Vandewalle (1993).

In this survey we shall only consider systems of ODEs, with some remarks on differential algebraic equations.

Practical problems are usually nonlinear, but it has been our working hypothesis that studying the linear case carefully, specifically introducing a clear notation and suitable concepts, might be what users really need. In applying the ideas to particular problems, including nonlinear mappings, it is often relatively easy to make intelligent guesses, if one has a good understanding of the nonlinear problem at hand and of the behaviour of the method on linear problems. On papers dealing with strongly nonlinear problems we mention Nevanlinna and Odeh (1987), partly because it was Farouk Odeh who introduced the second author to waveform relaxation in 1983.

The iterative method has many names. To call it waveform relaxation is natural when the application area is electronics. To call it Picard–Lindelöf iteration is historically motivated, although 'block Picard–Lindelöf iteration' would perhaps be more accurate, if cumbersome. The names Picard and Lindelöf also occur in the analysis of the iteration: our convergence theory on finite windows is based on the theory of entire functions, to which both Picard and Lindelöf made important contributions.

We shall not discuss implementation issues at all, not because they are unimportant, but because they are well described in the literature. On linear PDEs we refer to Lubich and Ostermann (1987) and Vandewalle (1992), mentioning that a combination of multigrid in space and waveform relaxation in time is fast and parallelizes reasonably well.

2. Finite windows

2.1. Basic estimates

Let A be a constant d by d complex matrix. We want to solve the initial value problem

$$\dot{x} + Ax = f, \quad x(0) = x_0, \tag{2.1}$$

where the forcing function f generally depends on time t. The matrix A is decomposed as $A = M - N$, where M would typically contain the diagonal blocks of A, and N the off-diagonal couplings. We consider the iteration

$$\dot{x}^k + Mx^k = Nx^{k-1} + f, \quad x^k(0) = x_0. \tag{2.2}$$

If nothing better is available, one can take $x^0(t) = x_0$. In practice, equation (2.2) would be solved by high-quality software, within a (perhaps k-dependent) tolerance. Here we assume it to be solved exactly.

Introducing the following *iteration* operator

$$\mathcal{K}u(t) := \int_0^t e^{-(t-s)M} Nu(s)\,\mathrm{d}s \tag{2.3}$$

we can write (2.2) in the form

$$x^k = \mathcal{K}x^{k-1} + g, \tag{2.4}$$

where

$$g(t) := e^{-tN}x_0 + \int_0^t e^{-(t-s)M} f(s)\,\mathrm{d}s. \tag{2.5}$$

In what follows we shall assume that f is defined for all $t > 0$ and is locally in L_1, that is

$$\int_0^T |f(t)|dt < \infty \quad \text{for all} \quad T < \infty. \tag{2.6}$$

So, we in fact replace (2.1) by the fixed-point problem

$$x = \mathcal{K}x + g, \tag{2.7}$$

which then has a unique solution within continuous functions on $[0, \infty)$.

Proposition 1   Let f be absolutely integrable on bounded subsets of $[0, \infty)$, and let x_0 be given. Then there exists exactly one continuous solution x on $[0, \infty)$ satisfying (2.7). In addition, x is absolutely continuous and satisfies (2.1) almost everywhere.

Proof.   To prove a result like this, one only has to show that $(1 - \mathcal{K})^{-1}$ is a bounded operator in $C[0, T]$ for all T. This is included in the estimates of the growth of the resolvent in Proposition 2. That x is absolutely continuous follows by differentiation. $\square$

We shall use $|\cdot|$ to denote the Euclidean norm, and its induced matrix norm, throughout the paper. In $C[0, T]$ we shall then use the uniform norm

$$|x|_T := \sup_{0 \le t \le T} |x(t)|. \tag{2.8}$$

We shall also use $|\cdot|_T$ to denote the induced operator norm

$$|\mathcal{K}|_T := \sup_{|x|_T = 1} : \mathcal{K}x|_T.$$

Theorem 1   For $k \ge 1$, we have the bound

$$|\mathcal{K}^k|_T \le e^{T|M|} \frac{(T|N|)^k}{k!}. \tag{2.9}$$

Proof.   The iterates $\mathcal{K}^k$ are integral operators whose kernels are k-fold convolutions of $e^{-sM}N$ satisfying

$$|(e^{-sM}N)^{*k}(t)| \le e^{t|M|}|N| \frac{(|N|t)^{k-1}}{(k-1)!}. \tag{2.10}$$

In fact, (2.10) is trivial for $k = 1$. For $k > 1$ we have

$$\mathcal{K}^k u(t) = \int_0^t e^{-(t-s)M} N \mathcal{K}^{k-1} u(s)\, ds, \tag{2.11}$$

and from $\mathcal{K}^{k-1}u(t) = (e^{-sM}N)^{*(k-1)} * u(t)$ we obtain the induction step needed to conclude (2.10). Then (2.9) follows from (2.10) as

$$|\mathcal{K}^k|_T \le \int_0^T |(e^{-sM}N)^{*k}(t)|\, dt. \tag{2.12}$$

$\square$

To obtain an explicit formula for the resolvent operator

$$R(\lambda, \mathcal{K}) := (\lambda - \mathcal{K})^{-1},$$

consider the following problem for $\lambda \neq 0$:

$$\lambda u - \mathcal{K} u = g. \tag{2.13}$$

Assuming g is smooth, differentiate (2.13) to obtain

$$\dot{u} + (M - \frac{1}{\lambda} N)u = \frac{1}{\lambda}(\dot{g} + Mg). \tag{2.14}$$

Thus the only solution of (2.13) is given by

$$u(t) = R(\lambda, \mathcal{K})g(t) = \frac{1}{\lambda}g(t) + \frac{1}{\lambda^2}\int_0^t e^{-(t-s)(M - \frac{1}{\lambda}N)} Ng(s)\,\mathrm{d}s. \tag{2.15}$$

Proposition 2 The resolvent $R(\lambda, \mathcal{K})$ mapping g to u is given by (2.15) for $\lambda \neq 0$ and it satisfies

$$|R(\lambda, \mathcal{K})|_T \leq \frac{1}{|\lambda|} + \frac{1}{|\lambda|^2} e^{T|M - \frac{1}{\lambda}N|} |N|T. \tag{2.16}$$

Proof. For smooth g, (2.16) follows in the same way as (2.9). As (2.15) only deals with values of g, the bound (2.16) holds as such for all $g \in C[0,T]$. $\square$

2.2. Quasinilpotency, order and type

Bounded operators with spectrum equalling the origin are called quasinilpotent. Their resolvents are *entire* functions in $1/\lambda$ whose growth can be used to bound the powers of the operators. From (2.16) we see that $R(\lambda, \mathcal{K})$ is an entire function in $1/\lambda$ and that it essentially grows like $\exp(T|N|/|\lambda|)$ as $\lambda \to 0$. This means that $R(\lambda, \mathcal{K})$ is of at most *order* 1, and if the order is 1, then the *type* satisfies $\tau \leq T|N|$. These concepts are important because the growth of the resolvent as $\lambda \to 0$ and the decay of the powers are intimately related.

Hadamard (1893) used the maximum modulus

$$M(r, f) := \sup_{|z|=r} |f(z)| \tag{2.17}$$

of an entire function f to define the order

$$\omega := \limsup_{r\to\infty} \frac{\log\log M(r, f)}{\log r}. \tag{2.18}$$

In our case $R(\lambda, \mathcal{K})$ is an operator valued entire function in $1/\lambda$ and likewise we set, following Miekkala and Nevanlinna (1992, page 207),

$$\omega := \limsup_{r\to 0} \frac{\log\log(\sup_{|\lambda|=r} |R(\lambda, \mathcal{K})|_T)}{\log \frac{1}{r}}. \tag{2.19}$$

In general ω could be any nonnegative number, but it follows immediately from (2.16) that $0 \leq \omega \leq 1$. In Miekkala and Nevanlinna (1992) we proved that ω must be a rational number with denominator not exceeding the dimension d of the vectors. Its value depends on the 'graph properties' of M and N only, and in particular is independent of the window size T.

By definition, the order ω is an *asymptotic* concept. Together with the order, one often talks about the type τ of an entire function. This is also an asymptotic concept, which here takes the following form.

If $R(\lambda, \mathcal{K})$ is of positive order ω in $1/\lambda$ then we say that it is of type τ where

$$\tau := \limsup_{r \to 0} r^{\omega} \log\left(\sup_{|\lambda|=r} |R(\lambda, \mathcal{K})|_T \right). \tag{2.20}$$

While the order ω is independent of T, the type is of the form $\tau = cT$, where c is a positive constant.

Thus, if $R(\lambda, \mathcal{K})$ is of order $\omega > 0$ and type cT, then

$$\sup_{|\lambda|=r} |R(\lambda, \mathcal{K})|_T \sim e^{cT/r^{\omega}}, \quad \text{as} \quad r \to 0, \tag{2.21}$$

and, in particular, we have for any $\varepsilon > 0$ a constant C such that

$$|R(\lambda, \mathcal{K})|_T \leq \frac{C}{|\lambda|} e^{(1+\varepsilon)cT/|\lambda|^{\omega}} \tag{2.22}$$

holds for all $\lambda \neq 0$. To see how the growth of the resolvent is connected with the decay of the powers of the operator we state the following result.

Theorem 2 Let $\mathcal{A}$ be a bounded linear operator on a Banach space. If $R(\lambda, \mathcal{A})$ is entire in $1/\lambda$ and satisfies

$$\sup_{|\lambda|=r} \|R(\lambda, \mathcal{A})\| \leq \frac{C}{r} e^{\tau/r^{\omega}} \tag{2.23}$$

for all $r > 0$, then

$$\|\mathcal{A}^k\| < C\left(\frac{\tau e \omega}{k}\right)^{k/\omega}, \quad k = 1, 2, 3, \ldots. \tag{2.24}$$

Conversely, if (2.24) holds, then for $0 < \alpha \leq 1/2$ and $r > 0$ we have

$$\sup_{|\lambda|=r} \|R(\lambda, \mathcal{A})\| \leq \frac{1}{r}\left(1 + \frac{13}{\alpha} C\omega e^{(1+\alpha)\tau/r^{\omega}}\right). \tag{2.25}$$

Proof. To obtain (2.24) from (2.23) write

$$\mathcal{A}^n = \frac{1}{2\pi i} \int_{|\lambda|=r} \lambda^n R(\lambda, \mathcal{A}) \, d\lambda \tag{2.26}$$

and substitute $r^{\omega} = \frac{\tau\omega}{n}$. The reverse direction is also standard in spirit but the actual constants needed in (2.25) require some care. Here we refer to the proof of Theorem 5.3.4 in Nevanlinna (1993). $\square$

2.3. Characteristic polynomial and computation of order and type

As the iteration operator is given by convolution with a matrix valued kernel, it is possible to analyse the growth properties of its resolvent using the Laplace transform.

Consider $u = \mathcal{K}g$ where, say, $|g(t)| \leq Ce^{\alpha t}$ for some positive constants C and α. Taking the Laplace transform we obtain

$$\hat{u}(z) = (z + M)^{-1} N \hat{g}(z) \tag{2.27}$$

and in particular $\hat{u}$ is analytic for sufficiently large $\operatorname{Re} z$. Here $(z + M)^{-1}N$ is the *symbol* of $\mathcal{K}$, denoted by $K(z)$. Analogously, the resolvent operator $R(\lambda, \mathcal{K})$ has the symbol

$$\frac{1}{\lambda}[1 + (z + M - \frac{1}{\lambda}N)^{-1}\frac{1}{\lambda}N]. \tag{2.28}$$

Definition 1   We shall call

$$P(z, \frac{1}{\lambda}) := \det(z + M - \frac{1}{\lambda}N) \tag{2.29}$$

the *characteristic polynomial* of the iteration operator $\mathcal{K}$.

In Section 3 we shall see how the zeros of P determine the spectrum of the operator $\mathcal{K}$ when considered on the infinite time interval $[0, \infty)$: one looks at the supremum of all roots $|\lambda|$ when z travels in a right half plane. Here the properties of $\mathcal{K}$ on the finite interval $[0, T]$ are explained in terms of growth of $|z|$ as λ decays to zero.

Expanding the determinant yields the following result.

Proposition 3   We have

$$P(z, \mu) = \sum_0^d q_j(\mu)z^j, \tag{2.30}$$

where q_j is a polynomial of degree at most $d - j$ and $q_d = 1$.

The equation $P(z(1/\lambda), 1/\lambda) = 0$ determines an algebraic function $z = z(1/\lambda)$ which is d-valued. We need to study the behaviour of $z(1/\lambda)$ as $\lambda \to 0$.

Let z_j denote the branches of z. If z_j is not independent of λ we define ω_j by

$$z_j(\frac{1}{\lambda}) = c_j(\frac{1}{\lambda})^{\omega_j} + o((\frac{1}{\lambda})^{\omega_j}) \quad \text{as} \quad \lambda \to 0 \tag{2.31}$$

(with $c_j \neq 0$). If z_j is independent of λ then we define $\omega_j = 0$. Further we set $\omega := \max \omega_j$.

Lemma 1

$$\omega \in \{\frac{m}{k} \; : \; k, m \quad \text{integers}, \quad 0 \leq m \leq k \leq d, k \neq 0\} \tag{2.32}$$

Proof. The ω_js are computed from the Newton diagram, which is explained in Section 2.5. Theorem 7 implies the claim. □

We can show that ω is the order of the iteration operator $\mathcal{K}$. Consider $\omega > 0$. Let $c := \max |c_j|$ where j runs over those indices for which $\omega_j = \omega$. Then we can formulate our result as follows.

Theorem 3 The iteration operator $\mathcal{K}$ of (2.3) is in $C[0, T]$ of order $\omega = \frac{m}{n}$ with some integers $0 \leq m \leq n \leq d$, independently of T. If $\omega > 0$ then there exists a positive c ($c = \max |c_j|$ as above) such that for all T the type is $\tau = cT$. If $\omega = 0$ then the operator is nilpotent with index $n \leq d$, independently of T. Furthermore, $\mathcal{K}$ is nilpotent if and only if the characteristic polynomial P is independent of λ.

Proof. This is Theorem 4.6 in Miekkala and Nevanlinna (1992).

Since ω can take only a finite number of rational values for a d-dimensional problem, it should not be surprising that ω depends on graph properties of M and N, but not on the values of their elements. This topic is discussed further in Section 2.5.

For $\omega < 1$ we have the following characterization.

Theorem 4 $R(\lambda, \mathcal{K})$ is of the order $\omega < 1$ in $C[0, T]$ if and only if N is nilpotent.

Proof. This is included in Section 2.5. □

Finally, if M and N commute, the analysis of convergence is easy.

Theorem 5 If M and N commute, then either N is nilpotent and then $\mathcal{K}$ is nilpotent too, or the order $\omega = 1$ and the type $\tau = \rho(N)T$.

Proof. For $z \notin \sigma(-M)$, we have

$$K(z)^k = (z + M)^{-k} N^k,$$

which gives

$$\rho(K(z)) \leq \frac{1}{\rho(z + M)} \rho(N).$$

On the other hand, from

$$N^k = (z + M)^k K(z)^k$$

we obtain

$$\rho(N) \leq \rho(z + M)\rho(K(z))$$

and thus

$$\rho(K(z)) = \frac{1}{\rho(z + M)}\rho(N) = (1 + o(1))\frac{1}{|z|}\rho(N)$$

as $z \to \infty$. The claim follows; see, for example, the proof of Theorem 6, equation (2.37). $\square$

2.4. Norm estimates

While the order and type are asymptotic concepts relating the decay of $|\mathcal{K}^n|_T$ to the behaviour of $z_j = z_j(1/\lambda)$ as $\lambda \to 0$, it is also possible to relate $|\mathcal{K}^n|_T$ to the decay of the symbol as $\operatorname{Re} z \to \infty$. Results of this nature are given in Nevanlinna (1989b), and here we present the following basic version.

As in the proof of Theorem 1 of Section 2.1 the claim takes a somewhat better form if formulated for the iterated kernels $(e^{-sM}N)^{*k}(t)$ pointwise in t rather than for the operator norm.

Let m, k be positive integers with $m \geq k$. Consider an estimate of the form

$$|(e^{-sM}N)^{*k}(t))| \leq Be^{\gamma t}\frac{(Bt)^{m-1}}{(m-1)!}, \quad \text{for} \quad t > 0. \tag{2.33}$$

Since $K(z)^k$ is the Laplace transform of $(e^{-sM}N)^{*k}$ we have

$$|K(z)^k| \leq B\int_o^\infty e^{-(\operatorname{Re} z - \gamma)t}\frac{(Bt)^{m-1}}{(m-1)!}\,dt = (\frac{B}{\operatorname{Re} z - \gamma})^m \quad \text{for} \quad \operatorname{Re} z > +\gamma. \tag{2.34}$$

Thus, an estimate for the iterated kernel implies an estimate for the power of the symbol. The nontrivial fact is that the reverse conclusion also holds.

Theorem 6 Suppose that there are positive integers k and m and positive constants B and γ such that (2.34) holds. Then, for all $j = 1, 2, \ldots$, we have

$$|\mathcal{K}^{jk}|_T \leq \frac{k}{m}de^{\gamma T}(\frac{BeT}{jm})^{jm}. \tag{2.35}$$

Proof. Theorem 2.4.1. in Nevanlinna (1989b) is slightly more general but formulated for the iterated kernels. Integrating the kernel estimate gives (2.35) but for a factor of 2. This can be dropped because Spijker (1991) has since proved a sharp version of a lemma by LeVeque and Trefethen. $\square$

Just for comparison, write $\omega := k/m$. Then for $n = jk$, $j = 1, 2, \ldots$ (2.35) reads

$$|\mathcal{K}^n|_T \leq d\omega e^{\gamma T}(\frac{Be\omega T}{n})^{n/\omega}, \tag{2.36}$$

which should be compared with (2.24) and with Theorem 3.

To make this connection explicit observe that (2.31) implies

$$\rho(K(z)) = (1 + o(1))(\frac{c}{|z|})^{\frac{1}{\omega}} \tag{2.37}$$

as $z \to \infty$. In fact, since

$$\lambda - K(z) = \lambda(z + M)^{-1}(z + M - \frac{1}{\lambda}N), \tag{2.38}$$

the eigenvalues of $K(z)$ are obtained from solving $P(z, 1/\lambda) = 0$ and (2.37) follows by 'inverting' (2.31). Thus (2.34) can be viewed as a 'norm' version of (2.37), with $k/m = \omega$ and $B \geq c$.

2.5. Computing order from the graph of A

It was explained in Section 2.3 that the order ω of the iteration operator $\mathcal{K}$ can be computed by solving $z = z(\lambda)$ from the characteristic polynomial $P(z, \frac{1}{\lambda}) = 0$ near $\lambda = 0$. The different branches are of the form (2.31) and the order ω is then the largest ω_j in (2.31). Before finding these solutions we need some background connecting graphs to matrices.

Let $G(B)$ be the directed graph associated with a $d \times d$-matrix B. $G(B)$ contains d vertices v_i. Each nonzero element B_{ij} of B corresponds to an edge of $G(B)$ with weight $B_{i,j}$ directed from v_j to v_i. By a *circuit* of $G(B)$ we mean a subgraph of $G(B)$ which consists of one or more nonintersecting loops. A circuit is denoted by $\mathcal{C}_j$ and its length (or number of edges) by $l(\mathcal{C}_j)$. Further, the product of the weights of the edges is called the *weight of the circuit* and is denoted by $w(\mathcal{C}_j, B)$. The second argument refers to $\mathcal{C}_j$ being a circuit in $G(B)$. Finally, j_{ev} means the number of components of even length in the circuit $\mathcal{C}_j$.

The coefficients b_i in the following expansion of the determinant by diagonal elements

$$\det(zI + B) = z^d + b_1 z^{d-1} + \ldots + b_d \tag{2.39}$$

can be linked to $G(B)$ by noticing first that each b_i is a sum of all principal minors of order i in $\det(B)$. From the definition of the determinant one can then show that these sums have the following graph interpretation (Chen 1976, Theorem 3.1):

$$b_i = \sum_{l(\mathcal{C}_j)=i} (-1)^{j_{ev}} w(\mathcal{C}_j, B), \quad i = 1, \ldots, d, \tag{2.40}$$

where the sum is taken over all circuits of length i in the digraph of B.

Now back to solving for $z(\lambda)$ from $P(z, 1/\lambda) = 0$, or rather from $p(\lambda, z) = 0$, where

$$p(\lambda, z) = \lambda^d P(z, \frac{1}{\lambda}) = \det(z\lambda I + \lambda M - N). \tag{2.41}$$

Now

$$p(\lambda, z) = \sum_{r=0}^{d} p_r(\lambda) z^r,$$

3. Infinite windows

3.1. Definition of spaces

In practice one would not usually iterate on $[0, \infty)$, but the infinitely long window is a natural setup for stiff problems and for DAEs: for the fast transients a finite window $[0, T]$ can be regarded as infinitely long.

The exceptional situation with stiff problems would appear if couplings are very small, that is, $T|N| = \mathcal{O}(1)$, then by the discussion of Section 2 we would have superlinear convergence. For $T|N| \gg 1$ we typically obtain only linear convergence, and one of the first interesting things is that the linear rate given by the spectral radius is very insensitive to the choice of norm.

Let X be a Banach space of functions $x : [0, \infty) \to \mathbb{C}^d$ such that the following conditions hold:

(i) $e^{\lambda t} c$ with $\operatorname{Re} \lambda > 0$ and $0 \neq c \in \mathbb{C}^d$ is not in X;

(ii) $e^{\lambda t} p(t)$, where p is a $\mathbb{C}^d$-valued polynomial and $\operatorname{Re} \lambda < 0$, is in X;

(iii) $x \mapsto \int_0^t e^{(s-t)B} C x(s) \, ds$, where B, C are constant matrices and the eigenvalues of B have positive real parts, are bounded operators in X;

(iv)   test functions C_0^∞ are dense in X.

Let $\| \cdot \|$ denote the norm in X. By (iii), $\sigma(M) \subset \mathbb{C}_+$ implies that $\mathcal{K}$ is a bounded operator in X. In order to formulate our results we recall the definition of the symbol

$$K(z) := (z + M)^{-1} N \tag{3.1}$$

The basic property of X is that it is 'unweighted' in exponential scales. However, such scaling is trivial and simply translates the imaginary axis: requirements on real parts being positive would become positive lower bounds on real parts.

The properties (i) and (iii) imply that if we try to solve our initial value problem in X we must require that all eigenvalues of A have positive real parts. Furthermore the following holds.

Theorem 9   If all eigenvalues of A have positive real parts, then $\mathcal{K}$ is a bounded operator in X if and only if the eigenvalues of M have positive real parts.

Proof.   The sufficiency part is of course obvious, while the necessity needs a small discussion. In Miekkala and Nevanlinna (1987*a*) the result is proved for L_p spaces.

Assume that $\mathcal{K}$ is bounded in X and that μ is an eigenvalue of M with nonpositive real part. Let J denote the Jordan block

$$J = \begin{pmatrix} \mu & 1 & & \\ & \mu & \ddots & \\ & & \ddots & 1 \\ & & & \mu \end{pmatrix},$$

associated with this eigenvalue, and let S be a similarity transform such that $\tilde{M} = S^{-1}MS$ is of the form

$$\tilde{M} = \begin{pmatrix} J & 0 \\ 0 & M_0 \end{pmatrix}.$$

Put $\tilde{N} = S^{-1}NS$ and let the corresponding operator be denoted by $\tilde{K}$. Since multiplication by a constant matrix is bounded in X, and $\tilde{K} = S^{-1}KS$, $\tilde{K}$ is bounded in X. Let the block structure induced by $\tilde{M}$ be denoted by

$$\tilde{N} = \begin{pmatrix} N_{11} & N_{12} \\ N_{21} & N_{22} \end{pmatrix}.$$

We claim that there exists $c \in \mathbb{C}^d$ such that

$$\tilde{N}c - \begin{pmatrix} b_1 \\ b_2 \end{pmatrix},$$

with $b_1 \neq 0$. Indeed, if $N_{11} \neq 0$, then this is trivial, while if $N_{11} = 0$, then the claim follows from $N_{12} \neq 0$. N_{11} and N_{12} cannot simultaneously vanish because every eigenvalue of A has positive real part.

Let c be as above and let λ be any complex number such that $\operatorname{Re}\lambda < 0$. Then $u := e^{\lambda t}c$ is an element of X, whence $\tilde{K}u \in X$. Let k be the largest index i for which $b_{1i} \neq 0$, where $b_1 = (b_{11}, b_{12}, \ldots)^T$. Then the kth component of the vector $\tilde{K}u$ satisfies

$$\left(\tilde{K}u\right)_k (t) = \int_0^t e^{\mu(s-t)} e^{\lambda s} \, ds \quad b_{1k}e_k,$$

where $e_k \in \mathbb{C}^d$ denotes the usual coordinate vector.

For any $f \in X$, define

$$\mathcal{L}f := \left(e_k, \tilde{K}f\right) c,$$

where $(\cdot, \cdot)$ denotes the usual inner product in $\mathbb{C}^d$. Since $\mathcal{L}f$ can also be written as $C\tilde{K}$, with $C - ce_k^T$, we see that $\mathcal{L}$ is bounded in X. By construction, $\mathcal{L}u = e^{-\mu t} * e^{\lambda t}c$ and, since $u = e^{\lambda t}c \in X$, we have $e^{-\mu t} * e^{\lambda t}c \in X$. This implies $e^{-\mu t}c \in X$, which, by (i), implies $\operatorname{Re}\mu \geq 0$. By assumption, $\operatorname{Re}\mu \leq 0$ and we conclude that $\mu = i\xi$ and $\nu := e^{-i\xi t}c \in X$.

It is easily checked that

$$\mathcal{L}^n v = \frac{t^n}{n!}v,$$

for all $n \geq 0$. But, since $\mathcal{L}$ is bounded, we may put

$$w := \left(\sum_{n \geq 0} \frac{(t/\|2\mathcal{L}\|)^n}{n!} \right) v = \exp(t/\|2\mathcal{L}\|)v$$

and obtain the contradiction

$$w = \exp \left((\|2\mathcal{L}\|^{-1} - i\xi)t \right) c \in X.$$

□

3.2. The spectrum and the spectral radius

From now on we assume that the eigenvalues of M have positive real parts. The main result here may be found in Nevanlinna (1990a).

Theorem 10   In every space X, we have $\sigma(\mathcal{K}) = \mathrm{cl} \bigcup_{\mathrm{Re}\, z \geq 0} \sigma(K(z))$.

We state some consequences before embarking on the proof.

Corollary 3 $\rho(\mathcal{K}) = \max\{K(i\xi) : \xi \in \mathbb{R}\}$.

Proof.   Since the eigenvalues of M have positive real parts, $K(z)$ is analytic in the closed right half plane. The claim follows from Theorem 10 using the maximum principle, on a Riemann surface corresponding to the algebraic function formed by the eigenvalues of $K(z)$, and the fact that all eigenvalues of $K(z)$ vanish at infinity. Alternatively, we may apply the maximum principle directly to the spectral radius of $K(z)$, because it is a subharmonic function; see Theorem 3.4.7 in Aupetit (1991). □

Corollary 4 $\sigma(\mathcal{K})$ is compact and connected, and $0 \in \sigma(\mathcal{K})$.

Proof.   All branches of the algebraic function vanish at infinity. Thus all components of $\sigma(\mathcal{K})$ contain the origin, which implies connectedness. Compactness is obvious, remembering that K is bounded by assumption (iii). □

Corollary 5 $\rho(\mathcal{K}) = 0$ if and only if there exists $m \leq d$ such that $\mathcal{K}^m = 0$.

Proof.   If $\rho(\mathcal{K}) = 0$ then $K(z)$ is nilpotent for all z in the right half plane. Thus there exists $m \leq d$ such that $K(z)^m \equiv 0$, for all z. Now $\mathcal{K}^m$ applied to, say, test functions can be written using the inverse Laplace transform in terms of $K(z)^m$. Since test functions are dense in X and $\mathcal{K}^m$ is continuous, $\mathcal{K}^m$ must vanish in all of X. The converse is trivial. □

Comparing this corollary with Theorem 3 in Section 2 we see that $\rho(\mathcal{K}) > 0$ in X if and only if $\mathcal{K}$ is of positive order in $C[0, T]$.

Proof of Theorem 10. The formula

$$R(\lambda, \mathcal{K})f(t) = \frac{1}{\lambda}f(t) + \frac{1}{\lambda^2}\int_0^t e^{-(t-s)(M-\lambda^{-1}N)}Nf(s)\,\mathrm{d}s \qquad (3.2)$$

is valid for all $t > 0$ and $\lambda \neq 0$, and at least for smooth f (see Section 2.1). This implies immediately that $R(\lambda, \mathcal{K})$ is bounded in X by (iii) and (iv), provided that all eigenvalues of $M - \lambda^{-1}N$ have positive real parts.

On the other hand, suppose that $M - \lambda^{-1}N$ has an eigenvalue, μ say, with negative real part. Denoting the corresponding eigenvector by b, we choose f to be the solution of $\dot{f} + Mf = 0$ with $f(0) = b$. Thus f is in X by (ii). However, from (2.14) we see that $u(t) = e^{-\mu t}b$, and thus $u \notin X$ by (i). Finally, suppose that $M - \lambda^{-1}N$ has a purely imaginary nonzero eigenvalue λ_0. Then, since $M - \lambda^{-1}N$ is analytic in λ, $M - \lambda^{-1}N$ would have at least one eigenvalue with negative real part near λ_0, unless the eigenvalue is constant. In that case M would also have a purely imaginary eigenvalue, contradicting our hypothesis. Since the spectrum is closed, such a λ_0 does belong to the spectrum, and we can conclude that $\lambda \in \sigma(\mathcal{K}) \setminus \{0\}$ if and only if $\det(M - \lambda^{-1}N \quad \mu)$ vanishes for some μ with nonpositive real part.

Writing $z = -\mu$ and recalling that $z + M$ is invertible for $\mathrm{Re}\,z \geq 0$, we deduce that $\det((\lambda - z + M)^{-1}N)$ vanishes for some z with non-negative real part.

Thus $0 \neq \lambda \in \sigma(\mathcal{K})$ if and only if there is a z, $\mathrm{Re}\,z \geq 0$, such that $\lambda \in \sigma(K(z))$. Since $K(z) \to 0$ as $z \to \infty$ and $\sigma(\mathcal{K})$ is closed, we deduce $0 \in \sigma(\mathcal{K})$, and the claim follows. $\square$

Since the spectral radius $\rho(\mathcal{K})$ is independent of the space X we may loosely say that the iteration converges on $[0, \infty)$ if and only if $\rho(\mathcal{K}) < 1$. Note that

$$\|\mathcal{K}^n\|^{1/n} \to \rho(\mathcal{K}), \qquad (3.3)$$

so that, for any $\varepsilon > 0$, there exists $C < \infty$ for which

$$\|\mathcal{K}^n\| \leq C(\rho(\mathcal{K}) + \varepsilon)^n, \qquad (3.4)$$

but that in general C depends on both ε and on the ambient norm.

The formula $\rho(\mathcal{K}) = \max_\xi \rho(K(i\xi))$ is very easy to use in practice. For example, in several special cases one has

$$\max_\xi \rho(K(i\xi)) = \rho(K(0)), \qquad (3.5)$$

which simply means that the convergence is dominated by the speed of convergence of the iteration

$$Mx^{k+1} = Nx^k + b \qquad (3.6)$$

for $Ax = b$. Such situations can occur, for instance in Jacobi splittings of linearized versions of parabolic equations. Results of this form have been discussed in Miekkala and Nevanlinna (1987a). This paper also contains results where $\rho(\mathcal{K}) > \rho(K(0))$. As an example we mention SOR for consistently ordered matrices. In this case the rate of convergence ($\rho(K(0))$) is known for iteration (3.6). It turns out that for consistently ordered matrices, we obtain $\rho(\mathcal{K}) = \rho(K(0))$ for small values of the overrelaxation parameter ω. However, when ω is close to 2, we have $\rho(\mathcal{K}) > \rho(K(0))$ and the iteration can diverge. For the precise result, see Theorem 4.1 in Miekkala and Nevanlinna (1987a).

In comparing two splittings it is important to notice that a splitting that looks favourable on $[0, \infty)$ may look inferior on $[0, T]$ and vice versa. For example, by Theorem 4 of Section 2, the order of $R(\lambda, \mathcal{K})$ is always less than 1 for Gauss–Seidel splitting on $C[0, T]$, whilst for overrelaxation splittings the order equals 1 if the diagonal does not vanish. On the other hand, for consistently ordered matrices, for instance, $\rho(\mathcal{K})$ initially decreases as the overrelaxation parameter increases from 1. So the Gauss–Seidel splitting provides ultimately the fastest convergence rate on finite windows, but on the infinite interval creates *propagating error waves*, which are best damped with a modest amount of overrelaxation. Overrelaxing too much will in turn cause growing error waves, making the process diverge on the infinite window.

3.3. On generalizing the theory for DAE systems

Let us change the model problem to

$$B\dot{x} + Ax = f \tag{3.7}$$

with consistent initial values for x, where B may be singular and f is sufficiently smooth (the required smoothness depends on the index of the system). The boundedness assumption for continuous solutions of (3.7) on the infinite time interval becomes

$$\det(zB + A) \neq 0, \quad \operatorname{Re} z \geq 0.$$

To see this and for the whole analysis of Miekkala (1989), one needs to use the Kronecker Canonical Form (KCF) of the DAE (Gantmacher 1959). Decompositions of the matrices $B = M_B - N_B$ and $A = M_A - N_A$ define the dynamic iteration for (3.7)

$$M_B\dot{x}^n + M_A x^n = N_B\dot{x}^{n-1} + N_A x^{n-1} + f, \quad n = 1, 2, \ldots \tag{3.8}$$

with consistent initial values for x. The iteration operator can now be written after transformation of (3.8) into KCF form, and constitutes two parts, one being an integral operator and the other a sum of matrix multiplication and differentiation operators. The basic difference to the ODE case is that, in order to guarantee boundedness of iteration (3.8), one needs to preserve the structure of the DAE while decomposing B and A in (3.7). Essentially, we mean that the index is preserved and the state variables and algebraic variables are preserved. The condition is formulated for the KCF of (3.8), but in Miekkala (1989) there is an error, corrected in Miekkala (1991). The space where x is iterated by (3.8) is chosen in Miekkala (1989) to be that of continuously differentiable functions with appropriate norm, but one could equally well use the space of continuous functions with the uniform norm. The smoothness requirement for f is essential since, for high index DAE systems, some components of the solution of (3.7) depend on derivatives of f. For index one (or zero) systems one might consider iteration (3.8) in L^p-space (both f and x^n in L^p), as in the ODE-case; the results of Miekkala (1989) still hold. For high index DAEs the space has to be modified so that the components corresponding to high index algebraic variables have different requirements from the other components. In general it would be difficult to recognize these components, but in applications it is sometimes possible. In Section 7 this kind of modified L^p-space formulation is used for the index two case.

In Miekkala (1989), assuming that the algebraic part of the iteration operator is bounded, the other results are analogous to the ODE case. For example,

$$\det(zM_B + M_A) \neq 0, \quad \operatorname{Re} z \geq 0, \tag{3.9}$$

is needed for boundedness of the iteration. The convergence rate is given by the 'Laplace transform' of (3.8),

$$\rho(\mathcal{K}_{DAE}) = \sup_{\operatorname{Re} z \geq 0} \rho((zM_B + M_A)^{-1}(zN_B + N_A)). \tag{3.10}$$

Convergence results, like those for consistently ordered matrices, can be generalized to special index one systems.

4. Acceleration techniques

We can accelerate waveform relaxation in two ways: we can try to get the error to decrease more rapidly per iteration, or spend less time integrating the early sweeps. The latter strategy is outlined in connection with the discretization, while here we address the former possibility.

4.1. The speed of optimal Krylov methods

Suppose the initial value problem has been transformed into the fixed-point problem

$$x = \mathcal{K}x + g. \tag{4.1}$$

Instead of iterating as usual, that is

$$x^{k+1} := \mathcal{K}x^k + g, \tag{4.2}$$

we could in principle keep all the vectors $\{x^k\}$ in memory and try to find as good a linear combination of these as possible.

We outline first the abstract Krylov subspace method approach; see Nevanlinna (1993). Let $\mathcal{A}$ be a bounded operator in a Banach space and b a vector in that space. Put

$$K_n(\mathcal{A}, b) := \text{span}\{\mathcal{A}^j b\}_0^{n-1} \tag{4.3}$$

and

$$K(\mathcal{A}, b) := \text{cl span}\{\mathcal{A}^j b\}_0^\infty. \tag{4.4}$$

Thus, $K(\mathcal{A}, b)$ is the smallest closed invariant subspace of $\mathcal{A}$ that contains b. In fact either $\dim K_n(\mathcal{A}, b) = n$ or there exists $m < n$ such that, for all $k > m$,

$$K_k(\mathcal{A}, b) = K_m(\mathcal{A}, b). \tag{4.5}$$

If we are given a fixed point problem

$$x = \mathcal{A}x + b, \tag{4.6}$$

such that $1 \notin \sigma(\mathcal{A})$, then clearly $x = (1 - \mathcal{A})^{-1} b$. Consider the following simple embedding:

$$x_\lambda = \frac{1}{\lambda} \mathcal{A} x_\lambda + b, \tag{4.7}$$

and assume that $\sigma(\mathcal{A})$ does not separate 1 from ∞. Then there exists a path $\lambda(s) : \lambda(1) = 1, \quad \lambda(\infty) = \infty$ such that (4.7) has a solution x_λ and clearly this solution is continuous along the path. Trivially, the Krylov subspace of $\lambda^{-1}\mathcal{A}$ equals that of $\mathcal{A}$ for nonzero λ. For $|\lambda| > \|\mathcal{A}\|$ we have

$$x_\lambda = \sum_0^\infty (\lambda^{-1} \mathcal{A})^k b, \tag{4.8}$$

which shows that $x_\lambda \in K(\mathcal{A}, b)$. By continuity, as $\lambda(s) \to 1$ and because $K(\mathcal{A}, b)$ is a closed set, we have $x \in K(\mathcal{A}, b)$, and $K(\mathcal{A}, b)$ is invariant for $(1 - \mathcal{A})^{-1}$ as well.

We assume in the following that $1 \notin \sigma(\mathcal{A})$ and that $\sigma(\mathcal{A})$ does not separate 1 from ∞. That the latter must be assumed is clear from the maximum principle, but can be understood immediately from the following example.

If $\mathcal{A} := \rho S$ where $\rho > 1$ and $S : e_j \mapsto e_{j+1}$ is the unitary shift in $\ell_2(\mathbb{Z})$, then $\sigma(\mathcal{A})$ is the circle centred at the origin of radius ρ and 1 is separated from ∞. If $b := e_0$, then $K(\mathcal{A}, e_0) = \text{cl span}\{e_j\}_0^\infty$ whilst the solution $x \notin K(\mathcal{A}, e_0)$. In fact, $x = \sum_{-\infty}^0 \rho^{1-j} e_j$.

Every vector in $K_n(\mathcal{A}, b)$ is of the form $q_{n-1}(\mathcal{A})b$ for some polynomial q_{n-1} of degree $n - 1$. Let us write

$$p(\lambda) := 1 - (1 - \lambda)q(\lambda), \tag{4.9}$$

where q is a given polynomial; then $q(\mathcal{A})$ approximates $(1 - \mathcal{A})^{-1}$ well if and only if $p(\mathcal{A})$ is small. In fact we have the following result (Nevanlinna 1993, Proposition 1.6.1).

Proposition 4 $\quad \dfrac{1}{\|1 - \mathcal{A}\|}\|p(\mathcal{A})\| \leq \|(1-\mathcal{A})^{-1}-q(\mathcal{A})\| \leq \|(1-\mathcal{A})^{-1}\|\|p(\mathcal{A})\|.$

Now, it is of interest to ask how small $p(\mathcal{A})$ can be. Therefore set $b_n(\mathcal{A}) := \inf \|p(\mathcal{A})\|$ where the infimum is taken over all polynomials of degree at most n, satisfying $p(1) = 1$ (see (4.9)).

Definition 3 (Nevanlinna 1990a, and Definition 3.3.1 in Nevanlinna 1993). Given a bounded $\mathcal{A}$, define

$$\eta(\mathcal{A}) := \inf_n b_n(\mathcal{A})^{1/n}.$$

We call $\eta(\mathcal{A})$ the *optimal reduction factor* of $\mathcal{A}$.

The main properties of $\eta(\mathcal{A})$ are collected in the following theorem.

Theorem 11

(i) $\eta(\mathcal{A}) < 1$ if and only if $1 \notin \sigma(\mathcal{A})$ and $\sigma(\mathcal{A})$ does not separate 1 from ∞;

(ii) if $\eta(\mathcal{A}) < 1$ then $\eta(\mathcal{A}) = 0$ if and only if $\mathrm{cap}(\sigma(\mathcal{A})) = 0$;

(iii) $0 < \eta(\mathcal{A}) < 1$, then the value of $\eta(\mathcal{A})$ only depends on $\sigma(\mathcal{A})$ and is given by $\eta(\mathcal{A}) = e^{-g(1)}$, where g is the (extended) Green's function, satisfying

- g is harmonic in the unbounded component G of $\mathbb{C} \backslash \sigma(\mathcal{A})$;
- $g(\lambda) = \log|\lambda| + \mathcal{O}(1)$ as $\lambda \to \infty$;
- $g(\lambda) \to 0$ as $\lambda \to \zeta$ from G, for every $\zeta \in \partial G \subset \partial \sigma(\mathcal{A})$.

Proof. These are covered by Theorem 3.3.4 and Theorem 3.4.9 in Nevanlinna (1993)

Operators $\mathcal{A}$ for which $\mathrm{cap}(\sigma(\mathcal{A})) = 0$ are *quasialgebraic*. In such a case $\sigma(\mathcal{A})$ cannot separate 1 from ∞, so that we can combine (i) and (ii) in the statement: The optimal reduction factor vanishes exactly for quasialgebraic operators with $1 \notin \sigma(\mathcal{A})$.

This is analogous to the vanishing of the spectral radius for quasinilpotent operators.

We shall say that $\mathcal{A}$ is *algebraic* if $q(\mathcal{A}) = 0$ for some polynomial q. Thus nilpotent operators form a subclass of algebraic operators.

4.2. Finite windows

Consider $\mathcal{K}$ in $C[0, T]$. From Section 2 we know that

$$|\mathcal{K}^n|_T \sim \left(\frac{cTe\omega}{n}\right)^{n/\omega} \tag{4.10}$$

as $n \to \infty$. We do not know sharp lower bounds for

$$b_n(\mathcal{K}) = \inf_{\deg p \leq n,\, p(1)=1} |p(\mathcal{K})|_T \tag{4.11}$$

for general $\mathcal{K}$, but we give an illustrative example instead. Consider the following operator $\mathcal{V}$,

$$\mathcal{V}u(t) := \int_0^t u(s)\, \mathrm{d}s \tag{4.12}$$

or $M = 0$ and $N = 1$. Then clearly

$$|\mathcal{V}^n|_T = \frac{T^n}{n!}, \tag{4.13}$$

so that $\omega = 1$, $\tau = T$. The following result shows that, when it is optimally accelerated, we obtain a speed of convergence in which the order is still 1 but the type is lowered from T to $T/4$.

Theorem 12 Let $\mathcal{V} = \int_0^t$ operate in $C[0, T]$. Then for $n \geq 2$

$$e^{-T}\frac{(T/4)^n}{n!} \leq b_n(\mathcal{V}) \leq 8(1 + T)e^T\frac{(T/4)^n}{(n-1)!} \tag{4.14}$$

Proof. This is proposition 5.2.5 in Nevanlinna (1993).

Thus, the speed can be accelerated, but not dramatically.

4.3. Infinite windows

Let X be any space considered in Section 3.1. The first result says that acceleration is possible, but then we shall see that the acceleration is often only of modest nature.

Theorem 13

(i) $\eta(\mathcal{K}) = 0$ only if $\rho(\mathcal{K}) = 0$.
(ii) If $0 < \eta(\mathcal{K}) < 1$ then $\eta(\mathcal{K}) < \rho(\mathcal{K})$.

Proof. This is Theorem 4 in Nevanlinna (1990a). □

Recall that $\rho(\mathcal{K}) = 0$ implies that $\mathcal{K}$ is nilpotent, so that the interesting case is (ii). By Theorem 11 $\eta(\mathcal{K}) < 1$ if and only if $1 \notin \sigma(\mathcal{K})$ and 1 is not separated from ∞ by $\sigma(\mathcal{K})$. In this setup, the case $1 \notin \sigma(\mathcal{K})$ can occur, so that the fixed point problem

$$x = \mathcal{K}x + g \tag{4.15}$$

would as such be well posed in X, but for all normalized polynomials p we would have

$$\|p(\mathcal{K})\| \geq 1. \tag{4.16}$$

In fact, since $\sigma(\mathcal{K})$ is connected and contains the origin, we require real M and N such that, if $\sigma(\mathcal{K})$ is symmetric over the real axis, then $1 \notin \sigma(\mathcal{K})$ but $\alpha \in \sigma(\mathcal{K})$ for some $\alpha > 1$.

To see how much smaller $\eta(\mathcal{K})$ can be compared with $\rho(\mathcal{K})$ consider the following simple example. Let

$$\mathcal{L}u(t) = \rho \int_0^t e^{-(t-s)} u(s)\, \mathrm{d}s, \tag{4.17}$$

with $\rho > 0$. Then

$$\sigma(\mathcal{L}) = \{\lambda : |\lambda - \rho/2| \leq \rho/2\}. \tag{4.18}$$

Thus $\rho(\mathcal{L}) = \rho$ while $\eta(\mathcal{L}) = \min\{\frac{\rho}{|2-\rho|}, 1\}$. On the other hand, for the operator $-\mathcal{L}$ we obtain

$$\rho(-\mathcal{L}) = \rho \quad \text{and} \quad \eta(-\mathcal{L}) = \frac{\rho}{2 + \rho}, \tag{4.19}$$

(Nevanlinna 1990a, page 155). In particular, if $\rho = 1 - \varepsilon$ with a small $\varepsilon > 0$ then $\eta(\mathcal{L}) \approx 1 - 2\varepsilon$, and this is only a 'modest' improvement, while $\eta(-\mathcal{L}) \approx \frac{1}{3}$, in which case we would speak about 'dramatic' improvement. More generally, if $\rho(\mathcal{K}) \in \sigma(\mathcal{K})$, with $\rho(\mathcal{K}) = 1 - \varepsilon$, then there cannot be any dramatic improvement for the following reason: the boundary of $\sigma(\mathcal{K})$ must be smooth near $\rho(\mathcal{K})$ (by Proposition 2 in Nevanlinna (1990a) and the Green's function $g(\lambda) \sim \mathcal{O}(\mathrm{dist}(\lambda, \sigma(\mathcal{K})))$ and $\eta(\mathcal{K}) = e^{-g(1)} = 1 - \mathcal{O}(\varepsilon)$. This should be contrasted with the situation for self-adjoint operators $\mathcal{A}$, for which the spectrum would be contained in an interval. Near the end point the corresponding Green's function would stretch the distance like the square root function and one would have $\eta(\mathcal{A}) = 1 - \mathcal{O}(\sqrt{\varepsilon})$, a well known effect of the conjugate gradient method. Finally, if $\rho(\mathcal{K}) \ll 1$, then it is possible to bound $\rho(\mathcal{K})$ in terms of $\eta(\mathcal{K})$. In fact, since $\sigma(\mathcal{K})$ is connected and contains both 0 and $\rho(\mathcal{K})e^{i\theta}$, for some θ, one has

$$\rho(\mathcal{K}) \geq \mathrm{cap}(\sigma(\mathcal{K})) \geq \tfrac{1}{4}\rho(\mathcal{K}). \tag{4.20}$$

This allows us to formulate the following theorem.

Theorem 14 For every $\mathcal{K}$ we have as $\varepsilon \to 0$,

$$\eta(\varepsilon\mathcal{K}) \geq (\tfrac{1}{4} + o(1))\rho(\varepsilon\mathcal{K}). \tag{4.21}$$

Proof. If g_ε is the Green's function for the outside of $\sigma(\varepsilon\mathcal{K})$, then

$$\eta(\varepsilon\mathcal{K}) = e^{-g_\varepsilon(1)} = (1 + o(1))\varepsilon\mathrm{cap}(\sigma(\mathcal{K})), \quad \text{as} \quad \varepsilon \to 0. \tag{4.22}$$

$\square$

To summarize: Krylov subspace acceleration is always possible, but dramatic improvement is obtained only if the distance between $\sigma(\mathcal{K})$ and 1 is essentially larger than $1 - \rho(\mathcal{K})$.

4.4. Time-dependent linear combinations

By a *subspace method* we mean any Krylov-subspace method that takes linear combinations of sweeps. To generalize this, we may think of processing the sweeps with some other operation. We outline here an approach of Lubich (1992). The basic special assumption here is that one decomposes $A = mI - (mI - A)$, so that the unaccelerated version would be

$$\dot{x}^{k+1} + m x^{k+1} = N x^k + f, \quad x^{k+1}(0) = x_0. \tag{4.23}$$

Observe that multiplication with m commutes with N.

The accelerated version is as follows. Given x^k, solve

$$\dot{u}^k + m u^k = N x^k + f, \quad u^k(0) = x_0, \tag{4.24}$$

set

$$v^k := u^k - x^k \tag{4.25}$$

and solve again for w^k from

$$\dot{w}^k + \lambda_k w^k = v^k, \quad w^k(0) = 0. \tag{4.26}$$

Finally, set

$$x^{k+1} := u^k + \alpha_k v^k + \beta_k w^k. \tag{4.27}$$

Note that all equations and substitutions (4.24)–(4.27) are on the component level, apart from the evaluation of $N x^k$ in (4.24) – in this sense the extra work is small compared with (4.23). The parameters α_k, β_k and λ_k can now be chosen so that the error reduction in $L_2(\mathbb{R}_+)$ is the same as that of Chebyshev acceleration of Richardson's iteration

$$x^{k+1} = x^k - \frac{1}{m} A x^k + \frac{1}{m} b$$

for the static linear system $Ax = b$. To see that this is possible, compute the Laplace transform of the iteration error, and require this to be the Chebyshev acceleration of the Laplace transform of the iteration error of the basic scheme (4.23) for every z.

Related ideas are also discussed in Skeel (1989) and Reichelt, White and Allen (1995).

4.5. Overlapping splittings

If M in the splitting $A = M - N$ is chosen to be a block diagonal of A then the iteration (2.2) can clearly be computed in parallel for each small subsystem corresponding to one block of A. This is known as block Jacobi iteration. If the order of the original system was d and we use s subsystems (blocks) we only need to solve systems of order d/s in parallel. The reduction of work (and time) is so large that one might as well increase the size of subsystems with a few components without losing this gain. The idea of overlapping was introduced by Jeltsch and Pohl (1995) in order to accelerate convergence of WR iteration. For block Jacobi iteration it can be best explained by an example.

Example 2   Let

$$A = \begin{pmatrix} 2 & -1 & & \\ -1 & 2 & -1 & \\ & -1 & 2 & -1 \\ & & -1 & 2 \end{pmatrix}, \quad \mathbf{x} = (x_1 \ x_2 \ x_3 \ x_4)^T$$

and we use two subsystems of the same size, so that

$$A = M - N = \begin{pmatrix} 2 & -1 & & \\ -1 & 2 & & \\ & & 2 & -1 \\ & & -1 & 2 \end{pmatrix} - \begin{pmatrix} 0 & 0 & & \\ 0 & 0 & 1 & \\ & 1 & 0 & 0 \\ & & 0 & 0 \end{pmatrix}.$$

Unknowns x_1, x_2 are solved from the first subsystem S_1 and x_3, x_4 from the second S_2. The idea of overlapping is that some components of the unknown vector are assigned to several subsystems, for instance x_3 in this example. Then, in (2.1), we obtain

$$\tilde{A} = \begin{pmatrix} 2 & -1 & & & \\ -1 & 2 & -1 & & \\ & -1 & 2 & & -1 \\ & -1 & & 2 & -1 \\ & & & -1 & 2 \end{pmatrix}, \quad \tilde{\mathbf{x}} = (x_1 \ x_2 \ x_{3.1} \ x_{3.2} \ x_4)^T$$

and

$$\tilde{M} - N = \begin{pmatrix} 2 & -1 & & & \\ -1 & 2 & -1 & & \\ & -1 & 2 & & \\ & & & 2 & -1 \\ & & & -1 & 2 \end{pmatrix} - \begin{pmatrix} 0 & 0 & 0 & & \\ 0 & 0 & 0 & & \\ 0 & 0 & 0 & 0 & 1 \\ & 1 & 0 & 0 & 0 \\ & & 0 & 0 \end{pmatrix}.$$

We use the first system to find $\{x_1, x_2, x_{3.1}\}$ and the second to calculate $\{x_{3.2}, x_4\}$. The value used for x_3 in the next iteration is taken as the linear combination $x_3 = \alpha x_{3.1} + (1 - \alpha)x_{3.2}$

The number of overlapping components between subsystems was first one, then two, in this example. This number is called the *overlap* and we denote it by o.

In general it is reasonable to assume that if we have $s > 2$ subsystems then

(A1): The overlapping components are assigned to at most two common subsystems.

The overlap o can be defined as the maximum number of overlapping components in the intersections $S_j \cap S_k$.

Jeltsch and Pohl (1995) formulated overlapping splittings for block Jacobi iteration at the subsystem level, and showed that a convergence analysis similar to that of basic WR can be carried out. Their numerical results suggested that overlapping accelerates the convergence of WR. In order to explain when and why this happens we describe the process for the whole system. Let us assume that the splitting $M - N$ corresponds to block Jacobi iteration. Thus the components of $\mathbf{x}$ corresponding to each subsystem must be numbered consecutively, and M must be block diagonal.

When we have chosen the overlapping components of $\mathbf{x}$ we modify (2.1) as follows:

- If x_i is copied from subsystem k_1 to subsystem k_2, then rename it $x_{i.k_1}$, and add a new component $x_{i.k_2}$ to subsystem k_2.
- Duplicate the ith row of system (2.1), $\dot{x}_i + \sum_{j=1}^d A_{ij} x_j = f_i$, and add the duplicated row to the row corresponding to index $i.k_2$.

Hence each overlapping component increments the dimension of A by one by duplicating a row and adjoining a new column to the new duplicated component. Between the integration sweeps, each overlapped component x_j of $\mathbf{x}$ is postprocessed by replacing both copies with a linear combination of the overlapped components. The effect for the whole solution can be viewed as a multiplication by a constant matrix E. From the iteration's perspective, the iteration matrix $(zI + M)^{-1}N$ is replaced by $(zI + M)^{-1}NE$. Since the overlapped components of $\mathbf{x}$ are in the nullspace of N, the graphs of NE and $G(N)$ are identical.

We shall show that overlapping can accelerate convergence by decreasing the order ω of the iteration operator $\mathcal{K}$. The order can be computed from the directed graph of the matrix A as stated in Section 2.5.

4.6. How overlapping decreases the order

The graph $G(\tilde{A})$ is formed from $G(A)$ by making the following modifications to $G(A)$.

- Duplicate the vertices of $G(A)$ corresponding to the overlapped components.
- Duplicate the edges coming into the vertices corresponding to the overlapped components.

The latter statement demands some explanation. If a vertex v is duplicated from subsystem k_1 to subsystem k_2, then for the copy in k_1, draw all edges incident to v in $G(A)$ not linked to subsystem k_2. Similarly, for the copy of v in subsystem k_2, draw the edges incident to v not intersecting subsystem k_1.

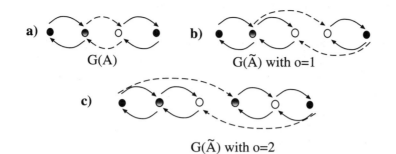

$G(A)$ $G(\tilde{A})$ with o=1

$G(\tilde{A})$ with o=2

Fig. 2. Directed graphs of Example 2. Edges belonging to $G(N)$ are denoted by dashed lines. Duplicated vertices are recognized by shading.

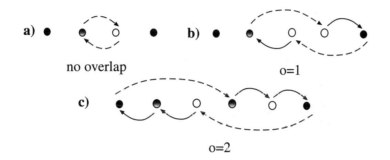

no overlap   o=1

o=2

Fig. 3. Critical cycles of Example 2. Edges belonging to $G(N)$ are denoted by dashed lines.

All these edges were also in $G(A)$. The new edges are copies of the incoming edges of v.

Example 2 continued. The directed graphs of A and $\tilde{A}$ are given in Fig. 2. The cycles giving $\max(n_j/l(\mathcal{C}_j))$ in $G(M - N)$ or $G(\tilde{M} - \tilde{N})$ are given in Figure 3.

This example is summarized in Table 1, and overlapping decreases ω_{graph}, and hence the order of the iteration operator.

Table 1.

o	ω_{graph}
0	1
1	$2/4 = 1/2$
2	$2/6 = 1/3$

Fig. 4. Graph structure corresponding to a band matrix with $b = 2$. The edges with arrows in both heads are abbreviations of two edges connecting the vertices both ways.

In general, assuming condition (A1) given in Section 4.5, and that only adjoining subsystems (S_l and S_{l+1}) overlap, we can derive our next result.

Theorem 15 Using overlapping in block Jacobi iteration never increases the ratio ω_{graph}.

The proof is based on showing that the only new circuits created in $G(\tilde{A})$ when compared to the original graph $G(A)$ are such that the maximum ratio of $n_j/l(\mathcal{C}_j)$ is smaller than in $G(A)$. Simultaneously, for the old circuits remaining also in $G(\tilde{A})$, the ratio $n_j/l(\mathcal{C}_j)$ may decrease to 0 if the overlapping is such that the circuit stays inside one subsystem in the new graph $G(\tilde{A})$. The detailed proof is given in Miekkala (1996); the following result is a direct consequence.

Corollary 6 If ω_{graph} is determined only by the cycle $\mathcal{C}_1$, and $\mathcal{C}_1$ can be contained into one of the subsystems using overlapping, then ω_{graph} decreases.

This result tells us how overlapping should be used to accelerate convergence of the iteration. Indeed, the cycle (or cycles) attaining $\max(n_j/l(\mathcal{C}_j))$ in $G(M - N)$ should first be located, and then the subsystems overlapped in such a way that this cycle stays inside one of the enlarged subsystems.

The matrix A in Example 2 was the so-called Laplacian matrix, a band matrix. We will now show how overlapping decreases the order for general band matrices. The band width is denoted by $2b + 1$, where b is the smallest integer for which $A_{ij} = 0$ whenever $|i - j| > b$.

Once again we need only study overlapping between two consecutive subsystems, say S_1 and S_2. In graph theoretic language, $b = 1$ means that every pair of adjacent vertices is connected by a loop of length two; for $b = n$, every pair of vertices at mutual distance at most n is connected by a loop of length two. Figure 2 shows the case $b = 1$ and Figure 4 case $b = 2$; the general case should be obvious (if too messy to draw).

We have already analysed overlapping vertices for $b = 1$, in Example 2. The interface between the subsystems has o overlapping vertices; thus the coupling edge entering one subsystem from another has to skip o vertices. From Figures 2 and 3, we conclude that the cycle between the subsystems satisfies $l(\mathcal{C}_j) = o + o + 2 = 2o + 2$ and $n_j = 2$. Therefore $\omega_{\text{graph}} = (o+1)^{-1}$.

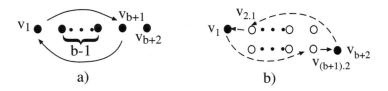

Fig. 5. a) Using overlap $o < b$ does not change the indicated loop between the subsystems. b) A critical circuit for $o = b$.

In the general case, the length of the critical circuit is $l(\mathcal{C}_j) = 2 + 2\lfloor o/b \rfloor$. Since the graph of a band matrix contains loops as in Figure 5a, it is clear that these loops remain in $G(\tilde{A})$ for $o \in \{1, \ldots, b-1\}$. Hence $\omega_{\text{graph}} = 1$ for these values of o. If $o = b$, then we duplicate b subsequent vertices as in Figure 5b and the critical circuit

$$v_1 \to v_{(b+1).2} \to v_{b+2} \to v_{2.1} \to v_1$$

has length 4. If overlap $o \subset \{b+1, \ldots, 2b-1\}$, then we still get a circuit of length 4 and $n_j = 2$, that is

$$v_1 \to v_{(b+1).2} \to v_{o+2} \to v_{(o-b+2).1} \to v_1.$$

When $o = 2b$, the second edge of this circuit cannot occur because $o + 2 - (b+1) > b$ and the length of the critical circuit increases to 6. The general result should now be obvious.

Theorem 16 Let A be a band matrix of band width $2b + 1$ and use block Jacobi iteration with overlap o in (2.2). Then

$$\omega \leq \frac{1}{\lfloor o/b \rfloor + 1}.$$

5. Discretized iterations

The results of the previous sections have analogues for discretized equations. We briefly discuss these analogues and then look at the new phenomena that arise when several grids are used during the calculation. Also, we mention some interesting step size control problems.

5.1. Discretization methods

The most natural approach to 'continuous time iteration' is simply to apply reliable software to integrate the associated equations. The process is sufficiently robust for results on the continuous version to describe what happens in practice, as long as the iteration errors are larger than the discretization errors. This robustness can be seen very well from an exact analysis of the discretized equations. Here we consider linear multistep methods with a constant time step h:

$$\frac{1}{h}\sum_{j=0}^{k}\alpha_j x_{n+j}^{\nu} + \sum_{j=0}^{k}\beta_j M x_{n+j}^{\nu} = \sum_{j=0}^{k}\beta_j(N x_{n+j}^{\nu-1} + f_{n+j}). \qquad (5.1)$$

As is customary, we use operator notation for the linear multistep methods. In order to avoid confusion with the spectral radius and the spectrum, we set

$$a(\zeta) := \sum_{j=0}^{k}\alpha_j\zeta^j, \quad b(\zeta) := \sum_{j=0}^{k}\beta_j\zeta^j. \qquad (5.2)$$

We normalize $b(1) = 1$, require that the order of consistency satisfies $p \geq 1$, and assume that $a(\zeta)$ and $b(\zeta)$ have no common factors. We abbreviate $\{\sum_{j=0}^{k}\alpha_j v_{n+j}\}$ to av.

In this notation, (5.1) reads

$$\frac{1}{h}ax^{\nu} + bMx^{\nu} = bNx^{\nu-1} + bf. \qquad (5.3)$$

As in the continuous case it is advantageous to introduce a linear operator $\mathcal{K}_h$ and write the solution of the difference equation (5.3) in the form

$$x^{\nu} = \mathcal{K}_h x^{\nu-1} + \varphi_h. \qquad (5.4)$$

Here $\mathcal{K}_h$ is well defined provided we understand the sequences to vanish for negative indices and

$$\frac{\alpha_k}{\beta_k} \notin \sigma(-hM). \qquad (5.5)$$

In what follows we shall always assume that (5.5) holds. The role of the Laplace transform is played by the 'ζ-transform'.

If X_h denotes $\mathbb{C}^d$-valued sequences, then we write

$$\tilde{v}(\zeta) := \sum_{n=0}^{\infty}\zeta^{-k}v_k, \qquad v \in X_h,$$

and this leads to the following expression for the *symbol* of $\mathcal{K}_h$:

$$K_h(\zeta) := (\frac{1}{h}a(\zeta) + b(\zeta)M)^{-1}b(\zeta)N. \qquad (5.6)$$

In particular, $K_h(\zeta) = K(a(\zeta)/hb(\zeta))$. It is also useful to write $v_n = v(nh)$ for $v \in X_h$. We shall need standard terminology to describe the stability properties of the method (a, b).

Definition 4 The *stability region* S consists of those $\mu \in \mathbb{C}\cup\{\infty\}$ for which the polynomial $a(\zeta) - \mu b(\zeta)$ (around ∞ consider $\mu^{-1}a - b$) satisfies the root condition. The method is called *strongly stable* if all roots of $a(\zeta)/(\zeta-1)$ are less than one in modulus. The method is called *A-stable* if S contains the closed left half plane.

5.2. Finite windows

Consider bounding $\mathcal{K}_h^\nu$ on the 'window' $0 \leq j \leq T/2$. We identify the vectors $v \in X_h$ by sequences indexed over $\mathbb{Z}$ with v vanishing for negative indices and

$$|v|_T = \max_{0 \leq h_j \leq T} |v_j|. \tag{5.7}$$

Theorem 17   If $\alpha_k/h\beta_k \notin \sigma(-M)$, then, in the window $0 \leq j \leq T/h$ $\mathcal{K}_h$ has the spectral radius

$$\rho(\mathcal{K}_h) = \rho(K(\alpha_k/h\beta_k)). \tag{5.8}$$

This is Theorem 4.1 in Nevanlinna (1989*c*). Comparing this with (5.6), observe that $z \to \infty$ corresponds to $\zeta \to \infty$ and $\lim_{\zeta \to \infty} a(\zeta)/hb(\zeta) = \alpha_k/h\beta_k$. It should be noticed that, unlike the infinite window case, we do *not* obtain a result of the form $\rho(\mathcal{K}_h) = \rho(\mathcal{K}) + \mathcal{O}(h^p)$ with p related to the accuracy of the discretization method. However, the following holds.

Corollary 7   Under the assumptions of Theorem 17 we have

$$\rho(\mathcal{K}_h) = (1 + o(1))(\frac{\beta_k}{\alpha_k}ch)^{1/\omega}, \quad \text{as } h \to 0, \tag{5.9}$$

if $\mathcal{K}$ is of order ω and of type $\tau = cT$.

Proof.   This follows immediately from the defining relation of ω and c (see Section 2.1, line (2.37)) by choosing $z = z(h) = \alpha_k/h\beta_k$. $\square$

As this corollary shows, the decay of $|\mathcal{K}_h^k|_T$ as $k \to \infty$ is again related to the order and type of the original resolvent operator. Upper bounds again hold, analogous to those in Section 2.4 for $|\mathcal{K}^k|_T$, but here we just refer to the original paper by Nevanlinna (1989*c*).

5.3. Infinite windows

We now look at the usual ℓ_2-space of square summable $\mathbb{C}^d$-valued sequences (but again the spectral radius of $\mathcal{K}_h$ would be the same for a large class of 'unscaled' norms). Now the local solvability condition $\alpha_k/h\beta_k \notin \sigma(-M)$ is still needed for $\mathcal{K}_h$ to be well defined, but another condition is needed to guarantee that $\mathcal{K}_h$ is bounded in ℓ_2. In the continuous case this was simply the condition $\sigma(-M) \subset \mathbb{C}_-$. Now the role of $\mathbb{C}_-$ is played by the stability region below.

Proposition 5   If $\sigma(-M) \subset h^{-1}\text{int}S$, then $\mathcal{K}$ is bounded in ℓ_2.

Proof.   This is Lemma 3.1 in Miekkala and Nevanlinna (1987*b*). $\square$

The correspondence of $\mathbb{C}_+$ with $h^{-1}(\mathbb{C} \setminus S)$, best seen in the fact $K_h(\zeta) = K(a(\zeta)/hb(\zeta))$, immediately explains this Proposition and related claims. For example, the boundaries of these sets osculate at the origin, displaying the *order of accuracy*. However, in order to obtain the exact formulation we need the following concept.

Definition 5   A multistep method (a, b) has *order of amplitude fitting q* if the principal root $\zeta_1(\mu)$ of $a(\zeta) - \mu b(\zeta) = 0$ (the zero for which $\zeta_1(\mu) - e^\mu = \mathcal{O}(\mu^{p+1})$) satisfies $|\zeta_1(it)| - 1 = \mathcal{O}(t^{q+1})$, for small real t.

Thus $q \geq p$ if p is the usual discretization order, and, with the trapezoidal rule for instance, we have $q = \infty$, while $p = 2$.

Theorem 18   Assume that the multistep method is of amplitude fitting order q and is strongly stable. Then, for all sufficiently small h, $\mathcal{K}_h$ is a bounded operator,

$$\|\mathcal{K}_h\| = \|\mathcal{K}\|[1 + \mathcal{O}(h^q)] \tag{5.10}$$

and

$$\rho(\mathcal{K}_h) = \rho(\mathcal{K})[1 + \mathcal{O}(h^q)]. \tag{5.11}$$

Theorem 19   Assume that the multistep method is A-stable. Then, for all $j \geq 1$,

$$\|\mathcal{K}_h^j\| \leq \|\mathcal{K}^j\| \tag{5.12}$$

and

$$\rho(\mathcal{K}_h) \leq \rho(\mathcal{K}). \tag{5.13}$$

Both Theorem (18) and (19) are proved in Nevanlinna (1990*b*), using results of Miekkala and Nevanlinna (1987*b*).

For Krylov acceleration it is interesting to know something of the spectrum.

Theorem 20   (Miekkala and Nevanlinna 1987*b*)

$$\sigma(\mathcal{K}_h) = \mathrm{cl} \bigcup_{|\zeta| \geq 1} \sigma(K(a(\zeta)/hb(\zeta))).$$

Corollary 8   For A-stable methods we have $\sigma(\mathcal{K}_h) \subset \sigma(\mathcal{K})$.

Proof.   By A-stability, $\bigcup_{|\zeta| \geq 1}\{a(\zeta)/hb(\zeta)\}$ is a subset of the closed right half plane. $\square$

Corollary 9 $\sigma(\mathcal{K}_h)$ consists of at most d components, each containing eigenvalues of $K(\alpha_k/h\beta_k)$.

Proof.   By letting $\zeta \to \infty$, we see that the eigenvalues of $K(\alpha_k/h\beta_k)$ belong to $\sigma(\mathcal{K}_h)$. Each eigenvalue, or, rather, branch of the algebraic function, can be continued to $|\zeta| \geq 1$, giving at most d components. $\square$

For example, if we take the implicit Euler method and relatively large step size, then $\sigma(\mathcal{K})$ and $\sigma(\mathcal{K}_h)$, and in particular the corresponding optimal reduction factors, may differ considerably. Lumsdaine and White (1995) give an example of this nature.

5.4. Multigrid in time

The effective use of 'multigrids' in our setup simply consists of balancing the iteration error and discretization error. Thus one moves only towards ever finer grids. The computational goal is to be able to compute or simulate the full system with an amount of work W which is a modest multiple of the work W_0, say, needed to compute the 'uncoupled' system

$$\dot{u} + Mu = Nx + f(t), \quad u(0) = x_0, \tag{5.14}$$

to the same tolerance, where x denotes the solution of

$$\dot{x} + Mx = Nx + f(t), \quad x(0) = x_0. \tag{5.15}$$

The ideas in setting up such a computational strategy, or 'tolerance game', have been discussed in Nevanlinna (1989c) and Nevanlinna (1990b). We shall not go into such a discussion here but rather concentrate on two issues that might cause difficulties if the implementation is careless. It may not be evident that it is possible to arrange for $W = \mathcal{O}(W_0)$ to hold. Two extremes are possible. First, the step size selection routine is extremely stupid and the step is constant on the grid. Second, the step size selection process is extremely clever and the step changes with the smoothness of the solution, so rapidly reducing the step size when the solution is rough, but increases the step size stably when the solution becomes smooth. The potential danger to be avoided is this: in solving stiff problems, it is to be expected that the solution at the end of the window is smooth. However, on the next window, say $[mT, (m + 1)T]$, the initial guess $x^0(t) \equiv x(mT)$ introduces an error which causes, in exact computation, a travelling error wave, which, however, has very small support. Roughly speaking, with fixed step strategy the error wave cannot be supported at all, while integration with automatic software has to be done with a good step size routine so that not too many time points are wasted at the thinly supported rough parts. Here we discuss the constant step case and in the next section the latter one.

For simplicity, let the grid at the ν_{th} iteration level be $\{jh_\nu\}_j$ where the time step $h_\nu = 2^{-n(\nu)}h_0$ and $n(\nu)$ is nondecreasing and unbounded. A detailed analysis of this is given in Section 3.2. of Nevanlinna (1990b). In the iteration process, whenever the grid is refined $(n(\nu) > n(\nu - 1))$, we need to be able to extend a grid function $v_{\nu-1} = \{v(jh_{\nu-1})\}$ to a grid function on $v_\nu = \{v(jh_\nu)\}$. Thus we have the prolongation operator

$$\mathcal{P}_\nu : \quad v_{\nu-1} \to v_\nu,$$

computed with accuracy matching the integration method, but the important point is that there is also a stability property to be satisfied. In fact, we want the overall process to decay with a rate essentially equalling $\rho(\mathcal{K})$, for all refinement sequences $\{n(\nu)\}$, and this is possible if the prolongation operators $\{\mathcal{P}_\alpha\}$ are stable: there exists a C such that

$$\|\Pi_1^n \mathcal{P}_{\alpha(j)}\| \leq C, \quad \mathcal{P}_{\alpha(j)} \in \{\mathcal{P}_\alpha\}.$$

The norms here are the naturally induced operator norms; grid functions $v_\nu = \{v(jh_\nu)\}$ are normed as follows:

$$\|v_\nu\| := \{h_\nu \sum_j |v(jh_\nu)|^2\}^{1/2}.$$

It turns out that there are arbitrary high-order stable prolongations but that the information should in general be collected from both sides of the grid points. A symbolic calculus for stepwise translation invariant prolongations was developed in Nevanlinna (1990b). The crucial dilation process here is quite similar to the subdivision algorithm in CAD or in wavelets and this eventually led Eirola (1992) to study the obtainable smoothness of wavelets.

For the error analysis the main result is the following theorem.

Let

$$B_{\mu\mu} := \quad \text{identity on grid functions on} \quad \{jh_\mu\}$$

$$B_{\nu\mu} := \mathcal{K}_{h_\nu} \mathcal{P}_\nu B_{\nu-1,\mu}, \quad \nu \geq \mu + 1.$$

Theorem 21   (Nevanlinna 1990b) Assume that the multistep method is strongly stable and we are given a stable set of prolongations. Let h_0 be small enough so that

$$\sigma(-M) \subset \frac{1}{h} \text{int } S$$

holds for all $h \leq h_0$. Given $\varepsilon > 0$ and $h_\nu = 2^{-n(\nu)} h_0$ with $n(\nu)$ nondecreasing and unbounded, there exists a C such that

$$\|B_{\nu\mu}\| \leq C(\rho(\mathcal{K}) + \varepsilon)^{\nu-\mu}, \quad \nu \geq \mu \geq 0.$$

This is the key result needed to show that the 'tolerance game' is possible.

5.5. A difficulty due to stiffness

Consider solving

$$\dot{x} + Ax = f \tag{5.16}$$

in a window where the solution is already smooth, that is, the transient has died out in the earlier window.

Usually one takes the initial function to be identically the initial value and it is no longer clear whether the iterates will stay smooth. A naive application of

Picard–Lindelöf iteration might then spend a lot of time in the early iterations, because any local error estimator would require tiny steps compared with the smoothness of the limit function.

We present here a model analysis of the smoothness of the iterates x^j, following Nevanlinna (1989a). We assume that the iterates x^j are computed exactly from

$$\dot{x}^{j+1} + Mx^{j+1} = Nx^j + f, \tag{5.17}$$
$$x^{j+1}(0) = x_0 \equiv x^0,$$

but we measure the 'cost of integration' as if we were using high quality software, based on first-order local error estimation: at time t the time step $h = h(t)$ would satisfy

$$h(t)|\ddot{x}^j(t)| = \epsilon_j. \tag{5.18}$$

This corresponds to the criterion of error per unit step; calculation for criterion of error per step is analogous.

Thus the relevant measure for the cost or for the total number of time points is proportional to $\sum \frac{1}{\epsilon_j} \int |\ddot{x}^j|$. An efficient implementation of Picard–Lindelöf iteration would gradually decrease the tolerance ϵ_j. Here we shall not discuss the choice of ϵ but focus on estimating $\int |\ddot{x}^j|$.

We put $T = 1$ and assume that x is so smooth that it can be represented as a convergent power series

$$x(t) = \sum t^i x_i. \tag{5.19}$$

Since we are interested in the second derivatives, we measure smoothness on the window [0,1] by

$$\|x\| := |x_0| + |x_1| + \sum_{i=2}^{\infty} i(i-1)|x_i|. \tag{5.20}$$

If $e^j := x - x^j$ denotes the iteration error, then

$$\dot{e}^{j+1} + Me^{j+1} = Ne^j, \quad e^j(0) = 0. \tag{5.21}$$

Introducing

$$k(t) := e^{-tM}N,$$

and setting

$$k^{*j} = k * k^{*(j-1)},$$

we have

$$e^j = k^{*j} * (x - x_0). \tag{5.22}$$

Substituting (5.19) into (5.22) yields

$$e^j(t) = \sum_{i=1}^{\infty} \int_0^t (t-s)^i k^{*j}(s) x_i ds, \tag{5.23}$$

and hence

$$\ddot{e}^j(t) = k^{*j}(t) x_1 + \sum_{i=2}^{\infty} i(i-1) \int_0^t (t-s)^{i-2} k^{*j}(s) x_i ds. \tag{5.24}$$

In order to estimate this we introduce the following bound:

$$C := \sup_j \sup_{|a|=1} \int_0^1 |k^{*j}(s)a| ds. \tag{5.25}$$

Theorem 22   We have

$$\int_0^1 |\ddot{x} - \ddot{x}^j| \le C\|x - x_0\|. \tag{5.26}$$

Furthermore, for any given splitting M, N there exists $x_0 \neq 0$ and f such that $x(t) = (1+t)x_0$, $\ddot{x} \equiv 0$ and for some j,

$$\int_0^1 |\ddot{x}^j| = C\|x - x_0\|. \tag{5.27}$$

Proof.   The definition of C immediately gives (5.26). Since $\int |k^{*j}|$ tends to zero and a in (5.25) runs over a compact set, there exist an integer j and a unit vector x_0 such that

$$C = \int_0^1 |k^{*j}(s)x_0| \, ds.$$

If $f(t) = x_0 + (1+t)Ax_0$ then $x(t) = (1+t)x_0$ is the solution of (5.16). Since $x_1 = x_0$, and $\ddot{e}^j = -\ddot{x}^j$, we obtain

$$\int_0^1 |\ddot{x}^j| = C = C\|x - x_0\|.$$

from (5.24). $\square$

We conclude from Theorem 22 that the important quantity $\int |\ddot{x}^j|$ stays small for smooth limit functions x if and only if C is of moderate size.

It is important to note that, even if C is of moderate size, the smallest 'steps' can be very small, since we may have $\sup_{[0,1]} |\ddot{x}^j| \gg C$, while $\int_0^1 |\ddot{x}^j| \le C$. Nevanlinna (1989a) contains an example of this.

Recall from Section 2 (proof of Theorem 5) that if

$$|K(z)| \le \frac{B}{\operatorname{Re} z - \gamma} \quad \operatorname{Re} z > \gamma,$$

then we obtain a bound for the norms of the iterated kernels and hence the upper bound:

$$C \le e^{\gamma} d \max_{j} (\frac{Be}{j})^j.$$

6. Periodic problems

We discuss briefly the iterative solution for periodic problems. We shall see that the speed of the basic iteration is related to the speed of the 'corresponding' initial value problem in the *infinite* window, while the speed after optimal Krylov acceleration is related to the speed obtained for initial value problems on the *finite* window.

6.1. The problem and the iteration operator

Consider solving the periodic boundary value problem

$$\dot{x} + Ax = f, \quad x(0) = x(T), \tag{6.1}$$

when f is a continuous function of period T. Splitting $A = M - N$ as usual leads to an iteration of the form

$$x^k = \mathcal{F}x^{k-1} + q \tag{6.2}$$

provided the *solvability condition* holds:

$$\frac{2\pi i n}{T} \notin \sigma(-M) \quad \text{for all } n \in \mathbb{Z}. \tag{6.3}$$

Here the integral operator $\mathcal{F}$ can be written in convolution form

$$\mathcal{F}x(t) = \int_0^T \varphi(t - s)x(s)\,ds, \tag{6.4}$$

where the kernel φ is periodic and

$$\varphi|_{[0,T)}(t) = e^{-tM}(1 - e^{-TM})^{-1}N.$$

For more details, see Vandewalle (1992).

6.2. Spectrum and consequences

Computation of the Fourier coefficients of φ gives $\hat{\varphi}(n) = K(\frac{2\pi i n}{T})$, where $K(z) = (z + M)^{-1}N$ is the symbol of the Volterra operator $\mathcal{K}$. This leads to the following result.

Theorem 23 (Vandewalle 1992) Let the solvability condition (6.3) hold. Then $\mathcal{F}$ is a compact operator in $C[0, T]$ with the spectrum

$$\sigma(\mathcal{F}) = \text{cl} \bigcup_{n \in \mathbb{Z}} \sigma(K(\frac{2\pi i n}{T})), \tag{6.5}$$

and spectral radius

$$\rho(\mathcal{F}) = \max_{n \in \mathbb{Z}} \rho(K(\frac{2\pi i n}{T})). \tag{6.6}$$

Corollary 10 Assume that eigenvalues of M have positive real parts. Then

$$\rho(\mathcal{F}) \leq \rho(\mathcal{K}) \leq \rho(\mathcal{F}) + \mathcal{O}\left(\frac{1}{T^2}\right). \tag{6.7}$$

Proof. Here $\mathcal{F}$ is considered in $C[0, T]$ while $\mathcal{K}$ is considered on the infinite window (any space X of Section 3). The first inequality is immediately verified because

$$\rho(\mathcal{F}) = \max_{n} \rho(K(\frac{2\pi i n}{T})) \leq \max_{\theta} \rho(K(i\theta)) = \rho(\mathcal{K}).$$

The second inequality follows from the fact that the boundary $\partial\sigma(\mathcal{K})$ is locally analytic at points $\lambda \in \sigma(\mathcal{K})$ where $|\lambda| = \rho(\mathcal{K})$; see Proposition 2 in Nevanlinna (1990a). Sampling this with density $\mathcal{O}(T^{-1})$ provides the maximum within tolerance $\mathcal{O}(T^{-2})$. $\square$

Observe in particular that even if the solvability condition holds (and in particular it holds for all T if M has eigenvalues of positive real parts) we can have $\rho(\mathcal{F}) > 1$, so that the iteration would diverge. However, Krylov acceleration would always work.

Corollary 11 Let the local solvability condition (6.3) hold and suppose $1 \notin \sigma(\mathcal{F})$. Then the optimal reduction factor vanishes: $\eta(\mathcal{F}) = 0$.

Proof. Since $\sigma(\mathcal{F})$ is countable, it is of zero capacity, and the claim follows from Theorem 11 in Section 4.1. $\square$

Since $\eta(\mathcal{F}) = 0$, we consider the superlinear decay of $b_n(\mathcal{F})$. Again the answer can be related to the corresponding initial value problem. Namely, Piirilä (1993) has shown that the order of decay for $b_n(\mathcal{F})$ equals that of $|\mathcal{K}^n|_T$, that is, if $R(\lambda, \mathcal{K})$ is of order ω, then

$$\limsup \frac{n \log n}{\log(1/b_n(\mathcal{F}))} = \omega.$$

7. A case study: linear RC-circuits

Linear RC circuits can be modelled in several ways. The sparse tableau model contains all the equations governing a circuit and results in a large DAE system. Nodal formulation results in a substantially smaller system and then equations are written for nodal voltages with the aid of so-called *stamps*. Using nodal formulation it was already shown in the earliest waveform relaxation paper of Lelarasmee et al. (1982) that waveform relaxation converges if the cutting is done only across such capacitors that there is a path connecting them to the ground involving only capacitors. A simple model problem where splitting is done across a capacitor not obeying this rule was considered by Miekkala, Nevanlinna and Ruehli (1990), showing that in this case waveform relaxation still converges, but convergence is sublinear. This result can be generalized to all splittings of linear RC circuits. waveform relaxation always converges, but convergence may be slow, like $\mathcal{O}(k^{-r})$, where k is the iteration index and r a small number. This result was proved by Nevanlinna (1991).

We consider here as a case study applying waveform relaxation for the sparse tableau formulation of linear RC circuits, following closely the treatment of Leimkuhler, Miekkala and Nevanlinna (1991). The system is a DAE of index one or two. We describe a splitting strategy that allows us to break the circuit into subcircuits only across resistors. This strategy leads to convergence that is shown using mainly Laplace transforms.

7.1. RC-circuit equations

The system of equations for an RC-circuit is

$$
\begin{pmatrix} C\dot{v}_C \\ 0 \\ 0 \\ 0 \\ 0 \end{pmatrix} + \begin{pmatrix} 0 & 0 & 0 & -I & 0 \\ 0 & -R & 0 & 0 & A_R \\ 0 & 0 & 0 & 0 & A_E \\ -I & 0 & 0 & 0 & A_C \\ 0 & A_R^T & A_E^T & A_C^T & 0 \end{pmatrix} \begin{pmatrix} v_C \\ i_R \\ i_E \\ i_C \\ v_N \end{pmatrix} = \begin{pmatrix} 0 \\ 0 \\ E(t) \\ 0 \\ 0 \end{pmatrix}. \quad (7.1)
$$

The unknown vector contains voltages across capacitors (v_C), currents through resistors (i_R), voltage sources (i_E) and capacitors (i_C), and nodal voltages (v_N). The matrices in (7.1) satisfy

$$
\begin{aligned}
R \quad &: \quad \text{a positive, diagonal } n_R \times n_R \text{ matrix;} \\
C \quad &: \quad \text{a positive, diagonal } n_C \times n_C \text{ matrix;} \\
A_R \quad &: \quad \text{an } n_R \times N \text{ incidence matrix;} \\
A_E \quad &: \quad \text{an } n_E \times N \text{ incidence matrix;} \\
A_C \quad &: \quad \text{an } n_C \times N \text{ incidence matrix.}
\end{aligned}
$$

Here an *incidence matrix* is a matrix whose elements belong to the set $\{-1, 0, 1\}$ and whose rows contain either two nonzeros $\{1, -1\}$ or one nonzero. The usual definition does not allow the latter case, which arises because we have eliminated the ground node from the circuit (which is a directed graph). So N is the number of nodes in the circuit after a reference (ground) node has been fixed and n_R, n_C and n_E are the number of resistors, capacitors and voltage sources in the circuit, respectively. The appropriate sizes of the variable vectors should be apparent.

The problem is well posed if

$$n_R + n_E + n_C \geq N + 1 \quad \text{and}$$

$$A := \begin{pmatrix} A_R \\ A_E \\ A_C \end{pmatrix} \quad \text{has full rank.} \tag{7.2}$$

Another basic assumption is that

$$A_E \quad \text{has linearly independent rows.} \tag{7.3}$$

This only means that there cannot be two voltage sources in parallel.

Proposition 6 Assume that assumptions (7.2) and (7.3) hold. Then the DAE (7.1) has index one if

$$\begin{pmatrix} A_E \\ A_C \end{pmatrix} \quad \text{has linearly independent rows.}$$

Otherwise it has index two.

Index two occurs if there are loops containing only capacitors and voltage sources. For the proof of the proposition and a discussion of the assumptions (7.2) and (7.3) see Manke et al. (1979).

We discuss the initial conditions for (7.1) after applying the Laplace transform to (7.1). The Laplace transform $\hat{x}$ of x is given by $\hat{x}(z) = \int_0^\infty e^{-zt} x(t) dt$. The transformed system becomes

$$\begin{aligned} zC\hat{v}_C &= \hat{i}_C + Cv_C(0), \\ R\hat{i}_R &= A_R\hat{v}_N, \\ A_E\hat{v}_N &= \hat{E}, \\ \hat{v}_C &= A_C\hat{v}_N, \\ \text{and} \quad A_R^T\hat{i}_R + A_E^T\hat{i}_E + A_C^T\hat{i}_C &= 0. \end{aligned} \tag{7.4}$$

Eliminating $\hat{i}_R$, $\hat{v}_C$ and $\hat{i}_C$ gives

$$\begin{pmatrix} 0 & A_E \\ A_E^T & A_R^T R^{-1} A_R + zA_C^T C A_C \end{pmatrix} \begin{pmatrix} \hat{i}_E \\ \hat{v}_N \end{pmatrix} = \begin{pmatrix} \hat{E} \\ A_C^T C A_C v_N(0) \end{pmatrix}, \tag{7.5}$$

where we have used $v_C(0) = A_C v_N(0)$. Equation (7.5) can be solved for $\begin{pmatrix} \hat{i}_E \\ \hat{v}_N \end{pmatrix}$ if the coefficient matrix is nonsingular. This can be shown by an indirect proof (Leimkuhler et al. 1991) for $\text{Re}\, z \geq 0$ and $z \neq 0$. If $z = 0$, (7.5) can still be solved if

$$\begin{pmatrix} A_R \\ A_E \end{pmatrix} \quad \text{has linearly independent columns.} \tag{7.6}$$

When solving the Laplace-transformed system (7.4), we find that if the transform of the input function $\hat{E}$ stays bounded as z grows, then $\hat{v}_C$, $\hat{i}_R$, $\hat{i}_E$, $\hat{i}_C$ and $\hat{v}_N$ are also bounded, if we have an index one DAE system. However, if index two occurs, then some components of $\hat{i}_C$ and $\hat{i}_E$ may grow linearly with z even when $\hat{E}$ is bounded (Leimkuhler et al. 1991).

Now let us discuss the initial values for (7.1). The form of the equation suggests that one can assign arbitrary initial values to the state variables v_{C_i}. However, if we study the Laplace transform of (7.1) we see from (7.5) that one may as well assume arbitrary initial values for all nodal voltages v_{N_i}. Not all of them will have any effect on the solution but only $A_C v_N(0)$, that is, those $v_{N_i}(0)$ corresponding to nodes adjacent to capacitors. Notice that although the solution for $\hat{v}_N$ is continuous for any initial values $v_N(0)$ there will, in general, be a discontinuity in the time domain solution because at $t > 0$ the algebraic equations in (7.1) determine $v_N(t)$, which may jump from the arbitrary $v_N(0)$. If one wants to avoid this discontinuity at the initial point, one should at least choose $v_N(0)$ consistent with $E(0)$ by the third equation of (7.1):

$$A_E v_N(0) = E(0).$$

Since A_E has independent rows by (7.3), the number of independent initial values that are used to obtain the solution of (7.5) is

$$\text{rank}\left(\begin{array}{c} A_E \\ A_C \end{array}\right) - n_E.$$

By Proposition 6, this equals n_C for index one and, for index two, it is at most n_C.

The results motivate us to assume all the bounded components are in an α-weighted L_2 space, but the 'index two' variables lie in a larger space, say Y_α. As shown above, the index two variables consist of some components of i_E and i_C. Since it is difficult to identify these particular components, we assume all components of i_E and i_C are elements of Y_α. The α-norm is defined by

$$|x|_\alpha^2 - \frac{1}{2\pi} \int_{-\infty}^{\infty} |\hat{x}(\alpha + i\xi)|^2 d\xi, \qquad \alpha > 0,$$

and the Y_α-norm by

$$|y|_{Y_\alpha}^2 = \frac{1}{2\pi} \int_{-\infty}^{\infty} |\hat{y}(\alpha + i\xi)|^2 \left(1 + \alpha^2 + |\xi|^2\right)^{-1} d\xi,$$

corresponding to the loss of one derivative. We may now take the space X_α to consist of elements $x^T = (v_C^T \; i_R^T \; i_E^T \; i_C^T \; v_N^T)$ where v_C, i_R and $v_N \in L_2^\alpha$ and i_E and $i_C \in Y_\alpha$. The norm in X_α is defined by

$$|x|_\alpha^2 = |v_C|_\alpha^2 + |i_R|_\alpha^2 + |i_E|_{Y_\alpha}^2 + |i_C|_{Y_\alpha}^2 + |v_N|_\alpha^2.$$

Remark 1 In the index one case we can simply take $X_\alpha = L_2^\alpha$ for all components of x, since the i_E and i_C have a special behaviour only in the index two case. Then one should replace $| \cdot |_{Y_\alpha}$ in the preceding norm definition by $| \cdot |_\alpha$.

From the input function $E(t)$ we assume

$$E \in L_2^\alpha. \tag{7.7}$$

Theorem 24 Assume (7.2), (7.3) and (7.7). Then (7.1) has a unique solution in X_α for all $\alpha > 0$.

Remark 2 In the classical treatment of DAEs, smoothness for the high index variables is guaranteed by requiring the input function ($E(t)$ in our application) to have as many derivatives as needed.

7.2. Splittings

For large-scale circuits it is sometimes natural to write (7.1) in the permuted form where the RC-circuit equations are repeated for each subcircuit. One tries to choose the subcircuits in such a way that there are as few connections, or couplings, to other subcircuits as possible. The resulting permuted form of (7.1) will have a block structure

$$\begin{pmatrix} \square & 0 & 0 \\ 0 & \ddots & 0 \\ 0 & 0 & \square \end{pmatrix} \frac{d}{dt} \begin{pmatrix} x_1 \\ \vdots \\ x_k \end{pmatrix} + \begin{pmatrix} \square & \star & \star \\ \star & \ddots & \star \\ \star & \star & \square \end{pmatrix} \begin{pmatrix} x_1 \\ \vdots \\ x_k \end{pmatrix} = \begin{pmatrix} f_1 \\ \vdots \\ f_k \end{pmatrix}, \tag{7.8}$$

where the coefficient matrix of $\dot{x}$ is still diagonal ≥ 0 and the coefficient matrix of x has nonzero elements mainly on its diagonal blocks, but also elsewhere because of the couplings. We will show that if the subcircuits are chosen in such a way that the *subcircuits are coupled solely through resistors*, then the waveform relaxation method converges linearly. One should duplicate each interface branch equation and assign the involved resistor current variable to both subsystems connected through this branch. Applying dynamic block Jacobi iteration to (7.8) after these modifications will always converge, as we will show in the next section.

As mentioned above, our rule is that when splitting (7.1) we only cut through resistors. So the equations that are possibly affected by this relaxation are those that contain i_R:

$$-Ri_R + A_R v_N = 0$$
$$\text{and} \quad A_R^T i_R + A_E^T i_E + A_C^T i_C = 0.$$

For one specific resistor r_i the first one is

$$-r_i i_{r_i} + v_{k_i} - v_{l_i} = 0, \tag{7.9}$$

where r_i is between the nodes k_i and l_i. If we cut through r_i it is not obvious to which subcircuit the above equation and variable i_{r_i} should be assigned. In order to preserve symmetry in the flow of information between subcircuits, it seems that both subcircuits should 'see' r_i in the same way; this means (7.9) should be assigned to both of them. To do that we have to duplicate equation (7.9) and also the variable i_{r_i}, in the sense that we associate equation (7.9) with $i_{r_i}^+$ to the first subcircuit and with $i_{r_i}^-$ to the second:

$$-r_i i_{r_i}^+ + v_{k_i} - v_{l_i} = 0 \quad \text{and} \quad -r_i i_{r_i}^- + v_{k_i} - v_{l_i} = 0 \qquad (7.10)$$

The relaxation is now defined by the iteration we apply to all pairs of equations involving 'cut resistors' as (7.10):

$$-r_i i_{r_i}^{+^k} + v_{k_i}^k - v_{l_i}^{k-1} = 0 \quad \text{and} \quad -r_i i_{r_i}^{-^k} + v_{k_i}^{k-1} - v_{l_i}^k = 0. \qquad (7.11)$$

All components of the unknown x other than those v_i occurring in the 'cut equations' (7.11) are treated at the new iteration index; thus the mentioned v_i are the only coupling terms.

There is of course no duplication of the KCL equations: the number of nodes does not change. The only change is that in the KCL equations corresponding to the nodes k_i and l_i (refer to (7.11)) we must use $i_{r_i}^+$ and $i_{r_i}^-$, respectively.

Next we want to describe the splitting process in equation form.

Let L_R be the set of indices of those resistors that are cut in the relaxation process. Then the resistor current variable i_R is modified so that each i_{r_i}, $i \in L_R$, is replaced by the pair of variables $i_{r_i}^+$ and $i_{r_i}^-$:

$$i_R = (i_{r_1} \ldots i_{r_i} \ldots i_{r_{n_R}})^T \mapsto \bar{i}_R = (i_{r_1} \ldots i_{r_i}^+ i_{r_i}^- \ldots i_{r_{n_R}})^T \text{ for all } i \in L_R.$$

Also, those rows $(A_R)_{i\cdot}$ of the incidence matrix A_R for which i is a member of L_R are duplicated, and the resulting new matrix $\bar{A}_R$ is split in the way suggested by (7.11):

$$\bar{A}_R = A_M - A_N,$$

where A_N has nonzero elements only on the pairs of rows corresponding to the duplicated equations. On those rows the splitting is

$$\bar{A}_R = \begin{pmatrix} \cdots & & \cdots \\ 1 & & -1 \\ 1 & & -1 \\ \cdots & & \cdots \end{pmatrix} = \begin{pmatrix} \cdots & & \cdots \\ 1 & & 0 \\ 0 & & -1 \\ \cdots & & \cdots \end{pmatrix} - \begin{pmatrix} \cdots & & \cdots \\ 0 & & 1 \\ 1 & & 0 \\ \cdots & & \cdots \end{pmatrix}$$

$$=: A_M - A_N.$$

If we modify the diagonal matrix of resistors R in the same way as i_R, that is

$$R = \operatorname{diag}(r_1, \ldots r_i, \ldots r_{n_R}) \mapsto \bar{R} = \operatorname{diag}(r_1, \ldots r_i, r_i, \ldots r_{n_R}),$$

then we obtain the iterative system

$$
\begin{pmatrix} C\frac{d}{dt}v_C^k \\ 0 \\ 0 \\ 0 \\ 0 \end{pmatrix} + \begin{pmatrix} 0 & 0 & 0 & -I & 0 \\ 0 & -\bar{R} & 0 & 0 & A_M \\ 0 & 0 & 0 & 0 & A_E \\ -I & 0 & 0 & 0 & A_C \\ 0 & A_M^T & A_E^T & A_C^T & 0 \end{pmatrix} \begin{pmatrix} v_C \\ \bar{i}_R \\ i_E \\ i_C \\ v_N \end{pmatrix}^k = \begin{pmatrix} 0 \\ A_N v_N^{k-1} \\ E(t) \\ 0 \\ 0 \end{pmatrix},
$$

$$(7.12)$$

with initial values $v_N^k(0) = v_N(0)$. The first observation of (7.12) is that its left-hand side has exactly the same symmetric structure as (7.1). In fact, if we can show that assumption (7.2) holds when A_R has been replaced by A_M, then we can immediately use the results of Section 7.1 to show that (7.12) can be solved in X_α for $\alpha > 0$.

The following lemma is proved in Leimkuhler et al. (1991).

Lemma 2

$$
\text{If} A = \begin{pmatrix} A_R \\ A_E \\ A_C \end{pmatrix} \quad \text{has full rank, then} \quad \begin{pmatrix} A_M \\ A_E \\ A_C \end{pmatrix} \quad \text{has full rank.}
$$

Let $\mathcal{K} : X_\alpha \to X_\alpha$ be the iteration operator of equation (7.12), and let

$$x^k = \mathcal{K}x^{k-1} + \varphi. \tag{7.13}$$

The Laplace transform of the iteration equation is then seen to be

$$\hat{x}^k = K(z)\hat{x}^{k-1} + \hat{\varphi},$$

where

$$
K(z) = \begin{pmatrix} Cz & 0 & 0 & -I & 0 \\ 0 & -\bar{R} & 0 & 0 & A_M \\ 0 & 0 & 0 & 0 & A_E \\ -I & 0 & 0 & 0 & A_C \\ 0 & A_M^T & A_E^T & A_C^T & 0 \end{pmatrix}^{-1} \begin{pmatrix} & & \\ & A_N & \\ & & \end{pmatrix} \tag{7.14}
$$

and $\hat{\varphi}$ is obvious from (7.12) because it does not depend on k.

The operator norm for $\mathcal{K}$ induced by $|\cdot|_\alpha$ is defined by

$$|\mathcal{K}|_\alpha = \sup_{|x|_\alpha=1} |\mathcal{K}x|_\alpha.$$

By Lemma 1 and the preceding analysis we now obtain

Theorem 25 Assume (7.2) and (7.3) and apply the described splitting process. Then $|\mathcal{K}|_\alpha$ is finite for all positive α.

7.3. Spectrum of $\mathcal{K}$

We can see from (7.12) that all the other components of x are in ker $\mathcal{K}$ than v_N. If we define the projection operators

$$\mathcal{P}\begin{pmatrix} v_C \\ \vdots \\ v_N \end{pmatrix} = \begin{pmatrix} 0 \\ \vdots \\ 0 \\ v_N \end{pmatrix} \quad \text{and} \quad P\begin{pmatrix} \hat{v}_C \\ \vdots \\ \hat{v}_N \end{pmatrix} = \begin{pmatrix} 0 \\ \vdots \\ 0 \\ \hat{v}_N \end{pmatrix},$$

we can deduce that the nontrivial part of the spectrum $\sigma(\mathcal{K})$ is actually $\sigma(\mathcal{P}\mathcal{K}\mathcal{P})$, which can be computed as in Leimkuhler et al. (1991):

$$\sigma(\mathcal{P}\mathcal{K}\mathcal{P}) = \mathrm{cl} \bigcup_{\mathrm{Re}\, z \geq \alpha} \sigma(PK(z)P).$$

As in Section 7.1 the equation $\hat{x}^k = K(z)\hat{x}^{k-1}$ can easily be manipulated to yield

$$\begin{pmatrix} 0 & A_E \\ A_E^T & B_M + zB_C \end{pmatrix}\begin{pmatrix} \hat{i}_E^k \\ \hat{v}_N^k \end{pmatrix} = \begin{pmatrix} 0 & 0 \\ 0 & B_N \end{pmatrix}\begin{pmatrix} 0 \\ \hat{v}_N^{k-1} \end{pmatrix}, \qquad (7.15)$$

where

$$B_M = A_M^T \bar{R}^{-1} A_M, \qquad B_C = A_C^T C A_C \quad \text{and} \quad B_N = A_M^T \bar{R}^{-1} A_N.$$

The solution of this equation clearly satisfies $\hat{v}_N^k \in \mathrm{Ker} A_E$. Because of the incidence matrix structure of A_E (each row has at most two nonzero elements ± 1) we can easily eliminate the $\hat{i}_E$ and n_E components of $\hat{v}_N$ from (7.15), ending up with the equation

$$(\tilde{B}_M + z\tilde{B}_C)\hat{v}_P^k = \tilde{B}_N \hat{v}_P^{k-1}, \qquad (7.16)$$

where $\hat{v}_P$ is a part of $\hat{v}_N$ with $N - n_E$ components.

The elimination described in Leimkuhler et al. (1991) is the same as 'shorting the edges' in graph theory: we short all edges containing voltage sources, and simultaneously the nodes adjacent to these edges are pairwise combined.

The computation of the spectrum relies on the following properties of the $\tilde{B}$-matrices. They are all symmetric, $\tilde{B}_N$ is nonnegative and $\tilde{B}_M$ and $\tilde{B}_C$ are positive semidefinite matrices with nonpositive off-diagonal elements. These facts imply that the splitting in iteration (7.16) is a regular splitting of the matrix $\tilde{B}_M + z\tilde{B}_C - \tilde{B}_N$, if $z \in (0, \infty)$. By the convergence theorem for regular splittings, iteration (7.16) then converges for $z \subset (0, \infty)$, since $\tilde{B}_M + z\tilde{B}_C - \tilde{B}_N$ is a nonsingular M-matrix. The spectral radius $\rho((\tilde{B}_M + z\tilde{B}_C)^{-1}\tilde{B}_N) = r < 1$ for $z \in \mathbb{R}_+$, and, by the Perron–Frobenius theorem, r is also an eigenvalue. For nonreal z with $\mathrm{Re}\, z > 0$, taking quadratic forms in the eigenvalue equation provides

$$(\tilde{B}_M + z\tilde{B}_C)^{-1}\tilde{B}_N x = \lambda x,$$

that is, following Leimkuhler et al. (1991),

$$|\text{Re}\,\frac{1}{\lambda}| \geq \inf_{|x|=1} \frac{x^*\tilde{B}_M x + \text{Re}\,zx^*\tilde{B}_C x}{|x^*\tilde{B}_N x|}.$$

The vector giving the minimum can be directly computed and is, of course, an eigenvector, so that

$$|\text{Re}\,\frac{1}{\lambda}| \geq \frac{1}{r} > 1.$$

This inequality can be restated as

$$(\text{Re}\,\lambda \mp \frac{r}{2})^2 + (\text{Im}\,\lambda)^2 \leq (\frac{r}{2})^2,$$

where the negative sign is used for $\text{Re}\,\lambda > 0$ and the positive sign for $\text{Re}\,\lambda < 0$. So the spectrum of $PK(z)P$ lies in the closed circles of Figure 6 for $\text{Re}\,z \geq \alpha > 0$.

Theorem 26 Let $\alpha > 0$. Assume (7.2), (7.3) and (7.7), and apply the described splitting process that only allows cutting through resistors. Then $\sigma(\mathcal{K})$ lies in the set $\mathcal{D}_\alpha$ of Figure 6. In particular, $\rho(\mathcal{K}) < 1$ and the iteration (7.12) converges in X_α.

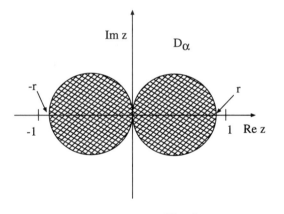

Fig. 6.

As mentioned in Section 7.1, for those circuits satisfying (7.6), the Laplace transform of (7.1) may also be boundedly solved for $z = 0$. Since

$$(7.6) \text{ implies } \begin{pmatrix} A_M \\ A_E \end{pmatrix} \text{ has linearly independent columns,}$$

this means that (7.12) can also be solved in the space X_0 without exponential weighting. In particular, for an index one system, $\mathcal{K}$ is a continuous operator in the ordinary L_2-space.

Theorem 27 Assume (7.2), (7.3), (7.6) and (7.7), and apply the described splitting process that only allows cutting through resistors. Then $\mathcal{K}$ is continuous in X_0 and $\rho(\mathcal{K}) < 1$, that is, the iteration (7.12) converges in X_0.

As stated in the beginning of Section 7.2 we can always permute the circuit equations and variables to a block form, where the blocks correspond to different subsystems. In that formulation, our iteration scheme is dynamic block Jacobi iteration and the corresponding iteration matrix clearly has zero trace. It has the same eigenvalues as $PK(z)P$ defined by (7.12), so we deduce that the trace of $K(z)$ vanishes.

REFERENCES

B. Aupetit (1991), *A Primer on Spectral Theory*, Springer, Berlin.

W. K. Chen (1976), *Applied Graph Theory, graphs and electrical networks*, North-Holland, Amsterdam.

T. Eirola (1992), 'Sobolev characterization of solutions of dilation equations', *SIAM J. Math. Anal.* **23**(4), 1015–1030.

F. R. Gantmacher (1959), *Matrizenrechnung II*, VEB Deutscher Verlag der Wissenschaften, Berlin.

J. Hadamard (1893), 'Etude sur les propriétés des fonctions entières et en particulier d'une fonction considérée par Riemann', *J. de Math. Pures et Appl.* **9**, 171–215.

R. Jeltsch and B. Pohl (1995), 'Waveform relaxation with overlapping splittings', *SIAM J. Sci. Comput.* **16**(1), 40–49.

F. Juang (1990), Waveform Methods for Ordinary Differential Equations, PhD thesis, University of Illinois at Urbana-Champaign, Dept of Computer Science. Report No. UIUCDCS-R-90-1563.

B. Leimkuhler, U. Miekkala and O. Nevanlinna (1991), 'Waveform relaxation for linear RC circuits', *Impact of Computing in Science and Engineering* **3**, 123–145.

E. Lelarasmee, A. Ruehli and A. Sangiovanni-Vincentelli (1982), 'The waveform relaxation method for time-domain analysis of large scale integrated circuits', *IEEE Trans. CAD* **1**(3), 131–145.

E. Lindelöf (1894), 'Sur l'application des méthodes d'approximations successives à l'Etude des intégrales réelles des Equations différentielles ordinaires', *J. de Math. Pures et Appl., 4e Série* **10**, 117–128.

C. Lubich (1992), 'Chebyshev acceleration of Picard–Lindelöf iteration', *BIT* **32**, 535–538.

C. Lubich and A. Ostermann (1987), 'Multigrid dynamic iteration for parabolic equations', *BIT* **27**, 216–234.

A. Lumsdaine and J. White (1995), 'Accelerating waveform relaxation methods with application to parallel semiconductor device simulation', *Numerical functional analysis and optimization* **16**(3,4), 395–414.

A. Lumsdaine and D. Wu (1995), Spectra and pseudospectra of waveform relaxation operators, Technical Report CSE-TR-95-14, Department of Computer Science and Engineering, University of Notre Dame, IN.

J. W. Manke, B. Dembart, M. A. Epton, A. M. Erisman, P. Lu, R. F. Sincovec and E. L. Yip (1979), *Solvability of Large Scale Descriptor Systems*, Boeing Computer Services Company, Seattle, WA.

U. Miekkala (1989), 'Dynamic iteration methods applied to linear DAE systems', *J. Comp. Appl. Math.* **25**, 133–151.

U. Miekkala (1991), Theory for iterative solution of large dynamical systems using parallel computations, PhD thesis, Helsinki University of Technology.

U. Miekkala (1996), Remarks on WR method with overlapping splittings. In preparation.

U. Miekkala and O. Nevanlinna (1987a), 'Convergence of dynamic iteration methods for initial value problems', *SIAM J. Sci. Stat. Comput.* **8**(4), 459–482.

U. Miekkala and O. Nevanlinna (1987b), 'Sets of convergence and stability regions', *BIT* **27**, 554–584.

U. Miekkala and O. Nevanlinna (1992), 'Quasinilpotency of the operators in Picard–Lindelöf iteration', *Numer. Funct. Anal. and Optimiz.* **13**(1,2), 203–221.

U. Miekkala, O. Nevanlinna and A. Ruehli (1990), Convergence and circuit partitioning aspects for waveform relaxation, in *Proceedings of the Fifth Distributed Memory Computing Conference, Charleston, South Carolina* (D. Walker and Q. Stout, eds), IEEE Computer Society Press, Los Alamitos, CA, pp. 605–611.

O. Nevanlinna (1989a), A note on Picard–Lindelöf iteration, in *Numerical Methods for Ordinary Differential Equations, Proceedings of the Workshop held in L'Aquila (Italy), Sept. 16-18, 1987. Vol. 1386 of Lecture Notes in Mathematics* (A. Bellen, C. W. Gear and E. Russo, eds), Springer.

O. Nevanlinna (1989b), 'Remarks on Picard–Lindelöf iteration, PART I', *BIT* **29**, 328–346.

O. Nevanlinna (1989c), 'Remarks on Picard Lindelöf iteration, PART II', *BIT* **29**, 535–562.

O. Nevanlinna (1990a), 'Linear acceleration of Picard–Lindelöf iteration', *Numer. Math.* **57**, 147–156.

O. Nevanlinna (1990b), 'Power bounded prolongations and Picard–Lindelöf iteration', *Numer. Math.* **58**, 479–501.

O. Nevanlinna (1991), Waveform relaxation always converges for RC-circuits, in *Proc. of NASECOD VII, held in April 8-12, 1991, Colorado*, Front Range Press, Colorado.

O. Nevanlinna (1993), *Convergence of iterations for linear equations*, Lectures in Mathematics, ETH Zürich, Birkhäuser, Basel.

O. Nevanlinna and F. Odeh (1987), 'Remarks on the convergence of waveform relaxation method', *Numer. Funct. Anal. and Optimiz.* **9**(3,4), 435–445.

E. Picard (1893), 'Sur l'application des méthodes d'approximations successives à l'étude de certaines équations différentielles ordinaires', *J. Math. Pures. Appl., 4e série* **9**, 217–271.

O. Piirilä (1993), 'Questions and notions related to quasialgebraicity in Banach algebras', *Annales Academia Scientiarum Fennica, Series A, Mathematica Dissertationes.* Helsinki.

M. Reichelt, J. White and J. Allen (1995), 'Optimal convolution SOR acceleration of waveform relaxation with application to parallel simulation of semiconductor devices', *SIAM J. Sci. Comput.* **16**(5), 1137–1158.

R. Skeel (1989), 'Waveform iteration and the shifted Picard splitting', *SIAM J. Sci. Stat. Comput.* **10**(4), 756–776.

M. N. Spijker (1991), 'On a conjecture by LeVeque and Trefethen related to the Kreiss matrix theorem', *BIT* **31**, 551–555.

L. N. Trefethen (1992), Pseudospectra and matrices, in *Numerical Analysis* (D. F. Griffiths and G. A. Watson, eds), Longman, Harlow, UK, pp. 234–266.

M. M. Vainberg and V. A. Trenogin (1974), *Theory of branching of solutions of non-linear equations*, Noordhoff International Publishing, Leyden.

S. Vandewalle (1992), The Parallel Solution of Parabolic Partial Differential Equations by Multigrid Waveform Relaxation Methods, PhD thesis, Katholieke Universiteit Leuven, Belgium.

S. Vandewalle (1993), *Parallel Multigrid Waveform Relaxation for Parabolic Problems*, B.G. Teubner, Stuttgart.

J. White and A. Sangiovanni-Vincentelli (1987), *Relaxation Techniques for the Simulation of VLSI Circuits*, Kluwer, Boston.

Acta Numerica (1996), *pp.* 309–395

Theory, algorithms, and applications of level set methods for propagating interfaces

James A. Sethian *

Department of Mathematics
University of California
Berkeley, CA 94720, USA
E-mail: sethian@math.berkeley.edu

We review recent work on level set methods for following the evolution of complex interfaces. These techniques are based on solving initial value partial differential equations for level set functions, using techniques borrowed from hyperbolic conservation laws. Topological changes, corner and cusp development, and accurate determination of geometric properties such as curvature and normal direction are naturally obtained in this setting. The methodology results in robust, accurate, and efficient numerical algorithms for propagating interfaces in highly complex settings. We review the basic theory and approximations, describe a hierarchy of fast methods, including an extremely fast marching level set scheme for monotonically advancing fronts, based on a stationary formulation of the problem, and discuss extensions to multiple interfaces and triple points. Finally, we demonstrate the technique applied to a series of examples from geometry, material science and computer vision, including mean curvature flow, minimal surfaces, grid generation, fluid mechanics, combustion, image processing, computer vision, and etching, deposition and lithography in the microfabrication of electronic components.

* Supported in part by the Applied Mathematics Subprogram of the Office of Energy Research under DE-AC03-76SF00098, and the National Science Foundation DARPA under grant DMS-8919074.

CONTENTS

1 Introduction   311

PART I: LEVEL SET FORMULATIONS
2 Theory of front evolution   312
3 The time-dependent level set formulation   321
4 The stationary level set formulation   325

PART II: NUMERICAL APPROXIMATION
5 Traditional techniques for tracking interfaces   326
6 A first attempt at constructing an approximation
   to the gradient   327
7 Schemes from hyperbolic conservation laws   330
8 Approximations to the time-dependent level set
   equation   331
9 First- and second-order schemes for convex speed
   functions   333
10 First- and second-order schemes for non-convex
   speed functions   335
11 Approximations to curvature and normals   336
12 Initialization and boundary conditions   337

PART III: EXTENSIONS
13 A hierarchy of fast level set methods   337
14 Additional complexities   343

PART IV: A NEW FAST METHOD
15 A fast marching level set method   351
16 Other speed functions: when does this method
   work?   356
17 Some clarifying comments   357

PART V: APPLICATIONS
18 Geometry   358
19 Combustion, crystal growth, and two-fluid flow   368
20 Two-phase flow   373
21 Constrained problems: minimal surfaces and shape
   recovery   375
22 Applications of the fast marching level set method   379
23 A final example: etching and deposition for the
   microfabrication of semiconductor devices   382
24 Other work   387
References   389

1. Introduction

Propagating interfaces occur in a wide variety of settings. As physical entities, they include ocean waves, burning flames, and material boundaries. Less obvious boundaries are equally important, and include shapes against backgrounds, hand-written characters, and iso-intensity contours in images.

The goal of this paper is to describe some recent work on level set methods, which attempts to unify these problems and provide a general framework for modelling the evolution of boundaries. Our aim is to review a collection of state-of-the-art details of computational techniques for tracking moving interfaces, and to give some sense of the flavour and breadth of applications.

Level set methods are numerical techniques that offer remarkably powerful tools for understanding, analysing, and computing interface motion in a host of settings. At their core, they rely on a fundamental shift in how one views moving boundaries; rethinking the Lagrangian geometric perspective and exchanging it for an Eulerian, initial value partial differential equation perspective. Five clear advantages result from this new view of propagating interfaces.

1   From a theoretical/mathematical point of view, the real complexities of front motion are illuminated, in particular, the role of singularities, weak solutions, shock formation, entropy conditions and topological change in the evolving interface.

2   From a numerical perspective, natural and accurate ways of computing delicate quantities emerge, including the ability to build high-order advection schemes, compute local curvature in two and three dimensions, track sharp corners and cusps, and handle subtle topological changes of merger and breakage.

3   From an implementation point of view, since the approach is based on an initial value partial differential equation, robust schemes result from numerical parameters set at the beginning of the computation. The error is thus controlled by

- the order of the numerical method
- the grid spacing Δh
- the time step Δt.

4   Computational adaptivity, both in meshing and in computational labour, is possible, as is a clear path to parallelism.

5   For monotonically advancing fronts obeying certain speed laws, we introduce exceptionally fast methods based on narrow band techniques and sorting algorithms.

In this paper, we survey an illustrative subset of past and current applications of level set methods. By no means is this an exhaustive review. A large body of work has been reluctantly skipped in the effort to keep this paper of reasonable length. The interested reader is referred to the many references given throughout the text.

PART I: LEVEL SET FORMULATIONS

2. Theory of front evolution

Consider a boundary, either a curve in two dimensions or a surface in three dimensions, separating one region from another. Imagine that this curve/surface moves in its normal direction with a known speed function F. Our goal is to track the motion of this interface as it evolves. We are only concerned with the motion of the interface in its normal direction: throughout, we shall ignore tangential motions of the interface.

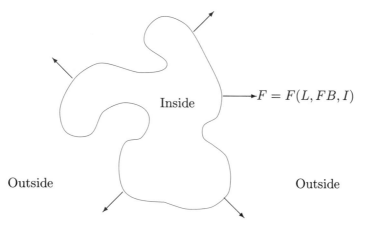

Fig. 1. Curve propagating with speed F in normal direction.

The speed function F can be thought of as depending on three types of arguments, namely

$$F = F(L, G, I), \qquad (2.1)$$

where

L = Local properties of the front are those determined by local geometric information, such as curvature and normal direction.

G = Global properties of the front such as integrals along the front and associated differential equations, are those whose solutions depend on the shape and position of the front. For example, suppose the interface is a source of heat affecting diffusion on either side of the interface, which in turn influences the motion of the interface. This would be characterized as global argument.

I = Independent properties are those that are independent of the shape of the front, such as an underlying fluid velocity which passively transports the front.

Much of the challenge in interface problems comes from producing an adequate model for the speed function F; this is a separate issue independent of the goal of an accurate scheme for advancing the interface based on the model for F. In this section, we assume that the speed function F is known. In Part V, we discuss the development of models for a collection of applications.

Our first goal is to develop the necessary theory to understand the interplay between the speed function F and the shape of the interface. For ease of discussion, we now turn to the simplest case of a closed curve propagating in the plane.

2.1. Fundamental formulation

Let γ be a smooth, closed initial curve in R^2, and let $\gamma(t)$ be the one-parameter family of curves generated by moving $\gamma = \gamma(t = 0)$ along its normal vector field with speed F. Here, F is the given scalar function. Thus, we have that $\vec{n} \cdot \vec{x}_t = F$, where $\vec{x}$ is the position vector of the curve, t is time, and $\vec{n}$ is the unit normal to the curve.

As a first attempt, a natural approach is to consider a parametrized description of the motion. We further restrict ourselves and imagine that the speed function F depends only on the local curvature κ of the curve, that is, $F = F(\kappa)$. Thus, we let the position vector $\vec{x}(s,t)$ parametrize γ at time t. Here, $0 \leq s \leq S$, and we assume periodic boundary conditions $\vec{x}(0,t) = \vec{x}(S,t)$. The curve is parametrized so that the interior is on the left in the direction of increasing s (see Fig. 2). Let $\vec{n}(s,t)$ be the parametrization of the outward normal and $\kappa(s,t)$ be the parametrization of the curvature. The equations of motion can then be written in terms of individual components $\vec{x} = (x,y)$ as

$$x_t = \frac{y_s F\left(\frac{y_{ss}x_s - x_{ss}y_s}{(x_s^2 + y_s^2)^{3/2}}\right)}{(x_s^2 + y_s^2)^{1/2}}, \qquad y_t = -\frac{x_s F\left(\frac{y_{ss}x_s - x_{ss}y_s}{(x_s^2 + y_s^2)^{3/2}}\right)}{(x_s^2 + y_s^2)^{1/2}}, \qquad (2.2)$$

where we have used the parametrized expression for the curvature $\kappa = (y_{ss}x_s - x_{ss}y_s)(x_s^2 + y_s^2)^{-3/2}$ inside the speed function $F(\kappa)$. This is a 'Lagrangian' representation because the range of $(x(s,t), y(s,t))$ describes the moving front.

2.2. Total variation: stability and the growth of oscillations

What happens to oscillations in the initial curve as it moves? We summarize the argument in Sethian (1985) showing that the decay of oscillations depends only on the sign of F_κ at $\kappa = 0$. The metric $g(s,t)$, which measures the

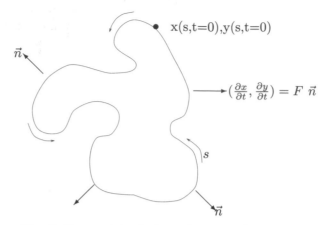

Fig. 2. Parametrized view of propagating curve.

'stretch' of the parametrization, is given by $g(s,t) = (x_s^2 + y_s^2)^{1/2}$. Define the total oscillation (also known as the total variation) in the front

$$Var(t) = \int_0^S |\kappa(s,t)|g(s,t)\,\mathrm{d}s. \tag{2.3}$$

This measures the amount of 'wrinkling'. Our goal is to find out if this wrinkling increases or decreases as the front evolves (see Fig. 3).

Differentiation of both the curvature and the metric with respect to time, together with substitution from equation (2.2) produces the corresponding evolution equations for the metric and curvature, namely

$$\kappa_t = -g^{-1}(F_s g^{-1})_s - \kappa^2 F \tag{2.4}$$

$$g_t = g\kappa F \tag{2.5}$$

(Here, g^{-1} is $1/g$, not the inverse). Now, suppose we have a non-convex initial curve moving with speed $F(\kappa)$, and suppose the moving curve stays smooth. By evaluating the time change of the total variation in the solution, we have the following (see Sethian (1985)).

Theorem　Consider a front moving along its normal vector field with speed $F(\kappa)$, as in equation (2.2). Assume that the initial curve $\gamma(0)$ is smooth and non-convex, so that $\kappa(s,0)$ changes sign. Assume that F is twice differentiable, and that $\kappa(s,t)$ is twice differentiable for $0 \le s \le S$ and $0 \le t \le T$. Then, for $0 \le t \le T$,

• 　if $F_\kappa \le 0$ $(F_\kappa \ge 0)$ wherever $\kappa = 0$, thcn

$$\frac{d\,\mathrm{Var}(t)}{dt} \le 0 \qquad (\frac{d\,\mathrm{Var}(t)}{dt} \ge 0); \tag{2.6}$$

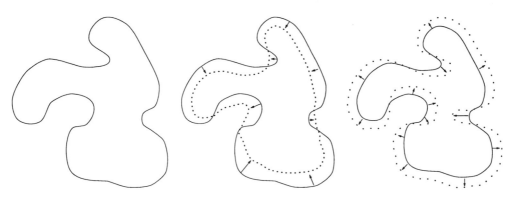

(a) Original curve   (b) Decrease in variation   (c) Increase in variation

Fig. 3. Change in variation.

- if $F_\kappa < 0$ $(F_\kappa > 0)$ and $\kappa_s \neq 0$ wherever $\kappa = 0$, then

$$\frac{d\,\mathrm{Var}(t)}{dt} < 0 \qquad (\frac{d\,\mathrm{Var}(t)}{dt} > 0). \qquad (2.7)$$

Remarks The theorem states that if $F_\kappa < 0$ wherever $\kappa = 0$, then the total variation decreases as the front moves and the front 'smooths out', that is, the energy of the front dissipates. The front is assumed to remain smooth in the interval $0 \leq t \leq T$ (the curvature is assumed to be twice differentiable). (In the next section, we discuss what happens if the front ceases to be smooth and develops a corner.) In the special case when $\gamma(t)$ is convex for all t, the theorem is trivial, since $\mathrm{Var}(t) = \int_0^S \kappa g \, ds = 2\pi$. The proof may be found in Sethian (1985).

Two important cases can be easily checked. A speed function $F(\kappa) = 1 - \epsilon\kappa$ for ϵ positive has derivative $F_\kappa = -\epsilon$, and hence the total variation decays. Conversely, a speed function of the form $F(\kappa) = 1 + \epsilon\kappa$ yields a positive speed derivative, and hence oscillations grow. We shall see that the sign of the curvature term in this case corresponds to the backwards heat equation, and hence must be unstable.

2.3. The role of entropy conditions and weak solutions

The above theorem assumes that the front stays smooth and differentiable. In many cases of evolving fronts, differentiability is soon lost. For example, consider the periodic initial cosine curve

$$\gamma(0) = (-s, [1 + \cos 2\pi s]/2) \qquad (2.8)$$

propagating with speed $F(\kappa) = 1$. The exact solution to this problem at time t may be easily constructed by advancing each point of the front in its normal

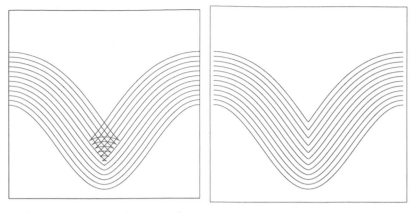

(a) Swallowtail ($F = 1.0$)   (b) Entropy solution ($F = 1.0$)

Fig. 4. Cosine curve propagating with unit speed.

direction a distance t. In fact, in terms of our parametrization of the front, the solution is given by

$$x(s,t) = \frac{y_s(s, t = 0)}{(x_s^2(s, t = 0) + y_s^2(s, t = 0))^{1/2}} t + x(s, t = 0), \qquad (2.9)$$

$$y(s,t) = -\frac{x_s(s, t = 0)}{(x_s^2(s, t = 0) + y_s^2(s, t = 0))^{1/2}} t + y(s, t = 0). \qquad (2.10)$$

In Fig. 4, the solution is given for this propagating cosine curve.

It is clear that the front develops a sharp corner, known as a *shock*, in finite time; however, once this corner develops, it is not at all clear how to construct the normal at the corner and continue the evolution, since the derivative is not defined there. Thus, beyond the formation of the discontinuity in the derivative, we need a *weak solution*, so-called because the solution weakly satisfies the definition of differentiability.

How shall we continue the solution beyond the formation of the singularity in the curvature corresponding to the corner in the front? The correct answer depends on the nature of the interface under discussion. If we regard the interface as a geometric curve evolving under the prescribed speed function, then one possible weak solution is the 'swallowtail' solution formed by letting the front pass through itself; this is the solution shown in Fig. 4a. We note that this solution is in fact the one given by equations (2.9), (2.10); the lack of differentiability at the centre point does not destroy the solution, since we have written the solution in terms of the initial data.

However, if we regard the moving curve as an interface separating two regions, the front at time t should consist of only the set of all points located a distance t from the initial curve. (This is known as the Huygens' principle

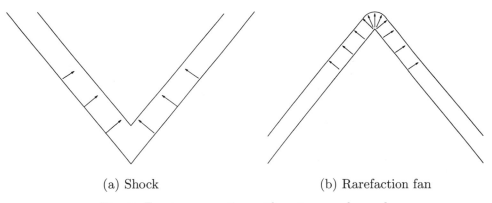

(a) Shock   (b) Rarefaction fan

Fig. 5. Front propagating with unit normal speed.

construction.) Roughly speaking, we want to remove the 'tail' from the 'swallowtail'. In Fig. 4b, we show this alternate weak solution. Another way to characterize this weak solution is via the following 'entropy condition' posed by Sethian (1982, 1985): if the front is viewed as a burning flame, then *once a particle is burnt it stays burnt*. Careful adherence to this stipulation produces the Huygens' principle construction.

What does this 'entropy condition' have to do with the notion of 'entropy'? An intuitive answer is that an entropy condition stipulates that no new information can be created during the evolution of the problem. Once an entropy condition is invoked, some information about the initial data is lost. Indeed, our entropy condition says that once a particle is burnt, it stays burnt, that is, once a corner has developed, the solution is no longer reversible. The problem cannot be run 'backwards' in time; if we try to do so, we will not retrieve the initial data. Thus, some information about the solution is forever lost.

As further illustration, we consider the case of a V-shaped front propagating normal to itself with unit speed ($F = 1$). In Fig. 5a, the point of the front is downwards: as the front moves inwards with unit speed, a shock develops as the front pinches off, and an entropy condition is required to select the correct solution to stop the solution from being multiple-valued. Conversely, in Fig. 5b, the point of the front is upwards: in this case the unit normal speed results in a circular fan, which connects the left state with slope $+1$ to the right state, which has slope -1.

It is important to summarize a key point in the above discussion. Our choice of weak solution given by our entropy condition rests on the perspective that the front separates two regions, and the assumption that we are interested in tracking the progress of one region into the other.

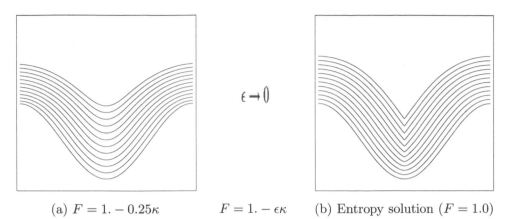

(a) $F = 1. - 0.25\kappa$ $F = 1. - \epsilon\kappa$ (b) Entropy solution ($F = 1.0$)

Fig. 6. Entropy solution is the limit of viscous solution.

2.4. Effects of curvature: the viscous limit and the link to hyperbolic conservation laws

Consider now a speed function of the form $F = 1 - \epsilon\kappa$, where ϵ is a constant. The modifying effects of curvature on the former constant speed law are profound, and in fact pave the way towards constructing accurate numerical schemes that adhere to the correct entropy condition.

Following Sethian (1985), the curvature evolution equation given by (2.4) can be rewritten in terms of arclength, namely

$$\kappa_t = \epsilon\kappa_{\alpha\alpha} + \epsilon\kappa^3 - \kappa^2, \tag{2.11}$$

where the second derivative of the curvature κ is taken with respect to arclength α. This is a reaction-diffusion equation; the drive toward singularities due to the reaction term ($\epsilon\kappa^3 - \kappa^2$) is balanced by the smoothing effect of the diffusion term ($\epsilon\kappa_{\alpha\alpha}$). Indeed, with $\epsilon = 0$, we have a pure reaction equation $\kappa_t = -\kappa^2$, and the developing corner can be seen in the exact solution $\kappa(s,t) = \kappa(s,0)/(1 + t\kappa(s,0))$, which is singular in finite t if the initial curvature is anywhere negative. Thus, as shown, corners can form in the moving curve when $\epsilon = 0$.

Consider again the cosine front given in equation (2.8) and the speed function $F(\kappa) = 1 - \epsilon\kappa$, $\epsilon > 0$. As the front moves, the troughs at $s = n + 1/2, n = 0, \pm1, \pm2, ...$ are sharpened by the negative reaction term (because $\kappa < 0$ at such points) and smoothed by the positive diffusion term (see Fig. 6a). For $\epsilon > 0$, it can be shown (see Sethian (1985) and Osher and Sethian (1988)) that the moving front stays C^∞. The entropy-satisfying solution to this problem when $F = 1$ from Fig. 4b is shown in Fig. 6b.

The central observation, and key to the level set approach, is the following link.

Consider the above propagating cosine curve and the two solutions:

- $X^\epsilon_{\text{curvature}}(t)$, obtained by evolving the initial front with $F_\epsilon = 1 - \epsilon\kappa$
- $X_{\text{constant}}(t)$, obtained with speed function $F = 1$ and the entropy condition.

Then, at any time T,

$$\lim_{\epsilon \to 0} X^\epsilon_{\text{curvature}}(T) = X_{\text{constant}}(T). \qquad (2.12)$$

Thus, the limit of motion with curvature, known as the 'viscous limit', is the entropy solution for the constant speed case, see Sethian (1985).

Why is this known as the 'viscous limit', or, more accurately, what does this have to do with viscosity? In order to see why viscosity is an appropriate name, we turn to the link between propagating fronts and hyperbolic conservation laws.

An equation of the form

$$u_t + |G(u)|_x = 0 \qquad (2.13)$$

is known as a hyperbolic conservation law. A simple example is Burger's equation, given by

$$u_t + uu_x = 0, \qquad (2.14)$$

which describes the motion of a compressible fluid in one dimension. The solution to this equation can develop discontinuities, known as 'shocks', where the fluid undergoes a sudden expansion or compression. These shocks (for example, a sonic boom) can arise from arbitrarily smooth initial data; they are a function of the equation itself. However, if one includes the effects of fluid viscosity, the equation includes a right-hand side, namely

$$u_t + uu_x = \epsilon u_{xx}. \qquad (2.15)$$

This second derivative on the right-hand side acts like a smoothing term and stops the development of such shocks; it can be shown that the solutions must remain smooth for all time.

What does this have to do with our propagating front equation? Consider the initial front given by the graph of $f(x)$, with f periodic on $[0, 1]$, and suppose that the propagating front remains a function for all time. Let ψ be the height of the propagating function at time t, thus $\psi(x, 0) = f(x)$. The tangent at (x, ψ) is $(1, \psi_x)$. Referring to Fig. 7, the change in height V in a unit time is related to the speed F in the normal direction by

$$\frac{V}{F} = \frac{(1 + \psi_x^2)^{1/2}}{1}, \qquad (2.16)$$

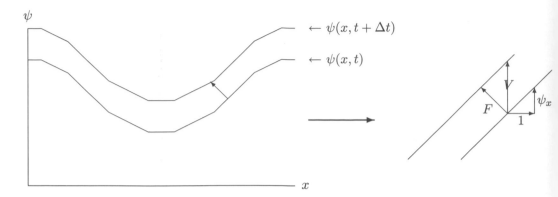

Fig. 7. Variables for propagating graph.

and thus the equation of motion becomes

$$\psi_t = F(1 + \psi_x^2)^{1/2}. \tag{2.17}$$

Using the speed function $F(\kappa) = 1 - \epsilon\kappa$ and the formula $\kappa = -\psi_{xx}/(1+\psi_x^2)^{3/2}$, we get

$$\psi_t - (1 + \psi_x^2)^{1/2} = \epsilon\frac{\psi_{xx}}{1 + \psi_x^2}. \tag{2.18}$$

We first note that this is a partial differential equation with a first-order time and space derivative on the left-hand side, and a parabolic second-order term on the right. Differentiating both sides of this equation yields an evolution equation for the slope $u = d\psi/dx$ of the propagating front, namely

$$u_t + [-(1 + u^2)^{1/2}]_x = \epsilon[\frac{u_x}{1 + u^2}]_x. \tag{2.19}$$

Thus, the derivative of our equation with parabolic right-hand side for the changing height ψ looks like a viscous hyperbolic conservation law with $G(u) = (1 + u^2)^{1/2}$ for the propagating slope u; see Sethian (1987). Hyperbolic conservation laws of the above form have a long history, in fact, our entropy condition is equivalent to the one for propagating shocks in hyperbolic conservation laws. Our goal will be to exploit the theory and technology of numerical solutions of hyperbolic conservation laws to devise accurate numerical schemes to solve the equation of motion for propagating fronts.

Before doing so, however, we have a technical problem. The equation of motion given by equation (2.17) only refers to fronts that remain the graph of a function as they move. The above ideas must be extended to include propagating fronts that are not easily written as functions. This is the time-dependent level set idea introduced by Osher and Sethian (1988).

3. The time-dependent level set formulation

3.1. Formulation

Given a closed $N - 1$ dimensional hypersurface $\Gamma(t)$, we now produce an *Eulerian* formulation for the motion of the hypersurface propagating along its normal direction with speed F, where F can be a function of various arguments, including the curvature, normal direction, *etc.* The main idea of the level set methodology is to embed this propagating interface as the zero level set of a higher-dimensional function ϕ. Let $\phi(x, t = 0)$, where x is a point in R^N, be defined by

$$\phi(x, t = 0) = \pm d, \tag{3.1}$$

where d is the distance from x to $\Gamma(t = 0)$, and the plus (minus) sign is chosen if the point x is outside (inside) the initial hypersurface $\Gamma(t = 0)$. Thus, we have an initial function $\phi(x, t = 0) : R^N \to R$ with the property that

$$\Gamma(t = 0) = (x|\phi(x, t = 0) = 0). \tag{3.2}$$

Our goal is to produce an equation for the evolving function $\phi(x, t)$ that contains the embedded motion of $\Gamma(t)$ as the level set $\phi = 0$. Let $x(t)$ be the path of a point on the propagating front. That is, $x(t = 0)$ is a point on the initial front $\Gamma(t = 0)$, and $x_t \cdot n = F(x(t))$ where n is the normal to the front at $x(t)$. Since we want the zero level set of the evolving function ϕ to always match the propagating hypersurface, we must have

$$\phi(x(t), t) = 0. \tag{3.3}$$

By the chain rule,

$$\phi_t + \nabla \phi(x(t), t) \cdot x'(t) = 0. \tag{3.4}$$

Since $n = \nabla \phi / |\nabla \phi|$, we have the evolution equation for ϕ, namely

$$\phi_t + F|\nabla \phi| = 0 \tag{3.5}$$

$$\phi(x, t = 0) \quad \text{given}. \tag{3.6}$$

This is our time-dependent level set equation. For certain forms of the speed function F, we obtain a standard Hamilton–Jacobi equation.

In Fig. 8, taken from Sethian (1994), we show the outward propagation of an initial curve and the accompanying motion of the level set function ϕ.

In Fig. 8a, we show the initial circle, and in Fig. 8c, we show the circle at a later time. In Fig. 8b, we show the initial position of the level set function ϕ, and in Fig. 8d, we show this function at a later time. We refer to this as an Eulerian formulation because the underlying coordinate system remains fixed.

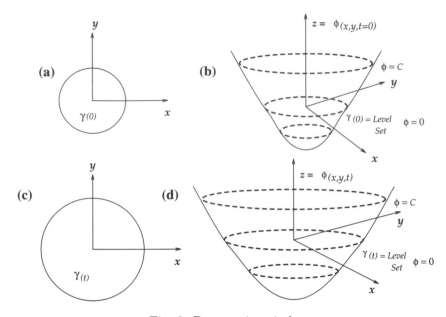

Fig. 8. Propagating circle.

3.2. Advantages

There are four major advantages to this Eulerian level set formulation.

1   The evolving function $\phi(x, t)$ always remains a function as long as F is smooth. However, the level surface $\phi = 0$, and hence the propagating hypersurface $\Gamma(t)$ may change topology, break, merge, and form sharp corners as the function ϕ evolves.

2   The second major advantage concerns numerical approximation. Because $\phi(x, t)$ remains a function as it evolves, we may use a discrete grid in the domain of x and substitute finite difference approximations for the spatial and temporal derivatives. For example, using a uniform mesh of spacing h, with grid nodes (i, j), and employing the standard notation that ϕ_{ij}^n is the approximation to the solution $\phi(ih, jh, n\Delta t)$, where Δt is the time step, we might write

$$\frac{\phi_{ij}^{n+1} - \phi_{ij}^n}{\Delta t} + (F)(\nabla_{ij}\phi_{ij}^n) = 0 \qquad (3.7)$$

Here, we have used forward differences in time, and let $\nabla_{ij}\phi_{ij}^n$ be some appropriate finite difference operator for the spatial derivative. Thus, an explicit finite difference approach is possible.

3   Intrinsic geometric properties of the front may be easily determined from the level set function ϕ. For example, at any point of the front,

the normal vector is given by

$$\vec{n} = \frac{\nabla\phi}{|\nabla\phi|}, \tag{3.8}$$

and the curvature of each level set is easily obtained from the divergence of the gradient of the unit normal vector to the front, that is,

$$\kappa = \nabla \cdot \frac{\nabla\phi}{|\nabla\phi|} = \frac{\phi_{xx}\phi_y^2 - 2\phi_x\phi_y\phi_{xy} + \phi_{yy}\phi_x^2}{(\phi_x^2 + \phi_y^2)^{3/2}}. \tag{3.9}$$

4   Finally, there are no significant changes required to follow fronts in three dimensions. By simply extending the array structures and gradient operators, propagating surfaces are easily handled.

As an example of the application of level set methods, consider once again the problem of a front propagating with speed $F(\kappa) = 1 - \epsilon\kappa$. In Fig. 9, we show two cases of a propagating initial triple sine curve. For ϵ small (Fig. 9a), the troughs sharpen up and result in transverse lines that come too close together. For ϵ large (Fig. 9b), parts of the boundary with high values of positive curvature can initially move downwards, and concave parts of the front can move quickly upwards.

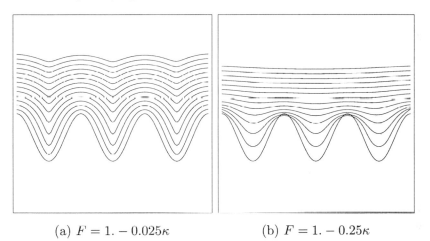

(a) $F = 1. - 0.025\kappa$ (b) $F = 1. - 0.25\kappa$

Fig. 9. Propagating triple sine curve.

3.3. Theoretical aspects of the level set formulation

Numerical techniques for approximating moving fronts have been the focus of much effort in computational physics. At the same time, the theoretical analysis of moving curves and surfaces has been a subject of considerable importance in its own right. The work of Gage (1984), Gage and Hamilton

(1986) and Grayson (1987), discussed later, provided some groundbreaking analysis of flow of curves under the curvature, and the seminal result that a closed curve shrinking under its curvature collapses smoothly to a point.

Along a very different approach, Brakke (1978) applied varifold theory to the problem of a hypersurface moving under its curvature, and in doing so provided a wide-ranging perspective for these problems. His was a general approach, and included problems in which the results were not necessarily smooth. His analysis provided a detailed look at surface curvature evolution problems.

There has been considerable theoretical analysis of the level set approach, its formulation, and its relation to other perspectives on front propagation. The model posed in Sethian (1982), which considered flame propagation as a function of curvature, and introduced an entropy condition for evolving fronts, served as the basis for theoretical analysis by Barles (1985). The full level set methodology of Osher and Sethian (1988) provides a view of surface evolution different from the one provided in the work of Gage, Grayson, and Brakke. First, the embedding of the front as a higher-dimensional function naturally accounts for some of the issues of topological change and corner formation. Second, and more importantly, the transformation of a geometry problem into an initial value partial differential equation means that available technology, including regularity of solutions, viscous solutions of Hamilton–Jacobi equations, and questions of existence and uniqueness, can be applied in this geometrical setting.

Using the above level set approach, Evans and Spruck (1991, 1992a, 1992b, 1995), and Chen, Giga and Goto (1991), Giga and Goto (1992) and Giga, Goto and Ishii (1992) performed detailed analysis of curvature flow in a series of papers. They exploited much of the work on viscosity solutions of partial differential equations developed over the past 15 years (see Lions (1982)), which itself was inspired by the corresponding work applied to hyperbolic conservation laws. These papers examined the regularity of curvature flow equations, pathological cases, and the link between the level set perspective and the varifold approach of Brakke. These papers opened up a series of investigations into further issues; we also refer the interested reader to Ilmanen (1992, 1994) and Evans, Soner and Souganidis (1992).

3.4. Summary

The discussion in Part I may be summarized as follows:

1   A front propagating at a constant speed can form corners as it evolves; at such points, the front is no longer differentiable and a weak solution must be constructed to continue the solution.

2   The correct weak solution, motivated by viewing the front as an evolving interface separating two regions, comes by means of an entropy condition.

3   A front propagating at a speed that depends on its curvature does not form corners and stays smooth for all time. Furthermore, the limit of this motion as the dependence on curvature vanishes is the entropy-satisfying solution obtained for the constant speed case.

4   If the propagating front remains a graph as it moves, there is a direct link between the equation of motion and one-dimensional hyperbolic conservation laws. The role of curvature in a propagating front is analogous to the role of viscosity in equations of viscous compressible fluid flow.

5   By embedding the motion of a curve as the zero level set of a higher-dimensional function, an initial value partial differential equation can be obtained which extends the above to include arbitrary curves and surfaces moving in two and three space dimensions.

4. The stationary level set formulation

In the above level set equation

$$\phi_t + F|\nabla\phi| = 0 \qquad (4.1)$$

the position of the front is given by the zero level set of ϕ at a time t. Suppose we now restrict ourselves to the particular case of a front propagating with a speed F that is either always positive or always negative. In this case, we can convert our level set formulation from a time-dependent partial differential equation to a stationary one, in which time has disappeared. We now describe a stationary level set formulation, which is common in control theory.

To explain this transformation, imagine the two-dimensional case in which the interface is a propagating curve, and suppose we graph the evolving zero level set above the xy plane. That is, let $T(x,y)$ be the time at which the curve crosses the point (x,y). The surface $T(x,y)$ then satisfies the equation

$$|\nabla T|F = 1. \qquad (4.2)$$

Equation (4.2) simply says that the gradient of arrival time surface is inversely proportional to the speed of the front. This is a Hamilton–Jacobi equation, and the recasting of the a front motion problem into a stationary one is common in a variety of applications; see Falcone (1994) and Falcone, Giorgi and Loretti (1994). In the case where the speed function F depends only on position, we get the well-known Eikonal equation. The requirement that the speed function always be positive[*] is so that the crossing time surface $T(x,y)$ is single-valued.

---

[*] or conversely, always negative

To summarize,

- In the time-dependent level set equation, the position of the front Γ at time t is given by the zero level set of ϕ at time t, that is $\Gamma(t) = \{(x, y) : \phi(x, t) = 0\}$.
- In the stationary level set equation, the position of the front Γ is given by the level set $\Gamma(t) = \{(x, y) : T(x, y) = t\}$.

That is, we wish to solve

Time-dependent formulation   **Stationary formulation**

$$\phi_t + F|\nabla\phi| = 0 \qquad\qquad |\nabla T|F = 1$$
$$\text{Front}= \Gamma(t) = \{(x, y) : \phi(x, t) = 0\} \quad \text{Front}= \Gamma(t) = \{(x, y) : T(x, y) = t\}$$
$$\text{Applies for arbitrary } F \qquad\qquad \text{Requires } F > 0.$$

$$(4.3)$$

In both cases, we require an 'entropy-satisfying' approximation to the gradient term. In the next section, we discuss appropriate approximations for this term, leading to schemes for both the time-dependent and stationary level set formulations. Our goal now is to turn to the issue of numerical approximations, and to develop the necessary theory and numerics to approximate accurately the level set initial value partial differential equation.

PART II: NUMERICAL APPROXIMATION

5. Traditional techniques for tracking interfaces

Before discussing the numerical approximation of these level set equations, it is instructive to review briefly more traditional techniques for computing the motion of interfaces.

Marker/string methods In these methods, a discrete parametrized version of the interface boundary is used. In two dimensions, marker particles are used; in three dimensions, a nodal triangularization of the interface is often developed. The positions of the nodes are then updated by determining front information about the normals and curvature from the marker representation. Such representations can be quite accurate; however, limitations exist for complex motions. Firstly, if corners and cusps develop in the evolving front, markers usually form 'swallowtail' solutions, which must be removed through delooping techniques which attempt to enforce an entropy condition inherent in such motion; see Sethian (1985). Second, topological changes are difficult to handle: when regions merge, some markers must be removed. Third, significant instabilities in the front can result, since the underlying marker particle motions represent a weakly ill-posed initial value problem; see Osher and Sethian (1988). Finally, extensions of such methods to three dimensions require additional work.

Cell-based methods In these methods, introduced by Noh and Woodward (1976), the computational domain is divided into a set of cells containing 'volume fractions'. These volume fractions are numbers between 0 and 1, and represent the fraction of each cell containing the physical material. At any time, the front can be reconstructed from these volume fractions. Since their introduction, many elaborate reconstruction techniques have been developed over the years to include pitched slopes and curved surfaces; see Chorin (1980), Hirt and Nicholls (1981) and Puckett (1991). The accompanying accuracy depends on the sophistication of the reconstruction and the so-called 'advection sweeps'. Some of the most elaborate and accurate versions of these schemes to date are due to Puckett; see Puckett (1991). Advantages of such techniques include the ability to easily handle topological changes, design adaptive mesh methods, and extend the results to three dimensions. However, determination of geometric quantities such as normals and curvature can be inaccurate.

Characteristic methods In these methods, 'ray-trace'-like techniques are used. The characteristic equations for the propagating interface are used, and the entropy condition at forming corners (see Sethian (1985)) is formally enforced by constructing the envelope of the evolving characteristics. Such methods handle the looping problems more naturally, but may be complex in three dimensions and require adaptive adding and removing rays, which can cause instabilities and/or oversmoothing.

6. A first attempt at constructing an approximation to the gradient

We now turn to our time-dependent level set equation itself, and attempt to construct a numerical method.

Recall that our goal is to solve the equation given in (3.5) by

$$\phi_t + F|\nabla\phi| = 0, \tag{6.1}$$

$$\phi(x, t = 0) \quad \text{given.} \tag{6.2}$$

The marker particle method discretizes the front. The volume-of-fluid (VOF) method divides the domain space into cells containing fractions of material. The level set method divides the domain up into grid points that discretize the values of the level set function ϕ. Thus, the grid values give the height of a surface above the domain, and if we slice this surface by the xy plane, we extract the zero level set corresponding to the front.

Another way to look at this is to say that each grid point contains the value of the level set function at that point. Thus, there is an entire family of contours, only one of which is the zero level set (see Fig. 10). Rather than move each of the contours in a Lagrangian fashion, we stand at each grid point and update its value to correspond to the motion of the surface, thus producing a new contour value at that grid point.

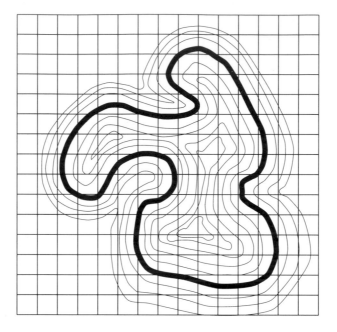

Fig. 10. Dark line is zero level set corresponding to front.

What is the right way to approximate this equation? We shall investigate the most straightforward numerical approach we can think of by studying the simpler case of an evolving curve whose position can always be described as the graph of a function. The equation for this case was given in (2.17) as shown in Fig. 7, namely

$$\psi_t = F(1 + \psi_x^2)^{1/2}. \tag{6.3}$$

Perhaps the most straightforward way of creating an algorithm to approximate the solution to this equation is to replace all spatial derivatives with central differences and the time derivative with a forward difference, just as we did in the Lagrangian case. However, it is easy to see that such an algorithm may not work. Let $F(\kappa) = 1$ and consider the initial value problem

$$\psi_t = (1 + \psi_x^2)^{1/2}, \qquad (6.4)$$

$$\psi(x,0) = f(x) = \left\{ \begin{array}{ll} 1/2 - x & x \le 1/2 \\ x - 1/2 & x > 1/2 \end{array} \right\}. \qquad (6.5)$$

The initial front is a 'V' formed by rays meeting at $(1/2, 0)$. By our entropy condition, the solution at any time t is the set of all points located a distance t from the initial 'V'. To construct a numerical scheme, divide the interval $[0, 1]$ into $2M - 1$ points, and form the central difference approximation to the spatial derivative ψ_x in equation (6.4), namely

$$\frac{\psi_i^{n+1} - \psi_i^n}{\Delta t} = [1 + [\frac{\psi_{i+1}^n - \psi_{i-1}^n}{2\Delta x}]^2]^{1/2} \qquad (6.6)$$

Since $x_M = 1/2$, by symmetry, $\psi_{M+1} = \psi_{M-1}$, thus $\psi_t(1/2, 0) = 1$. However, for all $x \ne 1/2$, ψ_t is correctly calculated to be $\sqrt{2}$, since the graph is linear on either side of the corner and thus the central difference approximation is exact. Note that this has nothing to do with the size of the space step Δx or the time step Δt. *No matter how small we take the numerical parameters, as long as we use an odd number of points, the approximation to ψ_t at $x = 1/2$ gets no better.* It is simply due to the way in which the derivative ψ_x is approximated. In Fig. 11, we show results using this scheme, with the time derivative ψ_t replaced by a forward difference scheme.

Exact solution   Central differences $\Delta t - .005$ Central differences $\Delta t = .0005$

Fig. 11. Central difference approximation to level set equation.

It is easy to see what has gone wrong. In the exact solution, $\psi_t = \sqrt{2}$ for all $x \ne 1/2$. This should also hold at $x = 1/2$ where the slope is not defined; the Huygens construction sets $\psi_t(x = 1/2, t)$ equal to $\lim_{x \to 1/2} \psi_t$. Unfortunately, the central difference approximation chooses a different (and, for our purpose, wrong) limiting solution. It sets the undefined slope ψ_x equal to the average of the left and right slopes. As the calculation progresses, this miscalculation of the slope propagates outwards from the spike as wild oscillations. Eventually, these oscillations cause blowup in the solution.

It is clear that some more care must be taken in formulating an algorithm. What we require are schemes that approximate the gradient term $|\nabla \phi|$ in a way that correctly accounts for the entropy condition. This is the topic of the next section.

7. Schemes from hyperbolic conservation laws

Our schemes are linked to those from hyperbolic conservation laws. As motivation, consider the single scalar hyperbolic conservation law

$$u_t + [G(u)]_x = 0. \tag{7.1}$$

It is well known that discontinuities known as shocks can develop in the solution, even with smooth initial data; see Lax (1970) and LeVeque (1992). These discontinuities occur because of the collision of characteristics, and an appropriate weak solution must be constructed to carry the solution beyond the collision time. The correct 'entropy solution' is obtained by considering the limit of the associated viscous conservation laws $u_t + [G(u)]_x = \epsilon u_{xx}$ as the viscous coefficient ϵ goes to zero.

From a numerical point of view, the equation can be approximated through the construction of appropriate numerical fluxes g such that

$$\frac{u_i^{n+1} - u_i^n}{\Delta t} = -\frac{g(u_i^n, u_{i+1}^n) - g(u_{i-1}^n, u_i^n)}{\Delta x}, \tag{7.2}$$

where we require that $g(u, u) = G(u)$. A wide collection of numerical flux functions are available, such as the Lax–Friedrichs flux, Godunox flux, and TVD schemes; see Colella and Puckett (1994) and LeVeque (1992). The goal in the construction of such flux functions is to make sure that the conservation form of the equation is preserved, make sure the entropy condition is satisfied, and try to give smooth (highly accurate) solutions away from the discontinuities. One of the most straightforward approximate numerical fluxes is the Engquist–Osher scheme (Engquist and Osher 1980), which is given by

$$g_{EO}(u_1, u_2) = G(u_1) + \int_{u_1}^{u_2} \min(\frac{\mathrm{d}G}{\mathrm{d}u}, 0)\, \mathrm{d}u. \tag{7.3}$$

For the Burger's equation in which $G(u) = u^2$, we have the particularly compact representation of this flux function as

$$g_{EO}(u_1, u_2) = (\max(u_1, 0)^2 + \min(u_2, 0)^2). \tag{7.4}$$

This flux function will serve as our core technique for approximating the level set equation.

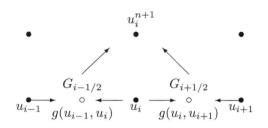

Fig. 12. Update of u through numerical flux function.

8. Approximations to the time-dependent level set equation

In this section, we develop schemes for the level set equation

$$\phi_t + F|\nabla\phi| = 0. \tag{8.1}$$

We begin by writing this equation with a general Hamiltonian H as

$$\phi_t + H(\phi_x, \phi_y, \phi_z) = 0, \tag{8.2}$$

where

$$H(u, v, w) = \sqrt{u^2 + v^2 + w^2}. \tag{8.3}$$

We begin with the one-dimensional version, that is, $\phi_t + H(\phi_x) = 0$, where $H(u) = \sqrt{u^2}$.

From the previous section, we have numerical flux functions for the conservation equation $u_t + [G(u)]_x = 0$, satisfying

$$\frac{u_i^{n+1} - u_i^n}{\Delta t} = -\frac{g(u_i^n, u_{i+1}^n) - g(u_{i-1}^n, u_i^n)}{\Delta x}. \tag{8.4}$$

In terms of our computational grid shown in Fig. 12, the value of G at the point $(i - 1/2)\Delta x$ (called $G_{i-1/2}$) is approximated by the numerical flux function g as

$$G_{i-1/2} = g(u_{i-1}^n, u_i^n). \tag{8.5}$$

Similarly, at the point $i + 1/2$, we have

$$G_{i+1/2} = g(u_i^n, u_{i+1}^n). \tag{8.6}$$

Then from Fig. 12, we see that the right-hand side of equation (7.2) is just the central difference operator applied to the numerical flux function g. As the grid size goes to zero, consistency requires that $g(u, u) = G(u)$.

We are now all set to build a scheme for our level set equation. Let $u = \phi_x$. Then we can write

$$\phi_t + H(u) = 0. \tag{8.7}$$

In terms of our computational grid in Fig. 13, in order to produce ϕ_i^{n+1} we

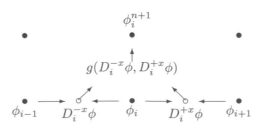

Fig. 13. Update of ϕ through numerical Hamiltonian.

need ϕ_i^n as well as a value for $H(u_i^n)$. Fortunately, an approximate value for $H(u_i^n)$ is *exactly* what is given by our numerical flux function, hence we have

$$H(u) \approx g(u_{i-1/2}, u_{i+1/2}). \tag{8.8}$$

All that remains is to construct values for u in the middle of our computational cells. Since $u = \phi_x$, we can use a central difference approximation in ϕ to construct those values. Thus (see Fig. 13), we have

$$\phi_i^{n+1} = \phi_i^n - \Delta t \; g \left(\frac{\phi_i^n - \phi_{i-1}^n}{\Delta x}, \frac{\phi_{i+1}^n - \phi_i^n}{\Delta x} \right), \tag{8.9}$$

where g is one of the numerical flux functions and, again, we have substituted forward and backward difference operators on ϕ for the values of u at the left and right states.

In the specific case of our one-dimensional level set equation with $H(u) = \sqrt{u^2}$, we can use the EO scheme given in the previous section and, for speed $F = 1$, write

$$\phi_i^{n+1} = \phi_i^n - \Delta t \left(\max(D_i^{-x}, 0)^2 + \min(D_i^{+x}, 0)^2 \right)^{1/2}. \tag{8.10}$$

This is the level set scheme given in Osher and Sethian (1988). As long as the Hamiltonian is symmetric in each of the space dimensions, the above can be replicated in each space variable to construct schemes for two- and three-dimensional front propagation problems.

In general we apply the following philosophy:

1   if the Hamiltonian H is convex, then we use the above scheme
2   if the Hamiltonian H is non-convex, then we use a variant on the Lax–Friedrichs scheme described below.

It is important to point out that far more sophisticated schemes exist than the ones presented here. In the applications of these schemes to hyperbolic problems and shock dynamics, high-order resolution schemes are often necessary (Colella and Puckett 1994), because of the differentiation of the numerical flux function g. However, in our case, because we are solving

(a) Exact solution   (b) Scheme with 20 points   (c) Scheme with 100 points

Fig. 14. Upwind, entropy-satisfying approximations to the level set equation.

$\phi_t + H(u) = 0$ rather than $u_t + [H(u)]_x$, the differentiation is not required. Thus, we have found that for almost all practical purposes, the first- and second-order schemes presented below are adequate.

Before constructing the general schemes, let's return to the example of the propagating curve. Earlier, we attempted to follow the propagation of a simple corner moving with speed $F = 1$. Our attempts to use a central difference approximation failed. In Fig. 14, we show what happens if we use the scheme given in equation (8.10). The exact answer is shown, together with two simulations. The first uses the entropy-satisfying scheme with only 20 points (Fig. 14b), the second (Fig. 14c) with 100 points. In the first approximation, the entropy condition is satisfied, but the corner is somewhat smoothed due to the small number of points used. In the more refined calculation, the corner remains sharp, and the exact solution is very closely approximated.

9. First- and second-order schemes for convex speed functions

Given a convex speed function F (that is, a speed function F such that the resulting Hamiltonian $H = F|\nabla\phi|$ is convex), we can produce the following schemes, first presented in Osher and Sethian (1988). Start with the equation

$$\phi_t + H(\phi_x, \phi_y, \phi_z) = 0, \tag{9.1}$$

and approximate it by

$$\phi_i^{n+1} = \phi_i^n - \Delta t \left(\frac{\phi_{ijk}^n - \phi_{i-1,j,k}^n}{\Delta x}, \frac{\phi_{i+1,j,k}^n - \phi_{i,j,k}^n}{\Delta x}, \right. \tag{9.2}$$
$$\frac{\phi_{ijk}^n - \phi_{i,j-1,k}^n}{\Delta y}, \frac{\phi_{i,j+1,k}^n - \phi_{i,j,k}^n}{\Delta y},$$
$$\left. \frac{\phi_{ijk}^n - \phi_{i,j,k-1}^n}{\Delta z}, \frac{\phi_{i,j,k+1}^n - \phi_{i,j,k}^n}{\Delta z} \right).$$

A multi-dimensional version of this scheme (Osher and Sethian 1988) is then given by

$$g_{EO}(u_1, u_2, v_1, v_2, w_1, w_2) = [\, \max(u_1, 0)^2 + \min(u_2, 0)^2 + \tag{9.3}$$
$$\max(v_1, 0)^2 + \min(v_2, 0)^2 +$$
$$\max(w_1, 0)^2 + \min(w_2, 0)^2]^{1/2}.$$

Thus we have

First-order space convex

$$\phi_{ijk}^{n+1} = \phi_{ijk}^{n} - \Delta t \left(\max(F_{ijk}, 0)\nabla^+ + \min(F_{ijk}, 0)\nabla^- \right) \tag{9.4}$$

where

$$\nabla^+ = \Big(\max(D_{ijk}^{-x}, 0)^2 + \min(D_{ijk}^{+x}, 0)^2 + \tag{9.5}$$
$$\max(D_{ijk}^{-y}, 0)^2 + \min(D_{ijk}^{+y}, 0)^2 +$$
$$\max(D_{ijk}^{-z}, 0)^2 + \min(D_{ijk}^{+z}, 0)^2 \Big)^{1/2},$$

$$\nabla^- = \Big(\max(D_{ijk}^{+x}, 0)^2 + \min(D_{ijk}^{-x}, 0)^2 + \tag{9.6}$$
$$\max(D_{ijk}^{+y}, 0)^2 + \min(D_{ijk}^{-y}, 0)^2 +$$
$$\max(D_{ijk}^{+z}, 0)^2 + \min(D_{ijk}^{-z}, 0)^2 \Big)^{1/2}.$$

Here, we have used a short-hand notation in which $D^{+x}\phi_i^n$ is rewritten as D_i^{+x}, etc.

Second-order space convex

The above schemes can be extended to higher order, using technology from Harten, Engquist, Osher and Chakravarthy (1987). The basic trick is to build a switch that turns itself off whenever a shock is detected; otherwise, it will use a higher-order polynomial approximation of minimal oscillations. These details will not be presented; see Osher and Sethian (1988) for details. The scheme is the same as the above; however, this time ∇^+ and ∇^- are given by

$$\nabla^+ = \Big(\max(A, 0)^2 + \min(B, 0)^2 + \tag{9.7}$$
$$\max(C, 0)^2 + \min(D, 0)^2 +$$
$$\max(E, 0)^2 + \min(F, 0)^2 \Big)^{1/2},$$

$$\nabla^- = \Big(\max(B, 0)^2 + \min(A, 0)^2 + \tag{9.8}$$
$$\max(D, 0)^2 + \min(C, 0)^2 +$$
$$\max(F, 0)^2 + \min(E, 0)^2 \Big)^{1/2},$$

where

$$A = D_{ijk}^{-x} + \frac{\Delta x}{2} m(D_{ijk}^{-x-x}, D_{ijk}^{+x-x}), \quad B = D_{ijk}^{+x} - \frac{\Delta x}{2} m(D_{ijk}^{+x+x}, D_{ijk}^{+x-x}),$$
(9.9)

$$C = D_{ijk}^{-y} + \frac{\Delta y}{2} m(D_{ijk}^{-y-y}, D_{ijk}^{+y-y}), \quad D = D_{ijk}^{+y} - \frac{\Delta y}{2} m(D_{ijk}^{+y+y}, D_{ijk}^{+y-y}),$$
(9.10)

$$E = D_{ijk}^{-z} + \frac{\Delta z}{2} m(D_{ijk}^{-z-z}, D_{ijk}^{+z-z}), \quad F = D_{ijk}^{+z} - \frac{\Delta z}{2} m(D_{ijk}^{+z+z}, D_{ijk}^{+z-z}),$$
(9.11)

and the switch function is given by

$$m(x, y) = \left\{ \begin{array}{ll} \left\{ \begin{array}{ll} x & \text{if } |x| \le |y| \\ y & \text{if } |x| > |y| \end{array} \right\} & xy \ge 0 \\ 0 & xy < 0 \end{array} \right\}.$$
(9.12)

10. First- and second-order schemes for non-convex speed functions

Given a non-convex speed function F (that is, a speed function F for which the resulting Hamiltonian $H = F|\nabla \psi|$ is non-convex), we can follow the work in Osher and Shu (1991), replacing the Hamiltonian $F|\nabla \phi|$ with the Lax–Friedrichs numerical flux function. This produces the following schemes.

First-order space non-convex

$$\phi_{ijk}^{n+1} = \phi_{ijk}^n - \Delta t [H(\frac{D_{ijk}^{-x} + D_{ijk}^{+x}}{2}, \frac{D_{ijk}^{-y} + D_{ijk}^{+y}}{2}, \frac{D_{ijk}^{-z} + D_{ijk}^{+z}}{2}) \quad (10.1)$$
$$\frac{1}{2} \alpha_u (D_{ijk}^{+x} - D_{ijk}^{-x}) - \frac{1}{2} \alpha_v (D_{ijk}^{+y} - D_{ijk}^{-y}) - \frac{1}{2} \alpha_w (D_{ijk}^{+z} - D_{ijk}^{-z})]$$

where α_u (α_v, α_w) is a bound on the partial derivative of the Hamiltonian with respect to the first (second, third) argument, and the non-convex Hamiltonian is a user-defined input function.

Second-order space non-convex

$$\phi_{ijk}^{n+1} = \phi_{ijk}^n - \Delta t [H(\frac{A+B}{2}, \frac{C+D}{2}, \frac{E+F}{2}) \quad (10.2)$$
$$- \frac{1}{2} \alpha_u (B - A) - \frac{1}{2} \alpha_v (D - C) - \frac{1}{2} \alpha_w (F - E)],$$

where A, B, C, D, E, and F are defined as above. For details, see Osher and Shu (1991) and Adalsteinsson and Sethian (1995b, 1995c).

11. Approximations to curvature and normals

As discussed above, one of the main advantages of level set formulations is that geometric qualities of the propagating interface, such as curvature and normal direction, are easily calculated. For example, consider the case of a curve propagating in the plane. The expression for the curvature of the zero level set assigned to the interface itself (as well as all other level sets) is given by

$$\kappa = \nabla \cdot \frac{\nabla \phi}{|\nabla \phi|} = \frac{\phi_{xx}\phi_y^2 - 2\phi_y\phi_x\phi_{xy} + \phi_{yy}\phi_x^2}{(\phi_x^2 + \phi_y^2)^{3/2}}. \tag{11.1}$$

In the case of a surface propagating in three space dimensions, one has many choices for the curvature of the front, including the mean curvature κ_M and the Gaussian curvature κ_G. Both may be conveniently expressed in terms of the level set function ϕ as

$$\kappa_M = \nabla \cdot \frac{\nabla \phi}{|\nabla \phi|} = \frac{\begin{array}{c}(\phi_{yy} + \phi_{zz})\phi_x^2 + (\phi_{xx} + \phi_{zz})\phi_y^2 + (\phi_{xx} + \phi_{yy})\phi_z^2 \\ -2\phi_x\phi_y\phi_{xy} - 2\phi_x\phi_z\phi_{xz} - 2\phi_y\phi_z\phi_{yz}\end{array}}{(\phi_x^2 + \phi_y^2 + \phi_z^2)^{3/2}} \tag{11.2}$$

$$\kappa_G = \frac{\begin{array}{c}\phi_x^2(\phi_{yy}\phi_{zz} - \phi_{yz}^2) + \phi_y^2(\phi_{xx}\phi_{zz} - \phi_{xz}^2) + \phi_z^2(\phi_{xx}\phi_{yy} - \phi_{xy}^2) \\ + 2[\phi_x\phi_y(\phi_{xz}\phi_{yz} - \phi_{xy}\phi_{zz}) + \phi_y\phi_z(\phi_{xy}\phi_{xz} - \phi_{yz}\phi_{xx}) \\ + \phi_x\phi_z(\phi_{xy}\phi_{yz} - \phi_{xz}\phi_{yy})]\end{array}}{(\phi_z^2 + \phi_y^2 + \phi_x^2)^2} \tag{11.3}$$

Construction of the normal itself requires a more sophisticated scheme than simply building the difference approximation to $\nabla \phi$. This is because at corners, the direction of the normal can undergo a jump. This suggests the following technique, introduced by Sethian and Strain (1992). First, the one-sided difference approximations to the unit normal in each possible direction are formed. All four limiting normals are then averaged to produce the approximate normal at the corner. Thus, the normal n_{ij} is formed by first letting

$$n_{ij}^* \equiv \frac{\phi_x, \phi_y}{(\phi_x^2 + \phi_y^2)^{1/2}} \tag{11.4}$$

$$= \frac{(D_{ij}^{+x}, D_{ij}^{+y})}{[(D_{ij}^{+x})^2 + (D_{ij}^{+y})^2]^{1/2}} + \frac{(D_{ij}^{-x}, D_{ij}^{+y})}{[(D_{ij}^{-x})^2 + (D_{ij}^{+y})^2]^{1/2}} \tag{11.5}$$
$$\frac{(D_{ij}^{+x}, D_{ij}^{-y})}{[(D_{ij}^{+x})^2 + (D_{ij}^{-y})^2]^{1/2}} + \frac{(D_{ij}^{-x}, D_{ij}^{-y})}{[(D_{ij}^{-x})^2 + (D_{ij}^{-y})^2]^{1/2}},$$

and then normalizing so that $n_{ij} \equiv n_{ij}^*/|n_{ij}^*|$. If any of the one-sided approximations to $|\nabla\phi|$ is zero, that term is not considered and the weights are adjusted accordingly.

12. Initialization and boundary conditions

The time-dependent level set approach requires an initial function $\phi(x, t = 0)$ with the property that the zero level set of that initial function corresponds to the position of the initial front. The original level set algorithm computed the signed distance from each grid point to the initial front, which is matched to the zero level set. This is an expensive technique. Many other initial functions are possible, including those that are essentially constant except in a narrow band around the front itself.

The use of a finite computational grid means that we must develop boundary conditions. If the speed function F causes the front to expand (such as in the case $F = 1$), upwind schemes will naturally default to outward-flowing one-sided differences at the boundary of the domain. However, in cases of more complex speed functions, mirror boundary conditions usually suffice.

PART III: EXTENSIONS

13. A hierarchy of fast level set methods

The above time-dependent level set method is easily programmed. However, it is not particularly fast, nor does it make efficient use of computational resources. In this section, we consider a sequence of more sophisticated versions of the basic scheme.

13.1. Parallel algorithms

The above method updates *all* the level sets, not just the zero level set corresponding to the front itself. The advantage of this approach is that the data structures and operations are extremely clear, and it is a good starting point for building level set codes.

There are a variety of circumstances in which this approach is desirable. If, in fact, all the level sets are themselves important (such as problems encountered in image processing, which are discussed below), then computation over the entire domain is required. A simple approach is to perform a parallel computation. Since each grid point is updated by a nearest neighbour stencil using only grid points on each side, this technique almost falls under the classification of 'embarrassingly parallel'. In Sethian (1989), a parallel version of the level set method was developed for the Connection Machine CM-2 and CM-5. In the CM-2, nodes are arranged in a hypercube fashion; in the CM-5, nodes are arranged in a fat-tree. The code was written in global CMFortran, and at each grid point CSHIFT operators were used to update the level set function. The operation count per time step reduces to $O(1)$, since in most cases the full grid can be placed into physical memory. A time-explicit second-order space method was used to update the level set equation.

13.2. Adaptive mesh refinement

As a first level of creating a more efficient level set method, an adaptive mesh refinement strategy can be pursued. This is the approach taken by Milne and Sethian (1995), motivated by the adaptive mesh refinement work in Berger and Colella (1989). Adaptivity may be desired in regions where level curves develop high curvature or where speed functions change rapidly. In Fig. 15a, we show mesh cells that are hierarchically refined in response to a parent–child relationship around the zero level set of ϕ. Calculations are performed on both the fine and coarse grid. Grid cell boundaries always lie along coordinate lines, and patches do not overlap; in the scheme presented in Milne and Sethian (1995), no attempt is made to align the refined cells with the front.

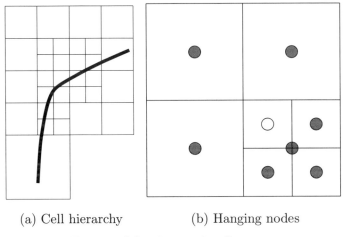

(a) Cell hierarchy   (b) Hanging nodes

Fig. 15. Adaptive mesh refinement.

The data structures for the adaptive mesh refinement are fairly straightforward. However, considerable care must be taken at the interfaces between coarse and fine cells; in particular, the update strategy for ϕ at so-called 'hanging nodes' is subtle. These are nodes at the boundary between two levels of refinement which do not have a full set of nearest neighbours required to update ϕ. To illustrate, in Fig. 15b, we show a two-dimensional adaptive mesh; the goal is to determine an accurate update strategy for the hanging node marked ○.

The strategy for updating ϕ at such points is as follows. Consider the archetypical speed function $F(\kappa) = 1 - \epsilon\kappa$.

- The advection term 1 leads to a hyperbolic equation; here, straightforward interpolation of the updated values of ϕ from the coarse cell grid is used to produce the new value of ϕ at $\circ$. The sophisticated technology in Berger and Colella (1989) is not required, since we are modelling the update according to the numerical flux function g, not the derivative of the numerical flux function as required for hyperbolic conservation laws.
- In the case of the curvature term $-\epsilon\kappa$, the situation is not as straightforward. The parabolic term cannot be approximated through simple interpolation. Interpolation from updated values on the coarse grid to the fine grid was shown to provide poor answers; if this procedure is employed, the boundary between the two levels of refinement acts as a source of noise, and significant error is generated at the boundary. Instead, values from both the coarse and refined grid next to the hanging node are used to construct a least squares solution for ϕ before the update. This solution surface is then formally differentiated to produce the various first and second derivatives in each component direction. These values are then used to produce the update value for ϕ similar to all other nodes. For details, see Milne and Sethian (1995).

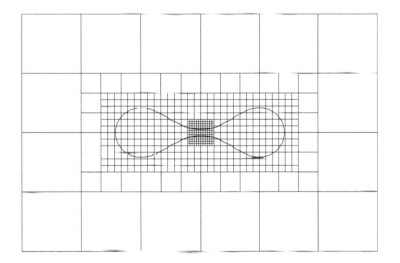

Fig. 16. Two-dimensional slice of adaptive mesh for propagating surface.

As illustration, in Fig. 16, we show a two-dimensional slice of a fully three-dimensional adaptive mesh calculation of a surface collapsing under its mean curvature. As discussed in a later section, the dumbbell neck pinches off under such a configuration, due to the high positive value of the principal axis of curvature.

13.3. Narrow banding and fast methods

An alternative to the above techniques is particularly valuable when one wants to track a specific front, namely the one associated with the zero level set. There are several disadvantages with the 'full-matrix' approach given above.

Speed Performing calculations over the entire computational domain requires $O(N^2)$ operations in two dimensions, and $O(N^3)$ operations in three dimensions, where N is the number of grid points along a side. As an alternative, an efficient modification is to perform work only in a neighbourhood of the zero level set; this is known as the *narrow band approach*. In this case, the operation count in three dimensions drops to $O(kN^2)$, where k is the number of cells in the width of the narrow band. This is a significant cost reduction. Typically, good results can be obtained with about six cells on either side of the zero level.

Calculating extension variables The time-dependent level set approach requires the extension of the speed function F in equation (3.5) to *all* of space; this then updates all of the level sets, not simply the zero level set on which the speed function is naturally defined. Recall that three types of arguments may influence the front speed F; local, global, and independent. Some of these variables may have meaning only on the front itself, and it may be both difficult and awkward to design a speed function that extrapolates the velocity away from the zero level set in a smooth fashion. Thus, another advantage of a narrow band approach is that this extension need only be done to points lying in the narrow band.

Choosing time step The full-matrix approach requires the choice of a time step that applies in response to the maximum velocity over the entire domain, not simply in response to the speed of the front itself. In a narrow band implementation, the time step can be adaptively chosen in response to the maximum velocity field only within the narrow band. This is of significance in problems in which the front speed changes substantially as it moves (for example, due to the local curvature or as determined by the underlying domain). In such problems, the CFL restriction for the velocity field for *all* the level sets may be much more stringent than the one for those sets within the narrow band.

The above 'narrow band' method was introduced by Chopp (1993), used in recovering shapes from images by Malladi, Sethian and Vemuri (1994), and analysed extensively by Adalsteinsson and Sethian (1995a).

In Fig. 17 we show the placement of a narrow band around an initial front. The entire two-dimensional grid of data is stored in a square array. A one-dimensional object is then used to keep track of the points in this array (dark grid points in Fig. 17 are located in a narrow band around the front of a user-

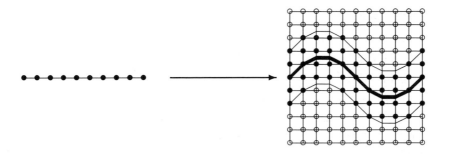

Fig. 17. Pointer array tags interior and boundary band points.

defined width). Only the values of ϕ at points within the tube are updated; values of ϕ at grid points on the boundary of the narrow band are frozen. When the front moves near the edge of the tube boundary, the calculation is stopped, and a new tube is built with the zero level set interface boundary at the centre. This rebuilding process is known as 're-initialization'.

Thus, the narrow band method consists of the following loop:

- tag 'alive' points in narrow band

- build 'land mines' to indicate nearby edge

- initialize 'far away' points outside(inside) narrow band with large positive(negative) values

- solve level set equation until a land mine is hit

- rebuild, loop.

Use of narrow bands leads to level set front advancement algorithms which are equivalent in complexity to traditional marker methods and cell techniques, while maintaining the features of topological merger, accuracy, and easy extension to multi-dimensions. Typically, the speed-up associated with the narrow band method is about ten times faster on a 160×160 grid than the full matrix method. Such a speed-up is substantial: in three-dimensional simulations, it can make the difference between computationally intensive problems and those that can be done with relative ease. Details on the accuracy, typical tube sizes, and number of times a tube must be rebuilt may be found in Adalsteinsson and Sethian (1995a).

This narrow banding technique requires a rebuilding and re-initialization of a new narrow band around the location of the front. There are several ways to perform this re-initialization, one of which leads to an efficient level set scheme for the particular case of a front propagating with a speed $F = F(x, y)$ where F is of one sign.

13.4. Re-initialization techniques: direct evaluation, iteration, Huygens' flowing

Direct evaluation

A straightforward approach is to find the zero level set by using a contour plotter and then recalculate the signed distance from each grid point to this zero level set to rebuild the band. This technique, first used in Chopp (1993), can be used to ensure that the level set function stays well behaved. However, this approach can be expensive, since the front must be explicitly constructed and distances must be calculated to neighbouring grid points.

Iteration

An alternative to this was given by Sussman, Smereka and Osher (1994), based on an observation of Morel. Its virtue is that one need not find the zero level set to re-initialize the level set function. Consider the partial differential equation

$$\phi_t = \text{sign}(\phi)(1 - |\nabla\phi|). \tag{13.1}$$

Given initial data for ϕ, solving the above equation to steady state provides a new value for ϕ with the property that $|\nabla\phi| = 1$, since convergence occurs when the right-hand side is zero. The sign function controls the flow of information in (13.1); if ϕ is negative, information flows one way and if ϕ is positive, then information flows the other way. The next effect is to 'straighten out' the level sets on either side of the zero level set and produce a ϕ function with $|\nabla\phi| = 1$, which in fact corresponds to the signed distance function. Thus, their approach is to periodically stop the level set calculation and solve (13.1) until convergence: if repeated sufficiently often, the initial data are often close to the signed distance function, and few iterations are required. One disadvantage of this technique is the relative crudeness of the switch function based on the sign of the level set equation; considerable motion of the zero level set can occur during the re-initialization, since the sign function does not accurately model the exact location of the front.

Huygens' principle flowing

An alternative technique is based on the idea of computing crossing times as discussed in Sethian (1994), and is related to the ideas given in Kimmel and Bruckstein (1993). Consider a particular value for the level set function $\phi_{\text{initial}}(x, t)$. With speed function $F = 1$, flow the level set function both forward and backwards in time and calculate crossing times (that is when ϕ changes sign) at each grid point. These crossing times (both positive and negative) are equal to the signed distance function by Huygens' principle. This approach has the advantage that one knows *a priori* how long to run the problem forward and backward to re-initialize grid points a given distance from the front. This calculation can be performed using a high-order scheme to produce accurate values for the crossing times.

This idea of computing crossing times is equivalent to converting the level set evolution problem into a stationary problem. This conversion can be used to develop an extremely fast marching level set scheme for the particular case of solving the level set equation for speed function $F = F(x, y)$, where F is always either positive or negative. This is discussed in the section on fast marching methods for the stationary formulation.

14. Additional complexities

Since their introduction, the capabilities and applicability of time-dependent level set methods have been considerably refined and extended. In this section, we discuss a few extensions that have proved to be useful in a variety of applications.

14.1. Masking and sources

Consider the problem of a front propagating with a speed F and subject to the constraint that the evolving interface cannot enter into a region Ω in the domain. This region Ω is referred to as a 'mask', since it inhibits all motion. There are several solutions to this problem, depending on the degree of accuracy required.

The simplest solution is to set the speed function F equal to zero for all grid points inside Ω. The location of all points inside Ω can be determined before any calculation is carried out. This technique assures that the front stops within one grid cell of the mask. In Fig. 18, we show a plane front propagating upwards with speed $F = 1$ in the upwards direction, with a rectangular block in the centre of the domain serving as a mask. In Fig. 18a, the speed function is reset to zero inside the mask region, and as the front propagates upwards it is stopped in the vicinity of the mask and is forced to bend around it.

The calculations in Fig. 18a are performed on a very crude 13×13 mesh in order to accentuate a problem with this approach, namely that the front can only be guaranteed to stop within one grid cell of the obstacle itself. This is because the level set method constructs an interpolated speed between grid points, and hence by setting the speed function to zero on and in the mask, the front slows down before it actually reaches the mask. Note that since this means one grid cell *normal* to the mask's boundary, a considerable amount of error can result.

A different fix, which eliminates much of this problem, comes from an alternate view. Given a mask area Ω, construct the signed distance function ϕ^Ω by taking the positive distance if inside Ω and the negative distance if outside (note that this is opposite sign choice from the one typically used). Then we limit motion into the masked region not by modifying the speed function, but instead by resetting the evolving level set function. Let $\phi^{(*)}$ be

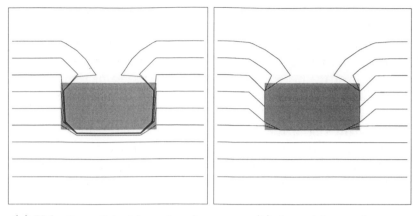

(a) Velocity = 0 inside rectangle   (b) ϕ reset by mask

Fig. 18. Front propagating upwards around masking block: 13×13 grid.

the value produced by advancing the level set ϕ^n one time step. Then let

$$\phi^{n+1} = \max(\phi^*, \phi^\Omega). \tag{14.1}$$

This resets the level set function so that penetration is not possible; of course, this is only accurate to the order of the grid. Results using this scheme are shown in Fig. 18b. Again, we have used a very coarse grid to accentuate the differences.

If we consider now the opposite problem, in which a region Ω acts as a source, then the solution is equally straightforward, and we obtain $\phi^{n+1} = \min(\phi^*, -\phi^\Omega)$; this is the technique used in Rhee, Talbot and Sethian (1995).

14.2. Discontinuous speed functions and sub-grid resolution

Let us generalize the above problem somewhat, and imagine that we want to solve an interface propagation problem in which there is a clear discontinuous speed function. For example, one may want to track the propagation of an interface through materials that correspond to differing media, across which propagation rates change quite sharply. As an example, consider again the evolution of the upwards propagating front, but this time the rectangular block halves the speed (that is, $F = 1$ outside Ω and $F = 0.5$ inside and on Ω). The standard level set will interpolate between these two speeds, and the results obtained will depend on the placement of the underlying grid; substantial variation in result will occur depending on whether a grid line lies directly on, below, or above the bottom edge of the rectangular block.

In order to solve this problem accurately, we need some sub-grid information about the speed function to construct correctly the speed function for those cells that lie only partially within Ω. Such a technique can be devised, motivated by the idea of the volume-of-fluid methods discussed earlier.

Given a region Ω, before any calculation proceeds we construct the cell fraction Vol_{ij}^{Ω}, which is a number between 0 and 1 for those cells that have at least one grid point in Ω and one outside Ω. This cell fraction corresponds to the amount of Ω material in the cell. These values are stored, and a list is kept of such boundary cells. We then proceed with the level set calculation, letting F be given by its value in the corresponding region. However, we modify the speed function for those cells that are marked as boundary cells. At the beginning of the time step, compute the volume fraction Vol_{ij}^{ϕ} for the zero level set in each cell; this may be approximately done without explicitly finding the zero level set through a least squares fit. This value is then compared with the stored value Vol_{ij}^{Ω} to provide an appropriate speed.

14.3. Multiple interfaces

As initially designed in Osher and Sethian (1988), the level set technique applies to problems in which there is a clear distinction between an 'inside' and 'outside'. This is because the interface is assigned the zero level value between the two regions. Extensions to multiple (more than two) interfaces have been made in some specific cases. In the case in which interfaces are passively transported and behave nicely, one may be able to use only one level set function and judiciously assign different values at the interfaces. For example, the zero level set may correspond to the boundary between two regions A and B, with the level set value 10 corresponding to the interface between two regions B and C. If A and C never touch, then this technique may be used to follow the interfaces in some cases.

However, in the more general case involving the emergence and motion of triple points, a different approach is required, because many different situations can occur; see, for example, Bronsard and Wetton (1995) and Taylor, Cahn and Handwerker (1992). Consider the following canonical example, as illustrated in Fig. 19. Regions A and B are both circular disks growing into region C with speed unity in the direction normal to the interface. At some point, the interfaces will touch and meet at a triple point, where a clear notion of 'inside' and 'outside' cannot be assigned in a consistent manner.

A level set approach to this problem has been proposed by Merriman, Bence and Osher (1994); they move each interface separately for one time step, find the interfaces of the various fronts, and then rebuild level set functions. This technique requires re-initializing the pairwise level set functions; any of the techniques described earlier can be used. Before describing this technique, we discuss a later set of techniques presented in Sethian (1995a, 1995b), applicable to many cases, and which do not require any such re-initialization. We will then return to the first algorithm in the case of motion driven by surface tension where some sort of re-initialization is required.

Fig. 19. Regions A and B expand into region C.

The key idea in each method lies in recasting the interface motion as the motion of one level set function for each material. In some sense, this is what was done in the re-ignition idea given in Rhee et al. (1995), where the front was a flame, which propagated downstream under a fluid flow. This front was re-ignited at each time step at a flame holder point by taking the minimum of the advancing flame and its original configuration around the flame holder, thus ensuring that the maximum burned fluid is achieved.

In general, imagine N separate regions and a full set of all possible pairwise speed functions F_{IJ} which describe the propagation speed of region I into region J: F is taken as zero if region I cannot penetrate J. The idea is to advance each interface to obtain a trial value for each interface with respect to motion into every other region, and then combine the trial values in such a way as to obtain the maximum possible motion of the interface.

In general then, proceed as follows. Given a region I, obtain $N - 1$ trial level set functions ϕ_{IJ}^* by moving region I into each possible region J, J=1,N $(J \neq I)$ with speed F_{IJ}. During the motion of region I into region J, assume that all other regions are impenetrable, that is, use the masking rule given by equation (14.1). We then test the penetrability of region J itself, leaving the value of ϕ_{IJ}^* unchanged if $F_{IJ} \neq 0$, otherwise modifying it with the maximum of itself and $-\phi_{JI}^*$. Finally, to allow region I to evolve as much as possible, we take the minimum over all possible motions to obtain the new position; this is the re-ignition idea described earlier. Complete details of the approach may be found in Sethian (1995a).

Three examples are shown to illustrate this approach. Given regions A, B, and C, the *influence matrix* describes the interaction of the various regions with each other. The interaction of each region with itself is null, hence the matrix has a dash on the diagonal. The interaction of any pair of regions is required to be zero in one of the two interactions.

In Fig. 20, regions A and B expand with unit speed into region C, but cannot penetrate each other. They advance and meet; the boundary between the two becomes a vertical straight line.

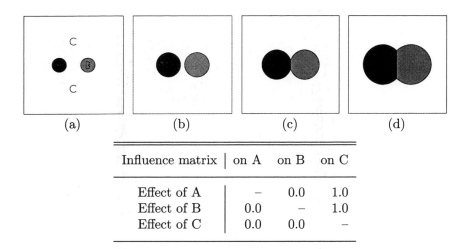

Influence matrix	on A	on B	on C
Effect of A	–	0.0	1.0
Effect of B	0.0	–	1.0
Effect of C	0.0	0.0	–

Fig. 20. A and B move into C with speed 1, stop at each other.

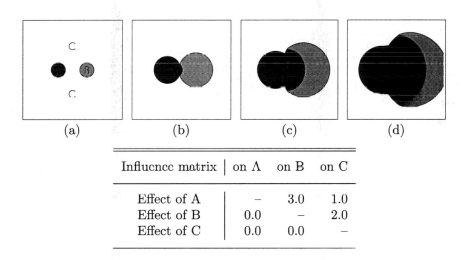

Influence matrix	on A	on B	on C
Effect of A	–	3.0	1.0
Effect of B	0.0	–	2.0
Effect of C	0.0	0.0	–

Fig. 21. A into C with speed 1, A into B with speed 3, B into C with speed 2.

Next, we consider a problem with different evolution rates. In Fig. 21, region A grows with speed 1 into region C (and region C grows with speed 0 into region A), and region B grows with speed 2 into region C. Once they come into contact, region A dominates region B with speed 3, thus region B grows through Fig. 21c, and then is 'eaten up' by the advancing region A. Note what happens: region A advances with speed 3 to the edge of region B, which is only advancing with speed 1 into region C. However, region A cannot pass region B, because *its* speed into region C is slower than that of region B.

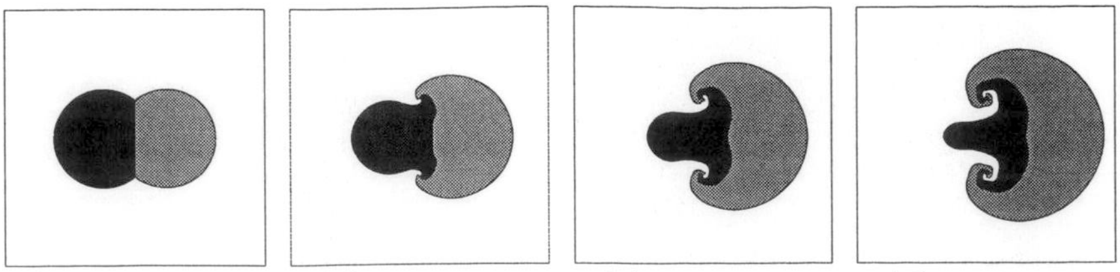

| (a) $T = 0.0$ | (b) $T = 2.0$ | (c) $T = 3.0$ | (d) $T = 5.0$ |

Influence matrix	on A	on B	on C
Effect of A	–	1.0	0.0
Effect of B	0.0	–	1.0
Effect of C	1.0	0.0	–

Fig. 22. Spiralling triple point: 98×98 grid.

Finally, in Fig. 22, the motion of a triple point between regions A, B, and C is shown. Assume that region A penetrates B with speed 1, B penetrates C with speed 1, and C penetrates A with speed 1. The exact solution is given by a spiral with no limiting tangent angle as the triple point is approached. The triple point does not move; instead, the regions spiral around it. In Fig. 22, results are shown from a calculation on a 98×98 grid. Starting from the initial configuration, the regions spiral around each other, with the leading tip of each spiral controlled by the grid size. In other words, we are unable to resolve spirals tighter than the grid size, and hence that controls the fine-scale description of the motion. However, we note that the triple point remains fixed. A series of additional calculations using this approach may be found in Sethian (1995*b*).

14.4. Triple points

Consider now the case of a triple point motion in which the speed of each interface is driven by curvature, which may correspond to surface tension. Imagine a triple point in which each of the three regions is attempting to move according to their own curvature. In Fig. 23a we show an initial configuration, and in Fig. 23b a final state, which consists of the three lines meeting in equal angles of 120 degrees.

If one attempts to apply the level set method for multiple interfaces described in the previous sections, a difficulty occurs, because each level function attempts to move away from the others, creating a gap. In Fig. 23c, we show this gap developing when a level set technique is applied to the final state.

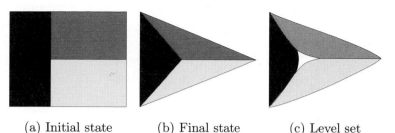

(a) Initial state   (b) Final state   (c) Level set

Fig. 23. Evolution of triple point under curvature.

Two different level set type algorithms were introduced in Merriman et al. (1994) for tackling this problem. The first can be viewed a 'fix' to the above problem; the second is a wholly different level set approach.

First, the problem with the above calculation is that the various level sets pull apart. A remedy is to reset the level set functions every time step to hold the triple point in place, that is,

$$\phi_i = \phi_i - \max \phi_j, \tag{14.2}$$

where the maximum is taken over all $j \neq i$. This keeps the triple point in place; however, the cost is that the level set functions can develop spontaneous zero crossings later in time. A remedy is to re-initialize all the level sets using any of the re-initialization techniques described in the previous section. With those two added steps in the algorithm, level set methods can easily handle some interesting problems concerning triple points.

A considerably different approach works by applying applying a reaction-diffusion type equation to a characteristic function assigned to each region, which is one inside the region and zero outside. This algorithm works by exploiting the link between curvature flow and a diffusion equation, along the lines of the material discussed earlier. The basic idea is that a diffusion term is applied, and then a sharpening term is executed, which sharpens up the solution. The net effect is to evolve the boundary line under curvature. This is a very clever algorithm, and can be applied to multiple interfaces. A series of fascinating calculations are presented in Merriman et al. (1994); for additional work on this topic, see Bronsard and Wetton (1995).

14.5. Building extension velocity fields

What happens when the speed of the moving front has meaning only on the front itself? This is a common occurrence in areas such as combustion, material science, and fluid mechanics, where the philosophy of embedding the front as the zero level set of a family of contours can be problematic. In fact, the most difficult part of level set methods is this 'extension' problem, and it will be a central focus of Part IV.

Recall the division of arguments in the speed function F given in equation (2.1). Front-based arguments are those that depend on geometric quantities of the front, such as curvature and normal vector, and have a clear meaning for all the level sets. Independent variables are equally straightforward, since their contribution to the speed makes no reference to particular information from the front itself.

The troublesome variables are the so-called 'global' variables, which can arise from solving differential equations on either side of the interface. Briefly, there are at least four ways to extend a velocity from the front to the grid points.

1 At each grid point, find the closest point on the front. This was the technique used in Malladi et al. (1994), and may be done efficiently in many cases by tracing backwards along the gradient given by $\nabla \phi$.

2 Evaluate the speed function off the front using an equation that only has meaning on the front itself. This is the technique used in the crystal growth/dendritic solidification calculations employed in Sethian and Strain (1992), where a boundary integral is evaluated both on and off the front.

3 Develop an evaluation technique that assigns artificial speeds to the level set going through any particular grid point. For example, in the etching/deposition simulations of Adalsteinsson and Sethian (1995b, 1995c, 1996), visibility of the zero level set must be evaluated away from the front itself.

4 Smearing the influence of the front. In the combustion calculations of Rhee et al. (1995), the influence of the front is mollified to neighbouring grid points on which an appropriate equation is solved.

PART IV: A NEW FAST METHOD

Consider the special case of a monotonically advancing front, that is, a front moving with speed F where F is always postive (or negative). Previously, we have produced a stationary level set equation, namely

$$|\nabla T|F = 1. \tag{14.3}$$

which is simply a static Hamilton–Jacobi equation; if F is only a function of position, this becomes the well-known Eikonal equation. A large body of research has been devoted to studying these types of equations; we refer the interested reader to Barles and Souganidis (1991), Lions (1982) and Souganidis (1985), to name just a few. At the same time, there are a wide collection of schemes to solve this problem; see, for example, Bardi and Falcone (1990),

Falcone et al. (1994), Rouy and Tourin (1992) and Osher and Rudin (1992). Here we introduce an entirely new and extremely fast method for solving this equation. It relies on a marriage between our narrow band technique and a fast heapsort algorithm, and can be viewed as an extreme one-cell version of our narrow band technique.

For ease of discussion, we limit ourselves to a two-dimensional problem inside a square from $[0, 1] \times [0, 1]$ and imagine that the initial front is along the line $y = 0$; furthermore, we assume that we are given a positive speed function $F(x, y)$ that is periodic in x. Thus, the front propagates upwards off the initial line, and the speed does not depend on the orientation of the front (it depends only on *independent variables*, using our earlier terminology). Using our approximation to the gradient, we are then looking for a solution in the unit box to the equation

$$F_{ij}^{-1} = \max(D_{ij}^{-x}T, 0)^2 + \min(D_{ij}^{+x}T, 0)^2 +$$
$$\max(D_{ij}^{-y}T, 0)^2 + \min(D_{ij}^{+y}T, 0)^2), \tag{14.4}$$

where $T(x, 0) = 0$.

Since equation (14.4) is in essence a quadratic equation for the value at each grid point (assuming the others are held fixed), we can iterate until convergence by solving the equation at each grid point, selecting the largest possible value as the solution in accordance with obtaining the correct viscosity solution. An iterative algorithm for computing the solution to this problem was introduced by Rouy and Tourin (1992). Typically, one iterates several times through the entire set of grid points until a converged solution is reached.

15. A fast marching level set method

The key to constructing a fast marching algorithm is the observation that the upwind difference structure of equation (14.4) means that information propagates 'one way', that is, from smaller values of T to larger values. Hence, our algorithm rests on 'solving' equation (14.4) by building the solution outwards from the smallest time value T. Our idea is to sweep the front ahead in an upwind fashion by considering a set of points in a narrow band around the existing front, and to march this narrow band forward, freezing the values of existing points and bringing new ones into the narrow band structure. The key is in the selection of *which* grid point in the narrow band to update. The technique is easiest to explain algorithmically; see Fig. 24. We imagine that we want to propagate a front through an N by N grid with speed F_{ij} giving the speed in the normal direction at each grid point. Here the set of grid points $j = 1$ corresponds to the y axis, and we assume that $F_{ij} > 0$.

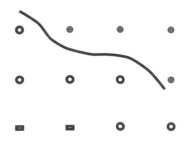

Fig. 24. Narrow band approach to marching level set method.

Algorithm

1   Initialize

   (a) (Alive points: grey disks): Let A be the set of all grid points $\{i, j = 1\}$; set $T_{i,1} = 0.0$ for all points in A.
   (b) (Narrow band points: black circles): Let *NarrowBand* be the set of all grid points $\{i, j = 2\}$, set $T_{i,1} = dy/F_{ij}$ for all points in *NarrowBand*.
   (c) (Far away points: black rectangles): Let *FarAway* be the set of all grid points $\{i, j > 2\}$, set $T_{i,j} = \infty$ for all points in *FarAway*.

2   Marching forwards

   (a) Begin loop: Let $(i_{\min}, j_{\min})$ be the point in *NarrowBand* with the smallest value for T.
   (b) Add the point $(i_{\min}, j_{\min})$ to A; remove it from *NarrowBand*.
   (c) Add to the narrow band list any neighbouring points $(i_{\min} - 1, j_{\min})$, $(i_{\min} + 1, j_{\min})$, $(i_{\min}, j_{\min} - 1)$, $(i_{\min}, j_{\min} + 1)$ that are not *Alive*. If the neighbour is in *FarAway*, remove it from that list.
   (d) Recompute the values of T at all neighbours according to equation (14.4), selecting the largest possible solution to the quadratic equation.
   (e) Return to top of loop.

We take periodic boundary conditions where required. Assuming for the moment that it takes no work to determine the member of the narrow band with the smallest value of T, the total work required to compute the solution at all grid points is $O(N^2)$, where calculation is performed on an N by N grid.

Why does the above algorithm work? Since we are always locating the smallest value in the narrow band, its value for T must be correct; other narrow band points or far away points with larger T values cannot affect it. The process of recomputing the T values at neighbouring points (that have not been previously accepted) cannot yield a value smaller than any of that

Fig. 25. Matrix of neighbouring values.

at any of the accepted points, since the correct viscosity solution is obtained by selecting the *largest* possible solution to the quadratic equation. Thus, we can march the solution outwards, always selecting the narrow band grid point with minimum trial value for T, and readjusting neighbours. Another way to look at this is that each minimum trial value begins an application of Huygens' principle, and the expanding wave front touches and updates all others.

15.1. Proof that the algorithm constructs a viable solution

Here, we prove that the above algorithm produces a solution that everywhere satisfies the discrete version of our equation, which is given by

$$f_{ij}^2 = \max\left(\max(D_{ij}^{-x}T, 0), -\min(D_{ij}^{+x}T, 0)\right)^2 +$$
$$\max\left(\max(D_{ij}^{-y}T, 0), -\min(D_{ij}^{+y}T, 0)\right)^2, \qquad (15.1)$$

where $f_{ij}^2 = 1/F_{ij}^2$. We shall give a constructive proof. Since the values of $T(x, y, z)$ are built by marching forwards from the smallest value to the largest, we need only show that whenever a 'trial' value is converted into an 'alive' value, none of the recomputed neighbours obtain new values less than the accepted value. If this is true, then we will always be marching ahead in time, and thus the correct 'upwind' nature of the differencing will be respected. We shall prove our result in two dimensions; the three-dimensional proof is the same.

Thus, consider the matrix of grid values given in Fig. 25. Our argument will follow the computation of the new value of T in the centre grid point to replace the value of ?, based on the neighbouring values. We will assume, without loss of generality, that the value A at the left grid point is the smallest of all 'trial' values, and prove that when we recompute the value at the centre grid point (called $T_{\text{recomputed from } A}$), it cannot be less than A. This will prove that the upwinding is respected, and that we need not go back and readjust previously set values. We shall consider the four cases that (1) none of the neighbours B, C, or D, are 'alive', (2) one of these neighbours is 'alive',

(3) two of the neighbours are 'alive', and (4) all three of these neighbours are 'alive'. †

A, B, C and D are 'trial', A is the smallest
In this case, all of the neighbours around the centre grid point are either 'trial' or set to *FarAway*. Since A is the smallest such value, we convert that value to 'alive' and recompute the value at the centre grid point. We now show that the recomputed value $A \leq T_{\text{recomputed from } A} \leq A + f$.

1   Suppose $A + f \leq \min(B, D)$. Then $T_{\text{recomputed from } A} = (A + f)$ is a solution to the problem, since only the difference operator to the left grid point is nonzero. ‡

2   Suppose $A + f \geq \min(B, D)$. Then, without loss of generality, assume that $B \leq D$. We can solve the quadratic equation

$$(T_{\text{recomputed from } A} - A)^2 + (T_{\text{recomputed from } A} - B)^2 = f^2. \qquad (15.2)$$

The discriminant is non-negative when $f \geq \frac{(B-A)}{\sqrt{2}}$, which must be true since we assumed that $A + f \geq B$ and hence $f \geq (B - A)$. Thus, a solution exists, and it is easy to check that this solution must then be greater than or equal to B and thus falls into the required range. Furthermore, we see that $T \leq A + f$, since the second term on the left is non-negative.

Thus, we have shown that $A \leq T_{\text{recomputed from } A} \leq A + f$, and therefore $T_{\text{recomputed from } A}$ cannot be less than the just converted value A.

    This case will act as template for the other cases.

B is 'alive', A, C and D are 'trial', A is the smallest of the trial values
In this case, A has just been converted, since it is the smallest of the trial values. We shall prove when we recalculate $T_{\text{recomputed from } A}$, its new value must still be greater than A. At some previous stage, when B was converted from trial to alive, the values of A, C and D were all trial values, and hence must have been larger. Then this means that when B was converted from trial to alive, we had the previous case above, and hence $B \leq T_{\text{recomputed from } B} \leq B + f$; furthermore, since the value at the centre was *not* chosen as the smallest trial value, we must have that $A \leq B + f$. By the above case, we then have that $B \leq A \leq T_{\text{recomputed from } A} \leq B + f$, and hence the recomputed value cannot be less than the just converted value of A.

---

† Recall that 'alive' means that their T values are less than A. Here, we are using the notation that the symbol A stands for both the grid point and its T value.

‡ We are absorbing the grid size Δx into the inverse speed function f.

C is 'alive', A, B and D are 'trial', A is the smallest of the trial values
In this case, due to the direction of the upwind differencing, the value at C is the contributor in the x direction, the acceptance of A does not affect the recomputation, and the case defaults into the first case above.

The remaining cases are all the same, since the differencing takes the smallest values in each coordinate direction. The proof in three dimensions is identical.

15.2. *Finding the smallest value*

The key to an efficient version of the above technique lies in a fast way of locating the grid point in the narrow band with the smallest value for T. We use a variation on a heapsort algorithm, see Press, Flannery, Teukolsky and Vetterling (1988) and Sedgewick (1988), with the additional feature of back pointers. In more detail, imagine that the list of narrow band points is initially sorted in a heapsort so that the smallest member can be easily located. We store the values of these points in the heapsort, together with their indices which give their location in the grid structure. We keep a companion array which points from the two-dimensional grid to the location of that grid point in the heapsort array. Finding the smallest value is easy. In order to find the neighbours of that point, we use the pointers from the grid array to the heapsort structure. The values of the neighbours are then recomputed, and then the results are bubbled upwards in the heapsort until they reach their correct locations, at the same time readjusting the pointers in the grid array. This results in an $O(\log N)$ algorithm for the total amount of work, where N is the number of points in the narrow band. For implementation details and further application of this technique, see Sethian (1995c, 1996) and Sethian, Adalsteinsson and Malladi (1996).

The above technique considered a flat initial interface for which trial values at the narrow band points could be easily initialized. Suppose we are given an arbitrary closed curve or surface as the initial location of the front. In this case, we use the original narrow band level set method to initialize the problem. First, label all grid points as 'far away' and assign them T values of ∞. Then, in a very small neighbourhood around the interface, construct the signed distance function from the initial hypersurface Γ. Propagate that surface both forwards and backwards in time until a layer of grid points is crossed in each direction, computing the signed crossing times as in Sethian (1994). Then collect the points with negative crossing times as 'alive' points with T value equal to the crossing time, and the points with positive crossing times as narrow band points with T value equal to the positive crossing times. Then begin the fast marching algorithm.

16. Other speed functions: when does this method work?

Our fast marching method as designed applied to speed functions F which depend only on position. In such cases, other forms for the gradient approximation can be used; for example, Rouy and Tourin (1992), namely

$$F_{ij}^{-1} = \max\left(\max(D_{ij}^{-x}T, 0), -\min(D_{ij}^{+x}T, 0)\right)^2 +$$
$$\max\left(\max(D_{ij}^{-y}T, 0), -\min(D_{ij}^{+y}T, 0)\right)^2. \qquad (16.1)$$

How general is our new technique? Suppose now we consider the more general case of stationary level set equation:

$$|\nabla T|F = 1. \qquad (16.2)$$

We begin by rewriting this in the standard form of a static Hamiltonian, namely

$$H(T_x, T_y, T_z) = 1. \qquad (16.3)$$

We already have a scheme for the case where $H = \nabla T$. Some variations on this Hamiltonian important in computer vision, such as

$$H = \max(|T_x|, |T_y|, |T_z|) \qquad H = |T_x| + |T_y| + |T_z|, \qquad (16.4)$$

may be approximated in a straightforward manner using any of the above entropy-satisfying approximations to the individual gradients. Our fast marching method will work in these cases. When the speed F depends in a subtle way on the value of ∇T (for example, in some problems in etching and deposition discussed in a later section), the situation is more delicate.

When will the technique work? Here, we present an intuitive perspective; complete details may be found in Sethian (1996) and Sethian et al. (1996). Suppose H is convex and always positive (or always negative), and suppose the approximation to H satisfies two properties:

1   the approximation scheme is consistent
2   at each grid point the scheme only makes use of smaller neighbouring values when updating the value at that point (this is the upwindness requirement), and cannot produce a new value which is less than any of the neighbours.

Then we can expect that our upwind sweeping method will work; searching for the smallest trial value will provide a consistent way of sweeping through the mesh and constructing the solution surface T. Complete details and many other schemes may be found in Sethian (1996) and Sethian et al. (1996).

17. Some clarifying comments

The time-dependent level set method and the stationary level set method each require careful construction of upwind, entropy-satisfying schemes, and make use of the dynamics and geometry of front propagation analysed in Sethian (1985). However, we note that the time-dependent level set method advances the front *simultaneously*, while the stationary method constructs 'scaffolding' to build the time solution surface T one grid point at a time. This means that the time at which the surface crosses a grid point (that is, its T value) may be found before other positions of that front at that time are determined. As such, there is *no* notion of a time step in the stationary method: one is simply constructing the stationary surface in an upwind fashion.

This means that if one is attempting to solve a problem in which the speed of a front depends on the current position of the front (such as in the case of visibility), or on subtle orientations in the front (such as in sputter yield problems), it is not clear how to use the stationary method, since the front is being constructed one grid point at a time.

The stationary method works because we were able to construct a simple approximation to the gradient. This was possible because the speed function F did *not* depend on the orientation of the front, nor on issues like visibility. Thus, returning to our earlier categorization of speed functions, our fast scheme works in cases where the speed F only depends on independent variables, such as in the case of photolithography development. Upwind entropy-satisfying schemes which can be transported to this fast stationary scheme for the case of more general speed functions F are more problematic, and discussed in detail in Sethian (1996).

To summarize,

- The stationary method is convenient for problems in which the front speed depends on independent variables, such as a photoresist rate function, and only applies if the speed function does not change sign.
- The time-dependent level set method is designed for more delicate speed functions, and can accurately evolve fronts under highly complex arguments.

In the next part, we discuss a variety of applications which employ both the time-dependent level set method and the fast marching method for monotonically advancing fronts.

PART V: APPLICATIONS

In this part, we present a series of applications of both the time-dependent level set method and the fast marching level set method to propagating interfaces. This is only a subset of the level set applications in the literature; throughout, we provide references to many other applications.

18. Geometry

In this section, we consider application of level set methods to problems in the geometric evolution of curves and surfaces. The motion will depend solely on local geometric properties such as normal direction and curvature; nonetheless, this is a fascinating and rich area.

18.1. Curvature flow

Suppose we are given a hypersurface in R^n propagating with some speed $F(\kappa)$. Previously, we have considered speed functions of the form $F(\kappa) = 1 - \epsilon\kappa$, where κ is the curvature. Let us now focus on a special speed function, namely $F = -\kappa$, where κ is the curvature. This corresponds to a geometric version of the heat equation; large oscillations are immediately smoothed out, and long-term solutions correspond to dissipation of all information about the initial state. As we shall see in later sections, curvature motion plays an important role in many applications such as a modelling term for surface tension in flexible membranes and a viscous term in physical phenomena.

The remarkable work of Gage and Grayson investigated the motion of a simple closed curve collapsing under its curvature. First, Gage showed that any convex curve moving under such a motion remains convex and must shrink to a point (Gage 1984, Gage and Hamilton 1986). Grayson (1987) followed this work with a stunning proof that *all* curves must shrink to a round point, regardless of their initial shape.

In Fig. 26, we take an odd-shaped initial curve and view this as the zero level set of a function defined in all of R^2. Here, for illustration, we have ϕ such that $\phi < 0$ as black and $\phi > 0$ as white, thus the zero level set is the boundary between the two. As the level curves flow under curvature, the ensuing motion carries each to a point, which then disappears. In the evolution of the front, one clearly sees that the large oscillations disappear quickly, and then, as the front becomes circular, motion slows, and the front eventually disappears.

In three dimensions, flow under mean curvature does not necessarily result in a collapse to a sphere. Huisken (1984) showed that convex shapes shrink to spheres as they collapse, analogous to the result of Gage. However, Grayson (1989) showed that non-convex shapes may in fact *not* shrink to a sphere,

Fig. 26. $F(\kappa) = -\kappa$.

and provided the counterexample of the dumbbell. A narrow handle of a dumbbell may have such a high inner radius that the mean curvature of the saddle point at the neck may still be positive, and hence the neck will pinch off.

As illustration, in Fig. 27, taken from Chopp and Sethian (1993), we show two connected dumbbells collapsing under mean curvature. As the intersection point collapses, the necks break off and leave a remaining 'pillow' region behind. This pillow region collapses as well, and eventually all five regions disappear.

Finally, what about self-similar shapes? In two dimensions, it is clear that a circle collapsing under its own curvature remains a circle; this can be seen by integrating the ordinary differential equation for the changing radius. In

Fig. 27. Collapse of two-handled dumbbell.

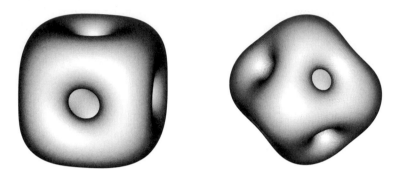

(a) Self-similar cube with holes (b) Self-similar octahedron with holes

Fig. 28. Self-similar shapes.

three dimensions, a sphere is self-similar under mean curvature flow, since its curvature is always constant. Angenent (1992) proved the existence of a self-similar torus that preserves the balance between the competing pulls towards a ring and a sphere.

In order to devise an algorithm to produce self-similar shapes, two things are required. First, since hypersurfaces get smaller as they move under their curvature, a mechanism is needed to 'rescale' their motion so that the evolution can be continued towards a possible self-similar shape. Second, a way of pushing the evolving fronts back towards self-similarity is required. Chopp has accomplished both in a clever numerical algorithm that produces a family of self-similar surfaces; see Chopp (1994). His family comes from taking a regular polyhedron (for example, a cube), and drilling holes in each face. The resulting figure then evolves according to auxiliary level set equations, which contain the re-scaling as part of the equation of motion. Two such self-similar surfaces are shown in Fig. 28.

18.2. Grid generation

Imagine that one is given a closed body, either as a curve in two space dimensions, or a surface in three space dimensions. In many situations, one wishes to generate a logically rectangular, body-fitted grid around or inside this body. By logically rectangular, we mean that each node of the grid has four neighbours (in two dimensions; in three dimensions, there are six neighbours). By body-fitted, we mean that the grid aligns itself with the body so that one set of coordinate lines matches the body itself. This grid generation problem is difficult in part because of the competing desires of uniformity in cell area and mesh orthogonality; we refer the interested reader to Knupp and Steinberg (1993).

Level set techniques offer an interesting technique for generating such grids. The idea, as presented in Sethian (1994), is to exploit the geometric nature of the problem and view the body itself as the initial position of an interface that must be advanced outwards away from the body. The initial position of the interface and its position at later times forms one set of grid lines; its orthogonal set forms the other. The body is propagated outwards with speed $F = 1 - \epsilon\kappa$; by finding the zero level set at discrete times, the set of coordinate lines that encircle the body is found. Construction of transverse lines normal to the body are obtained by following trajectories of $\nabla\phi$. Additional node adjustment is possible through application of additional smoothing operators. Node placement on the boundary and the ensuing exterior/interior grid can be automatically controlled. For details, see Sethian (1994).

This technique can almost be viewed as a hyperbolic solver. However, by solving the correct evolution equation for an advancing front, we avoid the difficulties of shock formation and colliding characteristics that plague most hyperbolic techniques. User intervention is kept to a minimum; for the most part, grids are generated automatically without the need to adjust parameters.

In Fig. 29, we show a variety of grids constructed using this level set approach, starting with relatively smooth grids and ending with a three-dimensional grid around an indented dumbbell. As can be seen, interior and exterior grids can be created, with the capability of handling significant corners and cusps. The grids are automatically created; there has been no adjustment of parameters in the creation of these different grids.

18.3. Image enhancement and noise removal

The previous sections concerned geometrical motion of a particular hypersurface of interest. Next, we turn to a level set problem in which all the level sets have meaning, and must be evolved.

The goal in this section is to apply some of the level set methodology to image enhancement and noise removal. To do so, we first need a few definitions. Define an *image* to be an intensity map $I(x, y)$ given at each point of a two-dimensional domain. The range of the function $I(x, y)$ depends on the type of image: for monochrome images, the range is $\{0, 1\}$; for greyscale images, $I(x, y)$ is a function mapping into $\{0, 1, \ldots, 255\}$; and for colour

An alternative approach is due to Rudin, Osher and Fatemi (1992). They take a total variation approach to the problem, which leads to a level set methodology and a very similar curvature-based speed function, namely $F(\kappa) = \kappa/|I|$; see also Osher and Rudin (1990). Following these works, a variation on these two approaches was produced by Sapiro and Tannenbaum (1993b). In that work, a speed function of the form $F(\kappa) = \kappa^{1/3}$ was employed. In each of these schemes, all information is eventually removed through continued application of the scheme. Thus, a 'stopping criterion' is required.

A recent level set scheme for noise removal and image enhancement was introduced by Malladi and Sethian (1995) and Malladi and Sethian (1996a). The scheme results from returning to the original ideas of curvature flow, and exploiting a 'min/max' function, which correctly selects the optimal motion to remove noise. It has two highly desirable features:

1   there is an intrinsic, adjustable definition of scale within the algorithm, such that all noise below that level is removed, and all features above that level are preserved

2   the algorithm stops automatically once the sub-scale noise is removed; continued application of the scheme produces no change.

These two features are quite powerful, and lead to a series of open questions about the morphology of shape and asymptotics of scale-removal; for details, see Malladi and Sethian (1996a).

The min/max flow
Consider the equation

$$\phi_t = \tilde{F}|\nabla\phi|. \tag{18.2}$$

A curve collapsing under its curvature will correspond to speed $\tilde{F} = \kappa$. Now, consider two variations on the basic curvature flow, namely

$$\tilde{F}(\kappa) = \min(\kappa, 0.0) \qquad \tilde{F}(\kappa) = \max(\kappa, 0.0).$$

Here, we have chosen the negative of the signed distance in the interior, and the positive sign in the exterior region. The effect of flow under $\tilde{F}(\kappa) = \min(\kappa, 0.0)$ is to allow the inward concave fingers to grow outwards, while suppressing the motion of the outward convex regions. Thus, the motion halts as soon as the convex hull is obtained. Conversely, the effect of flow under $\tilde{F}(\kappa) = \max(\kappa, 0.0)$ is to allow the outward regions to grow inwards while suppressing the motion of the inward concave regions. However, once the shape becomes fully convex, the curvature is always positive and the flow becomes the same as regular curvature flow.

Our goal is to select the correct choice of flow that smoothes out small oscillations, but maintains the essential properties of the shape. In order to do so, we discuss the idea of the min/max switch.

Consider the following speed function, introduced in Malladi and Sethian (1995) and refined considerably in Malladi and Sethian (1996a):

$$
\tilde{F}^{\text{Stencil}=k}_{\min/\max} = \begin{cases} \min(\kappa, 0) & \text{if Ave}^{R=kh}_{\phi(x,y)} < 0 \\ \max(\kappa, 0) & \text{if Ave}^{R=kh}_{\phi(x,y)} \geq 0, \end{cases} \tag{18.3}
$$

where $\text{Ave}^{R=kh}_{\phi(x,y)}$ is defined as the average value of ϕ in a disk of radius $R = kh$ centred around the point (x, y). Here, h is the step size of the grid. Thus, given a *StencilRadius* k, the above yields a speed function that depends on the value of ϕ at the point (x, y), the average value of ϕ in neighbourhood of a given size, and the value of the curvature of the level curve going through (x, y).

We can examine this speed function in some detail. For ease of exposition, consider a black region on a white background, chosen so that the interior has a negative value of ϕ and the exterior a positive value.

StencilRadius $k = 0$ If the radius $R = 0$ ($k = 0$), then choice of $\min(\kappa, 0)$ or $\max(\kappa, 0)$ depends only on the value of ϕ. All the level curves in the black region will attempt to form their convex hull, when seen from the black side, and all the level curves in the white region will attempt to form *their* convex hull. The net effect will be no motion of the zero level set itself, and the boundary will not move.

StencilRadius $k = 1$ If the average is taken over a stencil of radius kh, then some movement of the zero level corresponding to the boundary is possible. If there are some oscillations in the front boundary on the order of one or two pixels, then the average value of ϕ at the point (x, y) can have a different sign from the value at (x, y) itself. In this case, the flow will act as if it were selected from the 'other side', and some motion will be allowed until these first-order oscillations are removed, and a balance between the two sides is again reached. Once this balance is reached, further motion is suppressed.

StencilRadius $k > 1$ By taking averages over a larger stencil, larger amounts of smoothing are applied to the boundary. In other words, decisions about where features belong are based on larger and larger perspectives. Once features on the order of size k are removed from the boundary, balance is reached and the flow stops automatically. As an example, let $k = \infty$. Since the average will compute to a value close to the background colour, on this scale all structures are insignificant, and the max flow will be chosen everywhere, forcing the boundary to disappear.

(a) Initial boundary
'noisy' shape

(b) Min/max flow: stencil
radius $= 0; (T = \infty)$

(c) Min/max flow: stencil
radius $= 1; (T = \infty)$

(d) Continued min/max flow:
stencil radius $= 2; (T = \infty)$

Fig. 31. Motion of star-shaped region with noise under min/max flow at various
stencil levels.

To show the results of this hierarchical flow, we start with an initial shape
in Fig. 31a and first perform the min/max flow until steady state is reached
with stencil size zero in Fig. 31b. In this case, the steady state is achieved
immediately, and the final state is the same as the initial state. Min/max flow
is then performed until steady state is achieved with stencil size $k = 1$ in Fig.
31c, and then min/max flow is again applied with a larger stencil until steady
state is achieved in Fig. 31d.

These results can be summarized as follows:

- the min/max flow switch selects the correct motion to diffuse the small-
  scale pixel notches into the boundary

- the larger, global properties of the shape are maintained

- furthermore, and equally importantly, the flow stops once these notches are diffused into the main structure

- edge definition is maintained and, in some global sense, the area inside the boundary is preserved

- the noise removal capabilities of the min/max flow are scale-dependent, and can be hierarchically adjusted

- the scheme requires only a nearest neighbour stencil evaluation.

Extension of min/max scheme to grey-scale, texture, and colour images
The above technique applies to monochrome images. An extension to grey-scale images can be made by replacing the fixed threshold test value of 0 with a value that depends on the local neighbourhood. As designed in Malladi and Sethian (1995), let $T_{\text{threshold}}$ be the average value of the intensity obtained in the direction perpendicular to the gradient direction. Note that since the direction perpendicular to the gradient is tangent to the iso-intensity contour through (x, y), the two points used to compute are either in the same region, or the point (x, y) is an inflection point, in which the curvature is in fact zero and the min/max flow will always yield zero. By choosing a larger stencil we mean computing this tangential average over endpoints located further apart.

Formally then, our min/max scheme becomes:

$$\tilde{F}_{\min/\max} = \begin{cases} \max(\kappa, 0) & \text{if Average}(x, y) < T_{\text{threshold}} \\ \min(\kappa, 0) & \text{otherwise.} \end{cases} \qquad (18.4)$$

Further details about this scheme may be found in Malladi and Sethian (1996a). In that work, these techniques are applied to a wide range of images, including salt-and-pepper, multiplicative and Gaussian noise applied to monochrome, grey-scale, textured, and colour images.

Results
In this section, we provide a few examples of this min/max flow. In Fig. 32, 50% and 80% grey-scale noise is added to a monochrome image of a hand-written character. The noise is added as follows: $X\%$ noise means that at $X\%$ of the pixels, the given value is replaced with a number chosen with uniform distribution between 0 and 255. Here, the min/max switch function is taken relative to the value 127.5 rather than zero. The restored figures are converged. Continued application of the scheme yields almost no change in the results.

Next, salt-and-pepper grey-scale noise is removed from a grey-scale image. The results are obtained as follows. Begin with 40% noise in Fig. 33a. First, the min/max flow from equation (18.4) is applied until a steady state is reached (Fig. 33b). This removes most of the noise. The scheme is then continued with a larger threshold stencil for the threshold to remove further noise (Fig. 33c). For the larger stencil, we compute the average over a larger

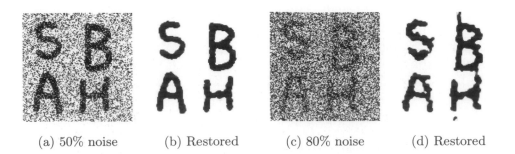

(a) 50% noise   (b) Restored   (c) 80% noise   (d) Restored

Fig. 32. Image restoration of binary images with grey-scale salt-and-pepper noise using min/max flow: restored shapes are final shape obtained ($T = \infty$).

(a) 40% noise   (b) Min/max flow   (c) Cont: larger stencil

Fig. 33. Min/max flow applied to grey-scale image.

disk and calculate the threshold value $T_{\text{threshold}}$ using a correspondingly longer tangent vector.

Finally, we show the effect of this scheme applied to an image upon which 100% Gaussian grey-scale noise has been superimposed; a random component drawn from a Gaussian distribution with mean zero is added to each (every) pixel. Fig. 34 shows the 'noisy' original together with the reconstructed min/max flow image.

19. Combustion, crystal growth, and two-fluid flow

In this section, we consider some applications of the level set methodology to a large and challenging class of interface problems. These problems are characterized by physical phenomena in which the front acts as a boundary condition to a partial differential equation, and the solution of this equation controls the motion of the front. In combustion problems, the interface is a flame, and both exothermic expansion along the front and flame-induced vorticity drive the underlying fluid mechanics. In crystal growth and dendritic

Original   Reconstructed

Fig. 34. Continuous Gaussian noise added to image.

formation, the interface is the solid/liquid boundary, and is driven by a jump condition related to heat release along the interface. In two-fluid problems, the interface represents the boundary between two immiscible fluids of significantly different densities and/or viscosities, and the surface tension along this interface plays a significant role in the motion of the fluids.

From an algorithmic perspective, the significant level set issue is that information about the speed of the front itself must be somehow transferred to the Eulerian framework that updates the level set function at the fixed grid points. This is a significant challenge for two reasons.

- In many situations, the interface velocity is determined by the interaction of local geometric quantities of the front itself (such as curvature) with global variables on either side of the interface (for example, jumps in velocity, heat, or concentration of species). If these global variables are calculated on a grid, it may be difficult to extend the values to the front itself where they are required to evaluate the speed function F. However, these quantities are only known at grid points, not at the front itself, where the relationship has meaning.

- It may be very difficult to extend this interface velocity back to the grid points (that is, to the other level sets). This problem, known as the *extension problem* (see Adalsteinsson and Sethian (1995b, 1995c)) must be solved in order for the level set method to work; some mechanism of updating the grid points in the neighbourhood of the zero level set is required.

In this section, we discuss three application areas where these problems have been solved. The common link in these sections is the presence of a term in the equation represented by a Dirac delta function along the interface. The interested reader is referred to the literature cited below for a detailed description of the algorithms and results.

19.1. Turbulent combustion of flames and vorticity, exothermicity, flame stretch and wrinkling

In Rhee et al. (1995), a flame is viewed as an infinitely thin reaction zone, separating two regions of different but constant densities. The hydrodynamic flow field is two dimensional and inviscid, and the Mach number is vanishingly small. This corresponds to the equations of zero Mach number combustion, introduced in Majda and Sethian (1984). The flame propagates into the unburnt gas at a prescribed flame speed S_u, which depends on the local curvature, due to the focusing/de-focusing of heating effects as a function of flame shape.

As the reactants are converted to products (that is, as the material makes the transition from 'unburnt' to 'burnt'), the local fluid undergoes volume increase known as exothermic expansion, associated with the density jump. At the same time, pressure gradients tangential to the flame cause different accelerations in the light and heavy gases. This causes a production of vorticity (known as baroclinic torque) across the flame, since the pressure gradient is not always aligned with the density gradient. Together, the burning of the flame acts as source of vorticity and volume for the underlying hydrodynamic field, both of which in turn affect the evolution of the flame interface.

This model presents a significant challenge for a level set method. The flame is tracked by identifying the flame interface as the zero level set of the level set function. The curvature is determined using the expression given in equation (11.1). The vortical field is represented by a collection of vortex blobs as in vortex method; see Chorin (1973) and Sethian (1991). The exothermic field is determined by solving a Poisson's equation on the underlying grid with right-hand side given by smearing the Dirac delta function to the neighbouring grid points. The no-flow boundary is satisfied by the addition of a potential flow that exactly cancels the existing flow field. Finally, the tangential stretch component is evaluated by tracing the values of the tangential velocity from the given position backwards along the normal to the front to evaluate the change in tangential velocity. For complete details, see Rhee et al. (1995).

Fig. 35, taken from Rhee et al. (1995), shows two results from this algorithm. We compare an anchored flame, with upstream turbulence imposed by a statistical distribution of positive and negative vortices. The goal here is to understand the effect of exothermicity and flame-induced vorticity on

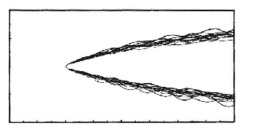
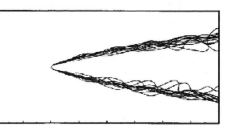

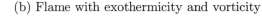

    (a) Flame with exothermicity   (b) Flame with exothermicity and vorticity

Fig. 35. Comparison of flame brush.

the flame wrinkling and stability. In Fig. 35a, we show an anchored flame in the oncoming turbulent field; here, different time snapshots are superimposed upon each other to show the flame 'brush'. In Fig. 35b, we turn on the effects of both volume expansion (a large density jump) and vorticity generation along the flame front. The resulting flow field generates a significantly wider flame brush, as the vorticity induces flame wrinkling and the exothermicity affects the surrounding flow field.

In the above application of the level method, the front acted as source of volume and vorticity. In the next application, developed in Sethian and Strain (1992), the interface motion is controlled by a complex jump condition.

19.2. Crystal growth and dendritic solidification

Imagine a container filled with a liquid such as water, which has been so smoothly and uniformly cooled below its freezing point that the liquid does not freeze. The system is now in a 'metastable' state, where a small disturbance such as dropping a tiny seed of the solid phase into the liquid will initiate a rapid and unstable process known as *dendritic solidification*. The solid phase will grow from the seed by sending out branching fingers into the distant cooler liquid nearer the undercooled wall. This growth process is *unstable* in the sense that small perturbations of the initial data can produce large changes in the time-dependent solid/liquid boundary.

Mathematically, this phenomenon can be modelled by a moving boundary problem. The temperature field satisfies a heat equation in each phase, coupled through two boundary conditions on the unknown moving solid/liquid boundary, as well as initial and boundary conditions. First, the normal velocity of the interface depends on the jump in the normal component of the heat flux across the interface. Second, the temperature at the interface itself depends on the local curvature and the velocity. Thus, the goal is to incorporate the influence of the front on the heat solvers across the interface. For further details, see Cahn and Hilliard (1958) and Mullins and Sekerka (1963).

A variety of techniques is possible to approximate numerically these equations of motion. One approach is to solve the heat equation in each phase and try to move the boundary so that the two boundary conditions are satisfied. Another approach is to recast the equations of motion as a single integral equation on the moving boundary and solve the integral equation numerically.

In Sethian and Strain (1992) a hybrid level set/boundary integral approach was developed, which includes the effects of undercooling, crystalline anisotropy, surface tension, molecular kinetics, and initial conditions. The central idea is to exploit a transformation due to Strain (1989), which converts the equations of motion into a single, history-dependent boundary integral equation on the solid/liquid boundary, given by

$$\epsilon_\kappa(n)\kappa + \epsilon_V(n)V + U + H \int_0^t \int_{\Gamma(t')} K(x, x', t-t')\, V(x', t')\, dx'\, dt' = 0 \quad (19.1)$$

for all x on the interface $\Gamma(t)$. Here, K is the heat kernel, ϵ_κ and ϵ_V are constants, U is the temperature, V is the normal velocity of the interface, and H is the latent heat of solidification. Note that the velocity V depends not only on the position of the front but also on its previous history. Thus, as shown in Strain (1989), information about the temperature off the front is stored in the previous history of the boundary. This can be evaluated by a combination of fast techniques; see Strain (1988, 1989, 1990) and Greengard and Strain (1990).

In Fig. 36, we show one example from Sethian and Strain (1992) in which the effect of changing the latent heat of solidification H is analysed. Since the latent heat controls the balance between the pure geometric effects and the solution of the history-dependent heat integral, increasing H puts more emphasis on the heat equation/jump conditions. Calculations are performed on a unit box, with a constant undercooling on the side walls of $u_B = -1$. The kinetic coefficient is $\epsilon_V = .001$; the surface tension coefficient is $\epsilon_\kappa = .001$; there is no crystalline anisotropy. The initial shape was a perturbed circle. A 96×96 mesh is used with time step $\Delta t = .00125$. The calculations are all plotted at the same time.

In the calculations shown, H is varied smoothly. In Fig. 36a, $H = .75$; the dominance of geometric motion serves to create a rapidly evolving boundary that is mostly smooth. H is increased in each successive figure, ending with $H = 1.0$ in Fig. 36d. As the latent heat of solidification is increased, the growing limbs expand outwards less smoothly, and instead develop flat ends. These flat ends are unstable and serve as precursors to tip splitting. We also note that the influence of the heat integral slows down the evolving boundary, as witnessed by the fact that all the plots are given at the same time. Presumably, increasing latent heat decreases the most unstable wavelength, as described by linear stability theory. The final shape shows side-branching, tip splitting, and the strong effects of the side walls.

Upper left: $H = .75$ Upper right: $H = .833$
Lower left: $H = .916$ Lower right: $H = 1.0$

Fig. 36. Effect of changing latent heat.

20. Two phase flow

In this section, we discuss level set methods applied to problems of two-phase flow.

Two early applications of fluid dynamical problems using level set methods to track the interface are the projection method calculations of compressible gas dynamics of Mulder, Osher and Sethian (1992) and the combustion calculations of Zhu and Sethian (1992). Each viewed the interface as the zero level set, and tracked this interface as a method of separating the two regions.

Firstly, in Mulder et al. (1992) the evolution of rising bubbles in compressible gas dynamics was studied. The level set equation for the evolving

interface separating two fluids of differing densities was incorporated inside
the conservation equations for the fluid dynamics. Both the Kelvin–Helmholtz
instability and the Rayleigh–Taylor instability were studied; the density ratio
was about 30 to 4, and both gases were treated as perfect gases. Considerable
discussion was devoted to the advantages and disadvantages of embedding the
level set equation as an additional conservation law.

Next, in the combustion calculations presented in Zhu and Sethian (1992),
the interface represented a flame propagating from the burnt region into the
unburnt region. Unlike the above calculations concerning flame stability in
flame holders, in these calculations the flame was viewed as a 'cold flame';
that is, the hydrodynamic flow field affected the position of the flame, but
the advancing flame did not in turn affect the hydrodynamic field. In these
calculations, the hydrodynamic field was computed using Chorin's projection
method (see Chorin (1968)), coupled to the level set approach. The problem
under study was the evolution of a flame inside a swirling two-dimensional
chamber; and the results showed the intermixing that can occur, the creating
of pockets of unburnt fuel surrounded by burnt pockets.

These two works were followed by the work of Chang, Hou, Merriman and
Osher (1994) and Sussman et al. (1994), using projection methods coupled to
the level set equation to study the motion of incompressible, immiscible fluids
where steep gradients in density and viscosity exist across the interface, and
the role of surface tension is crucial.

There are three key aspects of these fascinating calculations. First, using a formulation first developed by Brackbill, Kothe and Zemach (1992),
the surface tension generated by the level set curvature expression can be
smoothed using a mollified delta function to the neighbouring grid points;
this smoothing denotes a 'thickness' for the interface layer, and allows the
role of surface tension to be transported to the grid for inclusion in the projection method. Second, the contribution due to surface tension is converted
from a delta function to the Heaviside formulation, and incorporated as such
into the projection step. This eliminates the standard numerical instabilities
and oscillations that plague attempts to directly difference the delta function
itself. Third, the level set lines cease to correspond to the distance function
due to the significant variation in fluid velocities across the interface; to redistribute the level set contours, the re-initialization idea using iteration on
the sign of ϕ described in equation (13.1) was developed.

The calculations performed using these techniques show a wide range of
applications concerning falling drops, colliding drops, and the role of surface
tension, and open the door to a range of important fluid dynamics applications. We refer the reader to Sussman et al. (1994) and Chang et al. (1994)
for further details.

21. Constrained problems: minimal surfaces and shape recovery

Another set of applications include problems in which the motion of the front is constrained by external boundary conditions. In this section, we briefly review some work on the construction of minimal surfaces and shape recovery from images.

21.1. Minimal surfaces

The basic problem may be stated as follows. Consider a closed curve $\Gamma(s)$ in R^3. The goal is to construct a membrane with boundary Γ and mean curvature zero. In some cases, this can be achieved as follows. Given the bounding wire frame Γ, consider some initial surface $S(t = 0)$ whose boundary is Γ. Let $S(t)$ be the family of surfaces parametrized by t obtained by allowing the initial surface $S(t = 0)$ to evolve under mean curvature, with boundary always given by Γ. Defining the surface S by $S = \lim_{t\to\infty} S(t)$, one expects that the surface S will be a minimal surface for the boundary Γ. Several computational approaches exist to construct such minimal surfaces based on this approach, including Brakke's Surface Evolver program (Brakke 1990).

A level set approach to this problem rests on embedding the motion of the surface towards its minimal energy as the zero level set of a higher-dimensional function. Thus, given an initial surface $S(0)$ passing through Γ, construct a family of neighbouring surfaces by viewing $S(0)$ as the zero level set of some function ϕ over all of R^3. Using the level set equation (3.5), evolve ϕ according to the speed law $F(\kappa) = -\kappa$. Then a possible minimal surface S will be given by

$$S = \lim_{t\to\infty} \{x : \phi(x,t) = 0\}. \tag{21.1}$$

The difficult challenge with the above approach is to ensure that the evolving zero level set always remains attached to the boundary Γ. This is accomplished in Chopp (1993), by creating boundary conditions of grid points closest to the wire frame linking together the neighbouring values of ϕ, to force the level set $\phi = 0$ through Γ. Thus we obtain a constrained level set problem: we track mean curvature flow requiring that the evolving zero level set remains attached to the front.

In Fig. 37, taken from Chopp (1993), the minimal surface spanning two rings each of radius 0.5 and at positions $x = \pm.277259$ is computed. A cylinder spanning the two rings is taken as the initial level set $\phi = 0$. A $27 \times 47 \times 47$ mesh with space step 0.025 is used. The final shape is shown in Fig. 37.

Fig. 37. Minimal surface: catenoid.

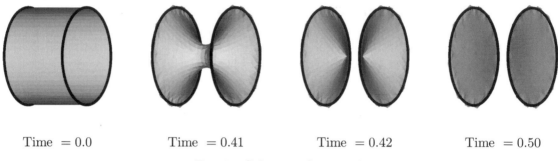

Time = 0.0   Time = 0.41   Time = 0.42   Time = 0.50

Fig. 38. Splitting of catenoid.

Next, in Fig. 38 (again taken from Chopp (1993)), this same problem is computed, but the rings are placed far enough apart so that a catenoid solution cannot exist. Starting with a cylinder as the initial surface, the evolution of this surface is computed as it collapses under mean curvature while remaining attached to the two wire frames. As the surface evolves, the middle pinches off and the surface splits into two surfaces, each of which quickly collapses into a disk. The final shape of a disk spanning each ring is indeed a minimal surface for this problem. This example illustrates one of the virtues of the level set approach. No special cutting or *ad hoc* decisions are employed to decide when to break the surface. Instead, viewing the zero level set as but one member of a family of flowing surfaces allows this smooth transition. Further results may be found in Chopp (1993).

21.2. Shape detection/recovery

Imagine that one is given an image. The goal in *shape detection/recovery* is to extract a particular shape from that image; here, 'extract' means to produce a mathematical description of the shape, which can be used in a variety of forms. The work on level set techniques applied to shape recovery described here was first presented in Malladi et al. (1994); further work using the level set scheme in the context of shape recovery may be found in Malladi, Sethian and Vemuri (1995*b*), Malladi and Sethian (1994), Malladi, Adalsteinsson and Sethian (1995*a*) and Caselles, Catte, Coll and Dibos (n.d.). We refer the interested reader to those papers for motivation, details, and a large number of examples.

Imagine that we are given an image, with the goal of isolating a shape within the image. Our approach (see Malladi et al. (1994)) is motivated by the active force contour/snake approach to shape recovery given by Kass, Witkin and Terzopoulos (1988). Consider a speed function of the form $1 - \epsilon\kappa$ $(1 + \epsilon\kappa)$, where ϵ is a constant. As discussed earlier, the constant acts as an advection term, and is independent of the moving front's geometry. The front uniformly expands (contracts) with speed 1 (-1) depending on the sign, and is analogous to the inflation force defined in Cohen (1991). The diffusive second term ϵK depends on the geometry of the front and smooths out the high curvature regions of the front. It has the same regularizing effect on the front as the internal deformation energy term in thin-plate-membrane splines (Kass et al. 1988).

Our goal now is to define a speed function from the image data that acts as a halting criterion for this speed function. We multiply the above speed function by the term

$$k_I(x,y) = \frac{1}{1 + \mid \nabla G_\sigma * I(x,y) \mid}, \qquad (21.2)$$

where the expression $G_\sigma * I$ denotes the image convolved with a Gaussian smoothing filter whose characteristic width is σ. The term $|\nabla G_\sigma * I(x,y)|$ is essentially zero except where the image gradient changes rapidly, in which case the value becomes large. Thus, the filter $k_I(x,y)$ is close to unity away from boundaries, and drops to zero near sharp changes in the image gradient, which presumably corresponds to the edge of the desired shape. In other words, the filter function anticipates steep changes in the image gradient, and stops the evolving front from passing out of the desired region.

Thus the algorithm works as follows. A small front (typically a circle) is started inside the desired region. This front then grows outwards and is stopped at the shape boundary by the filter term, which drops the value of the speed function F to zero.

There are several desirable aspects of this approach:

- the initial front can consist of many fronts; due to the topological capabilities of the level set method, these fronts will merge into a single front as it grows into the particular shape
- the front can follow intricate twists and turns in the desired boundary
- use of narrow band techniques makes the algorithm very fast
- the technique can be used to extract three-dimensional shapes as well by initializing in a ball inside the desired region

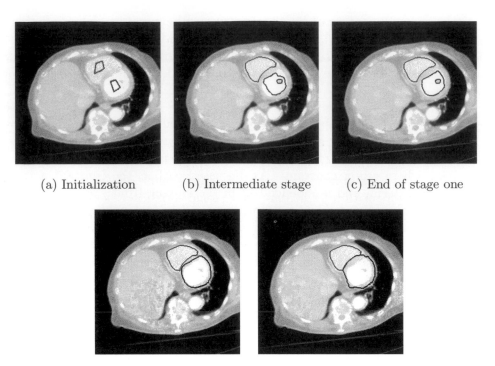

(a) Initialization   (b) Intermediate stage   (c) End of stage one

(d) Intermediate stage two   (e) End of stage two

Fig. 39. Shape extraction from heart data.

- small isolated spots of noise where the image gradient changes substantially are ignored; the front propagates *around* these points and closes back in on itself and then disappears.

As a demonstration, level set shape recovery techniques are applied to the difficult problem of extracting images of the left and right ventricles of the heart. In these calculations, taken from Malladi and Sethian (1996*b*), the problem is initialized by simultaneously tagging both the left and the right ventricle; the right ventricle is found by the evolving front, as is the left ventricle. Note that in the evolution of the right ventricle front, the papillary muscle is also found; see Fig. 39. This feature is obtained by initializing with a single contour, enclosing the papillary muscle and separating into an inner ring and outer ring. After the outer walls of the left and right ventricles are recovered, the outer wall of the right ventricle is extracted; this is done by temporarily relaxing the stopping criterion, and allowing the front to move past the inner wall of the right ventricle. Once this occurs, the stopping criterion is turned back on, and the front expands until the outer wall is found.

This technique for shape detection/recovery can be performed in three dimensions if three-dimensional data is available. For details of this and related work, see Malladi et al. (1994), Malladi et al. (1995b) and Malladi and Sethian (1996b).

22. Applications of the fast marching level set method

In the case of a monotonically advancing front whose speed in the normal direction depends only on position, we have previously seen that this can be converted into a stationary time problem. Furthermore, we have developed a fast marching algorithm for solving the Eikonal equation associated with this problem. Here, we show two applications of this technique.

22.1. Shape-from-shading

Suppose we illuminate a non-self-shadowing surface from a single point light source. At each point of the surface, one can define the brightness map I which depends on the reflectivity of the surface and the angle between the incoming light ray and the surface normal. Points of the surface where the normal is parallel to the incoming beam are brightest; those where the normal is almost orthogonal are darkest (again, we rule out surfaces that are self shadowing). The goal of *shape-from-shading* is to reconstruct the surface from its brightness function I.

We point out right away that the problem as posed does not have a unique solution. For example, imagine a beam coming straight down; it is impossible to differentiate a surface from its mirror image from the brightness function. That is, a deep valley could also be a mountain peak. Other ambiguities can exist, we refer the reader to Rouy and Tourin (1992) and Kimmel and Bruckstein (1992). Nonetheless, in its simplest form the shape-from-shading problem provides a simple example of an Eikonal equation that can be solved using our fast marching level set method.

We begin by considering a surface $T(x, y)$; the surface normal is then given by

$$n = \frac{(-T_x, -T_y, 1)}{(|\nabla T|^2 + 1)^{1/2}}. \tag{22.1}$$

Let (α, β, γ) be the direction from the light source. In the simplest case of a Lambertian surface, the brightness map is given in a very simple form by

$$I(x, y) = (\alpha, \beta, \gamma) \cdot n. \tag{22.2}$$

Thus, the shape-from-shading problem is to reconstruct the surface $T(x, y)$ given the brightness map $I(x, y)$.

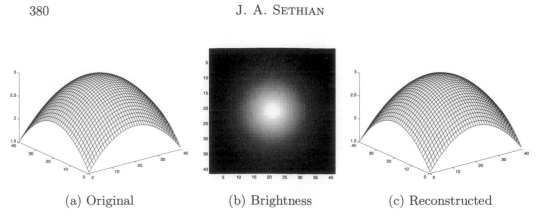

(a) Original   (b) Brightness   (c) Reconstructed

Fig. 40. Shape-from-shading reconstruction of paraboloid surface.

Consider the simplest case, namely that in which the light comes from straight down. Then the light source vector is $(0, 0, 1)$, and we have

$$I(x, y) = \frac{1.}{(|\nabla T|^2 + 1)^{1/2}}. \tag{22.3}$$

Rearranging terms, we then have an Eikonal equation for the surface, namely

$$|\nabla T| = \sqrt{\frac{1}{I^2} - 1}, \tag{22.4}$$

where n is the normal to the surface. We still need initial conditions for this problem. Let us imagine that at extrema of T we know the values of T. Then we can construct a viable solution surface using our fast marching method.

To demonstrate, we start with a given surface, first compute the brightness map I, and then reconstruct the surface by solving the above Eikonal equation. In Fig. 40, we show a paraboloid surface of the form $T = 3. - 3(x^2 + y^2)$. In Fig. 40a we show the original surface, in Fig. 40b we show the brightness map $I(x, y)$, and in Fig. 40c we show the reconstructed surface. This surface is 'built' by setting $T = 3$ at the point where the maximum is obtained, and then solving the Eikonal equation.

As a more complex example, we use a double Gaussian function of the form

$$T(x, y) = 3e^{-(x^2 + y^2)} - 2e^{-20((x - .05)^2 + (y - .05)^2)}. \tag{22.5}$$

Once again, we compute the brightness map and then reconstruct the surface; see Fig. 41.

We have barely touched the topic of shape-from-shading; in the case of multiple extrema and non-vertical light sources, more care must be taken, and we refer the interested reader to the above sources. Nonetheless, the fast marching level set algorithm is extremely effective for these problems.

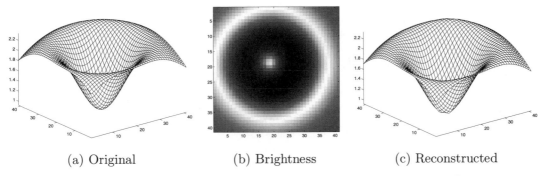

(a) Original   (b) Brightness   (c) Reconstructed

Fig. 41. Shape-from-shading reconstruction of double Gaussian surface.

22.2. *Photolithography development*

One component process in the manufacturing of microchips is the stage of *lithography development*; see Section 17. In this process, the resist properties of a material have been altered due to exposure to a beam which has been partially blocked by a pattern mask. The material is then 'developed', which means the material with less resistivity is etched away. While the process is discussed in more detail in the next section, at this point we simply note that the problem reduces to that of following an initially plane interface propagating downwards in three dimensions, where the speed in the normal direction is given as a supplied rate function at each point. The speed $F = F(x, y, z)$ depends only on position; however, it may change extremely rapidly. The goal in lithography development is to track this evolving front. In order to develop realistic structures in three-dimensional development profiles, a grid of size $300 \times 300 \times 100$ is not unreasonable; hence a fast algorithm is required to perform the development step.

Start with a flat profile at height $z = 1$ in the unit cube centred at $(.5, .5, .5)$ and follow the evolution of the interface downwards with speed given by the model Gaussian rate function

$$F(x, y, z) = e^{-64(r^2)}(\cos^2(12z) + .01), \qquad (22.6)$$

where $r = \sqrt{(x - .5)^2 + (y - .5)^2}$. This rate function F corresponds to the effect of standing waves which change the resist properties of the material, and causes sharp undulations and turns in the evolving profile. In Fig. 42, we show the profile etched out by such an initial state; the calculation is carried out until $T = 10$.

In Fig. 43, we give timings for a parameter study on a Sparc10 for the speed function $F = e^{-64(r^2)}(\cos^2(6z) + .01)$. We note that loading the file containing the model Gaussian rate function F is a significant proportion of the total compute time.

Fig. 42. Lithographic development on $50 \times 50 \times 50$ grid.

Grid size	Time to load rate file	Time to propagate front	Total time
$50 \times 50 \times 50$	0.1 secs	0.7 secs	0.8 secs
$100 \times 100 \times 100$	1.2 secs	8.2 secs	9.4 secs
$150 \times 150 \times 150$	3.9 secs	37.8 secs	41.7 secs
$200 \times 200 \times 200$	9.0 secs	80.0 secs	89 secs

Fig. 43. Timings for development to T=10: Sparc 10.

Further details of the application of our fast marching level set method may be found in Sethian (1995*c*, 1996) and Sethian et al. (1996).

23. A final example: etching and deposition for the microfabrication of semiconductor devices

23.1. Background

A goal of numerical simulations in microfabrication of semiconductor devices is to model the process by which silicon devices are manufactured. Here, we briefly summarize the stages involved. First, a single crystal ingot of silicon is extracted from molten pure silicon. This silicon ingot is then sliced into several hundred thin wafers, each of which is then polished to a smooth finish. A thin crystalline layer is then oxidized, a light sensitive 'photoresist' is applied, and then the wafer is covered with a pattern mask that shields part of the photoresist. This pattern mask contains the layout of the circuit itself. Under exposure to a light or an electron beam, the exposed photoresist polymerizes and hardens, leaving an unexposed material which is then etched away in a dry etch process, revealing a bare silicon dioxide layer. Ionized impurity atoms such as boron, phosphorus and argon are then implanted into the pattern of the exposed silicon wafer, and silicon dioxide is deposited at reduced pressure in a plasma discharge from gas mixtures at a low temperature. Finally, thin films such as aluminium are deposited by processes such as plasma sputtering, and contacts to the electrical components and component interconnections are established. The result is a device that carries the desired electrical properties.

The above processes produce considerable change in the surface profile as it undergoes the stages of etching, deposition, and photolithography. This problem is known as the 'surface topography problem' in microfabrication, and is controlled by a large collection of physical effects, including the visibility of the etching/deposition source at each point of the evolving profile, surface diffusion along the front, non-convex sputter laws that produce faceting, shocks and rarefactions, material-dependent discontinuous etch rates, and masking profiles.

The underlying physical effects involved in etching, deposition and lithography are quite complex; excellent reviews are due to Neureuther and his group: see Helmsen (1994), Scheckler (1991), Scheckler, Toh, Hoffstetter and Neureuther (1991), Toh (1990) and Toh and Neureuther (1991), as well as Cale and Raupp (1990a, 1990b, 1990c), McVittie, Rey, Bariya et al. (1991) and Rey, Cheng, McVittie and Saraswat (1991). The effects may be summarized briefly as follows.

Deposition Particles are deposited on the surface, which causes build-up in the profile. The particles may either isotropically condense from the surroundings (known as chemical or 'wet' deposition), or be deposited from a source. In the latter case, we envision particles leaving the source and depositing on the surface; the main advantage of this approach is increased control over the directionality of surface deposition. The rate of deposition, and hence growth of the layer, may depend on source masking, visibility effects between the source and surface point, angle-dependent flux distribution of source particles, the angle of incidence of the particles relative to the surface normal direction, reflection of deposited particles, and surface diffusion effects.

Etching Particles remove material from the evolving profile boundary. The material may be isotropically removed, as in chemical or 'wet' etching, or chipped away through reactive ion etching, also known as 'ion milling'. Similar to deposition, the main advantage of reactive ion etching is en-

hanced directionality, which becomes increasingly important as device
sizes decrease substantially and etching must proceed in vertical directions without affecting adjacent features. The total etch rate consists of
an ion-assisted rate and a purely chemical etch rate due to etching by
neutral radicals, which may still have a directional component. As in
the above, the total etch rate due to wet and directional milling effects
can depend on source masking, visibility effects between the source and
surface point, angle-dependent flux distributions of source particles, the
angle of incidence of the particles relative to the surface, reflection/re-
emission of particles, and surface diffusion effects.

Lithography As discussed earlier, the underlying material is treated by an
electromagnetic wave that alters the resist property of the material. The
aerial image is found, which then determines the amount of crosslinking
at each point in the material, which then produces the etch/resist rate
at each point of the material. A profile is then etched into the material,
where the speed of the profile in its normal direction at any point is
given by the underlying etch rate. The key factors that determine the
evolving shape are the etch/resist profile and masking effects.

In the final analysis, the above reduces to our familiar problem of tracking the boundary of a moving interface moving under a speed function F.
Abstractly, we may write

$$F = F_{\text{Deposition/Etching}} + F_{\text{Lithography}}. \tag{23.1}$$

Of course, all effects do not take place at once; however, the design of the numerical algorithm allows various combinations of terms to be 'turned on' during any time step of the surface advancement. For details and additional calculations of level set methods applied to microfabrication, see Adalsteinsson
and Sethian (1995b, 1995c, 1996).

23.2. Results

Etching/deposition
We begin in Fig. 44 with a deposition source above a trench, where deposition
material is emitted from a line source from the solid line above the trench. In
this experiment, the deposition rate is the same in all directions. The effects of
shadowing are considered. Fig. 44a shows results for 40 computational cells
along the width of the computed region (between the two vertical dashed
lines); Fig. 44b has 80 cells, and Fig. 44c has 160 cells. The time step for
all three calculations is $\Delta t = .00625$. The calculations are performed with
a narrow band tube width of 6 cells on either side of the front. There is
little change between the calculation with 80 cells and the one with 160 cells,
indicating that convergence has been achieved. As the walls pinch toward
each other, the seen visible angle decreases and the speed diminishes.

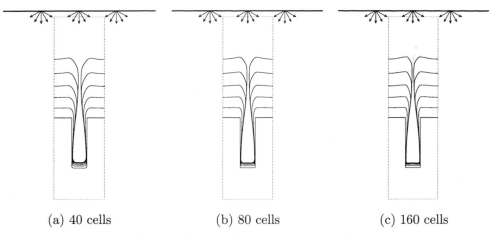

(a) 40 cells   (b) 80 cells   (c) 160 cells

Fig. 44. Source deposition into trench.

Ion-milling: non-convex sputter laws

A more sophisticated set of examples arises in simulations (for example, of ion milling) in which the normal speed of the profile depends on the angle of incidence between the surface normal and the incoming beam. This yield function is often empirically fit from experiment, and has been observed to cause such effects as faceting at corners; see Leon, Tazawa, Saito, Yoshi and Scharfetter (1993) and Katardjiev, Carter and Nobes (1988). As shown in Adalsteinsson and Sethian (1995*b* and 1995*c*), such yield functions can often give rise to non-convex Hamiltonians, in which case alternative schemes must be used. To study this phenomenon, in Fig. 45 we consider several front motions and their effects on corners. We envision an etching beam coming down in the vertical direction. In the cases under study here, the angle θ refers to the angle between the surface normal and the positive vertical. For this set of calculations, in order to focus on the geometry of sputter effects on shocks/rarefaction fan development, visibility effects are ignored. The calculations are made using the schemes for non-convex Hamiltonians described earlier. Following our usual notation, let $F(\theta)$ be the speed of the front in direction normal to the surface.

In column A, the effects of purely isotropic motion are shown; thus the yield function is $F = 1$. Located above the yield graph are the motions of a downwards square wave. In column B, the effects of directional motion are shown; thus the yield function is $F = \cos(\theta)$. In this case, the horizontal components on the profile do not move, and vertical components move with unit speed. In column C, the effects of a yield function of the form $F = [1 + 4\sin^2(\theta)]\cos(\theta)$ are shown.

The results of these calculations are given in Fig. 45. The results show that the effects of angle dependent yield functions are pronounced. In column A

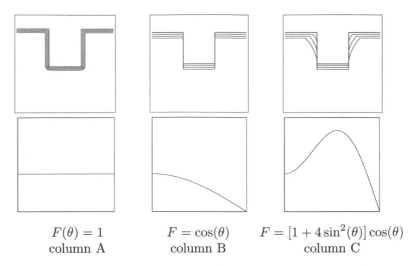

$$F(\theta) = 1 \qquad F = \cos(\theta) \qquad F = [1 + 4\sin^2(\theta)]\cos(\theta)$$
$$\text{column A} \qquad\qquad \text{column B} \qquad\qquad \text{column C}$$

Fig. 45. Effect of different yield functions: non-convex scheme.

the isotropic rate produces smooth corners, correctly building the necessary rarefaction fans in outward corners and entropy satisfying shocks in inward corners, as discussed and analysed in Sethian (1982 and 1985). In column B, the directional rate causes the front to be essentially translated upwards, with minimal rounding of the corners. In column C, the yield function results in faceting of inward corners where shocks form and sharp corners in the construction of rarefaction fans.

Discontinuous etch rates
Next, we study the effects of etching through different materials. In this example, the etch rates are discontinuous, and hence sharp corners develop in the propagating profile. The results of these calculations are shown in Fig. 46. A top material masks a lower material, and the profile etches through the lower material first and underneath the upper material. The profile depends on the ratio of the etch rates. In Fig. 46a, the two materials have the same etch rate, and hence the front simply propagates in its normal direction with unit speed, regardless of which material it is passing through. In Fig. 46b, the bottom material etches four times faster than the top; in Fig. 46c, the ratio is ten to one. Finally, in Fig. 46d, the ratio is forty to one, in which case the top material almost acts like a mask.

Simultaneous etching and deposition
Next, a parameter study of simultaneous etching and deposition is taken from Adalsteinsson and Sethian (1996). The speed function is

$$F = (1 - \alpha)F_{\text{etch}} + \alpha F_{\text{Deposition}}, \qquad (23.2)$$

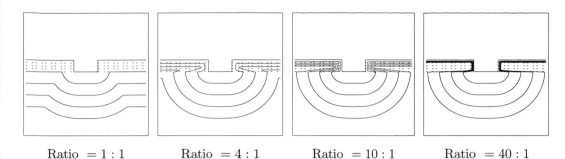

Ratio $= 1:1$ Ratio $= 4:1$ Ratio $= 10:1$ Ratio $= 40:1$

Fig. 46. Etch ratio = bottom material rate to top material rate.

where

$$F_{\text{etch}} = (5.2249\cos\theta - 5.5914\cos^2\theta + 1.3665\cos^4\theta),$$
$$F_{\text{Deposition}} = \beta F_{\text{Isotropic}} + (1-\beta)F_{\text{Source}}. \tag{23.3}$$

Visibility effects are considered in all terms except isotropic deposition. Fig. 47 shows the results of varying α and β between 0 and 1.

23.3. Three-dimensional simulations

Finally, a three-dimensional example of a non-convex sputter yield law is applied to an indented saddle, which gives rise to faceting as shown in Fig. 48. Complete details of the above and a large variety of simulations of etching, deposition, and lithography development may be found in Adalsteinsson and Sethian (1995b, 1995c, 1996)

24. Other work

The range of level set techniques extends far beyond the work covered here. Here, we point the reader to some additional topics.

On the theoretical side, considerable analysis of level set methods has been performed in recent years; see, for example, Brakke (1978), Ecker and Huisman (1991), Evans and Spruck (1991, 1992a, 1992b, 1995), Chen et al. (1991), Giga and Goto (1992), Giga et al. (1992) and Ambrosio and Soner (1994). This work has concentrated on many aspects, including questions of existence and uniqueness, pathological cases, extensions of these ideas to fronts of co-dimension greater than one (such as evolving curves in three dimensions), coupling with diffusion equations, links between the level set technique and Brakke's groundbreaking original varifold approach.

On the theoretical/numerical analysis side, level set techniques exploit the considerable technology developed in the area of viscous solutions to Hamilton–Jacobi equations; see the work in Barles (1993), Crandall, Evans

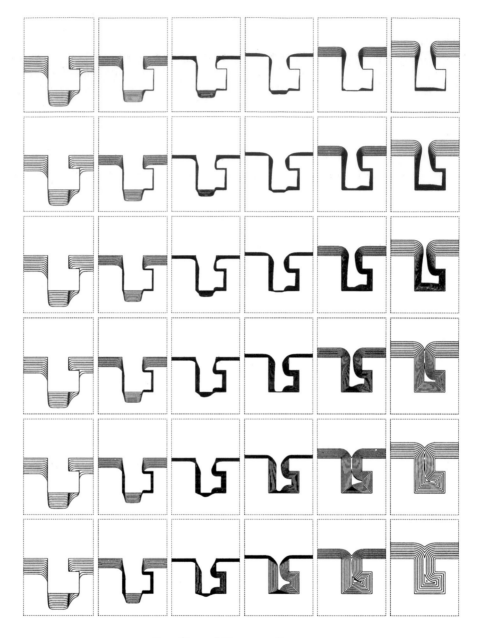

$$F = (1 - \alpha)F_{\text{etch}} + \alpha F_{\text{Deposition}}$$

$$F_{\text{etch}} = (5.2249\cos\theta - 5.5914\cos^2\theta + 1.3665\cos^4\theta)\cos\theta$$

$$F_{\text{Deposition}} = \beta F_{\text{Isotropic}} + (1 - \beta)F_{\text{Source}}$$
α increases from left to right
β increases from bottom to top

Fig. 47. Simultaneous etching and deposition.

Initial shape: $T = 0$ $F = [1 + 4\sin^2(\theta)]\cos(\theta)$, $T = 8$ Final rotated

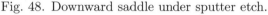

Fig. 48. Downward saddle under sputter etch.

and Lions (1984), Crandall, Ishii and Lions (1992), Crandall and Lions (1983) and Lions (1982).

A wide range of applications relates to level set methods, including work on minimal arrival times by Falcone (1994), flame propagation work by Zhu and Ronney (1995), a wide collection of applications from computer vision by Kimmel (1995), gradient flows applied to geometric active contour models (Kichenassamy, Kumar, Olver, Tannenbaum and Yezzi 1995), work on affine invariant scale space (Sapiro and Tannenbaum 1993a), and some related work on the scalar wave equation (Fatemi, Engquist and Osher 1995). We also refer the reader to the collection of papers from the International Conference on Mean Curvature Flow (Buttazzo and Visitin 1994).

Acknowledgements All calculations were performed at the University of California at Berkeley and the Lawrence Berkeley Laboratory. The detailed applications of the level set schemes discussed in this work are joint with D. Adalsteinsson, D. Chopp, R. Malladi, B. Milne, C. Rhee, J. Strain, L. Talbot and J. Zhu.

REFERENCES

D. Adalsteinsson and J. A. Sethian (1995a), 'A fast level set method for propagating interfaces', *J. Comp. Phys.* **118**, 269–277.

D. Adalsteinsson and J. A. Sethian (1995b), 'A unified level set approach to etching, deposition and lithography I: Algorithms and two-dimensional simulations', *J. Comp. Phys.* **120**, 128–144.

D. Adalsteinsson and J. A. Sethian (1995c), 'A unified level set approach to etching, deposition and lithography II: Three-dimensional simulations', *J. Comp. Phys.* **122**, 348–366.

D. Adalsteinsson and J. A. Sethian (1996), 'A unified level set approach to etching, deposition and lithography III: Complex simulations and multiple effects', *J. Comp. Phys.* To be submitted.

L. Alvarez, P.-L. Lions and M. Morel (1992), 'Image selective smoothing and edge detection by nonlinear diffusion. II', *SIAM J. Num. Anal.* **29**, 845–866.

L. Ambrosio and H. M. Soner (1994), Level set approach to mean curvature flow in arbitrary codimension. Preprint.

S. Angenent (1992), Shrinking doughnuts, in *Proceedings of Nonlinear Diffusion Equations and Their Equilibrium States, 3* (N. G. Lloyd et al., eds), Birkhäuser, Boston.

M. Bardi and M. Falcone (1990), 'An approximation scheme for the minimum time function', *SIAM J. Control Optim.* **28**, 950–965.

G. Barles (1985), Remarks on a flame propagation model, report 464, INRIA.

G. Barles (1993), 'Discontinuous viscosity solutions of first order Hamilton–Jacobi equations: A guided visit', *Non-linear Analysis: Theory, Methods, and Applications* **20**, 1123–1134.

G. Barles and P. E. Souganidis (1991), 'Convergence of approximation schemes for fully non-linear second order equations', *Asymptotic Anal.* **4**, 271–283.

J. B. Bell, P. Colella and H. M. Glaz (1989), 'A second-order projection method for the incompressible Navier–Stokes equations', *J. Comp. Phys.* **85**, 257–283.

M. Berger and P. Colella (1989), 'Local adaptive mesh refinement for shock hydrodynamics', *J. Comp. Phys.* **82**, 62–84.

J. U. Brackbill, D. B. Kothe and C. Zemach (1992), 'A continuum method for modeling surface tension', *J. Comp. Phys.* **100**, 335–353.

K. A. Brakke (1978), *The Motion of a Surface by Its Mean Curvature*, Princeton University Press, Princeton University.

K. A. Brakke (1990), Surface evolver program, Technical Report GCC 17, University of Minnesota. Geometry Supercomputer Project.

L. Bronsard and B. Wetton (1995), 'A numerical method for tracking curve networks moving with curvature motion', *J. Comp. Phys.* **120**, 66–87.

G. Buttazzo and A. Visitin (1994), Motion by mean curvature and related topics, in *Proceedings of the International Conference at Trento, 1992*, Walter de Gruyter, New York.

J. E. Cahn and J. E. Hilliard (1958), 'Free energy of a non-uniform system. I. Interfacial Free Energy', *J. Chem. Phys.* **28**, 258–267.

T. S. Cale and G. B. Raupp (1990a), 'Free molecular transport and deposition in cylindrical features', *J. Vac. Sci. Tech., B* **8**, 649–655.

T. S. Cale and G. B. Raupp (1990b), 'Free molecular transport and deposition in long rectangular trenches', *J. Appl. Phys.* **68**, 3645–8652.

T. S. Cale and G. B. Raupp (1990c), 'A unified line-of-sight model of deposition in rectangular trenches', *J. Vac. Sci. Tech., B* **8**, 1242–1248.

V. Caselles, F. Catte, T. Coll and F. Dibos (n.d.), A geometric model for active contours in image processing, Technical Report 9210, CEREMADE, Université de Paris-Dauphiné, France. Internal report.

J. E. Castillo (1991), *Mathematical Aspects of Grid Generation*, Frontiers in Applied Mathematics 8, SIAM Publications.

Y. C. Chang, T. Y. Hou, B. Merriman and S. J. Osher (1994), 'A level set formulation of Eulerian interface capturing methods for incompressible fluid flows', *J. Comp. Phys.* Submitted.

Y. Chen, Y. Giga and S. Goto (1991), 'Uniqueness and existence of viscosity solutions of generalized mean curvature flow equations', *J. Diff. Geom.* **33**, 749–786.

D. L. Chopp (1993), 'Computing minimal surfaces via level set curvature flow', *J. Comp. Phys.* **106**, 77–91.

D. L. Chopp (1994), 'Numerical computation of self-similar solutions for mean curvature flow', *J. Exper. Math.* **3**, 1–15.

D. L. Chopp and J. A. Sethian (1993), 'Flow under curvature: Singularity formation, minimal surfaces, and geodesics', *J. Exper. Math.* **2**, 235–255.

A. J. Chorin (1968), 'Numerical solution of the Navier–Stokes equations', *Math. Comp.* **22**, 745.

A. J. Chorin (1973), 'Numerical study of slightly viscous flow', *J. Fluid Mech.* **57**, 785–796.

A. J. Chorin (1980), 'Flame advection and propagation algorithms', *J. Comp. Phys.* **35**, 1–11.

L. D. Cohen (1991), 'On active contour models and balloons', *Computer Vision, Graphics, and Image Processing* **53**, 211–218.

P. Colella and E. G. Puckett (1994), *Modern Numerical Methods for Fluid Flow*, Lecture Notes, Department of Mechanical Engineering, University of California, Berkeley.

M. G. Crandall and P.-L. Lions (1983), 'Viscosity solutions of Hamilton–Jacobi equations', *Tran. AMS* **277**, 1 43.

M. G. Crandall, L. C. Evans and P.-L. Lions (1984), 'Some properties of viscosity solutions of Hamilton–Jacobi equations', *Tran. AMS* **282**, 487–502.

M. G. Crandall, H. Ishii and P.-L. Lions (1992), 'User's guide to viscosity solutions of second order partial differential equations', *Bull. AMS* **27**, 1–67.

K. Ecker and G. Huisman (1991), 'Interior estimates for hypersurfaces moving by mean curvature', *Inventiones Mathematica* **105**, 547–569.

B. Engquist and S. J. Osher (1980), 'Stable and entropy-satisfying approximations for transonic flow calculations', *Math. Comp.* **34**, 45.

L. C. Evans and J. Spruck (1991), 'Motion of level sets by mean curvature I', *J. Diff. Geom.* **33**, 635–681.

L. C. Evans and J. Spruck (1992a), 'Motion of level sets by mean curvature II', *Trans. AMS* **330**, 321–332.

L. C. Evans and J. Spruck (1992b), 'Motion of level sets by mean curvature III', *J. Geom. Anal.* **2**, 121–150.

L. C. Evans and J. Spruck (1995), 'Motion of level sets by mean curvature IV', *J. Geom. Anal.* **5**, 77–114.

L. C. Evans, H. M. Soner and P. E. Souganidis (1992), 'Phase transitions and generalized motion by mean curvature', *Comm. Pure Appl. Math.* **45**, 1097–1123.

M. Falcone (1994), The minimum time problem and its applications to front propagation, in *Motion by Mean Curvature and Related Topics. Proceedings of the International Conference at Trento, 1992*, Walter de Gruyter, New York.

M. Falcone, T. Giorgi and P. Loretti (1994), 'Level sets of viscosity solutions: Some applications to fronts and rendez-vous problems', *SIAM J. Appl. Math.* **54**, 1335–1354.

E. Fatemi, B. Engquist and S. J. Osher (1995), 'Numerical solution of the high frequency asymptotic wave equation for the scalar wave equation', *J. Comp. Phys.* **120**, 145–155.

M. Gage (1984), 'Curve shortening makes convex curves circular', *Inventiones Mathematica* **76**, 357.

M. Gage and R. Hamilton (1986), 'The heat-equation shrinking convex plane-curves', *J. Diff. Geom.* **23**, 69–96.

Y. Giga and S. Goto (1992), 'Motion of hypersurfaces and geometric equations', *J. Math. Soc. Japan* **44**, 99–111.

Y. Giga, S. Goto and H. Ishii (1992), 'Global existence of weak solutions for interface equations coupled with diffusion equations', *SIAM J. Math. Anal.* **23**, 821–835.

M. A. Grayson (1987), 'The heat equation shrinks embedded plane curves to round points', *J. Diff. Geom.* **26**, 285–314.

M. A. Grayson (1989), 'A short note on the evolution of a surface by its mean-curvature', *Duke Math. J.* **58**, 555–558.

L. Greengard and J. Strain (1990), 'A fast algorithm for evaluating heat potentials', *Comm. Pure Appl. Math.* **43**, 949–963.

A. Harten, B. Engquist, S. Osher and S. R. Chakravarthy (1987), 'Uniformly high order accurate essentially non-oscillatory schemes. III', *J. Comp. Phys.* **71**, 231–303.

J. J. Helmsen (1994), A Comparison of Three-Dimensional Photolithography Development Methods, PhD thesis, University of California, Berkeley.

C. W. Hirt and B. D. Nicholls (1981), 'Volume of fluid (COF) method for dynamics of free boundaries', *J. Comp. Phys.* **39**, 201–225.

G. Huisken (1984), 'Flow by mean curvature of convex surfaces into spheres', *J. Diff. Geom.* **20**, 237–266.

T. Ilmanen (1992), 'Generalized flow of sets by mean curvature on a manifold', *Indiana University Mathematics Journal* **41**, 671–705.

T. Ilmanen (1994), *Elliptic Regularization and Partial Regularity for Motion by Mean Curvature*, Memoirs of the American Mathematical Society, 108.

M. Kass, A. Witkin and D. Terzopoulos (1988), 'Snakes: Active contour models', *Int. J. Comp. Vision* pp. 321–331.

I. V. Katardjiev, G. Carter and M. J. Nobes (1988), 'Precision modeling of the mask-substrate evolution during ion etching', *J. Vac. Sci. Tech. A* **6**, 2443–2450.

S. Kichenassamy, A. Kumar, P. Olver, A. Tannenbaum and A. Yezzi (1995), *Gradient Flows and Geometric Active Contours, 1994*, ICCV.

R. Kimmel (1995), Curve Evolution on Surfaces, PhD thesis, Dept. of Electrical Engineering, Technion, Israel.

R. Kimmel and A. Bruckstein (1992), Shape from shading via level sets, Technical Report 9209, Technion, Israel. Center for Intelligent Systems.

R. Kimmel and A. Bruckstein (1993), 'Shape offsets via level sets', *Computer-Aided Design* **25**, 154–161.

P. Knupp and S. Steinberg (1993), The fundamentals of grid generation. Preprint.

P. D. Lax (1970), *Hyperbolic Systems of Conservation Laws and the Mathematical Theory of Shock Waves*, SIAM Reg. Conf. Series, Lectures in Applied Math, 11, SIAM, Philadelphia.

F. A. Leon, S. Tazawa, K. Saito, A. Yoshi and D. L. Scharfetter (1993), Numerical algorithms for precise calculation of surface movement in 3-D topography simulation, in *1993 International Workshop on VLSI Process and Device Modeling.*

R. J. LeVeque (1992), *Numerical Methods for Conservation Laws*, Birkhäuser, Basel.

P.-L. Lions (1982), *Generalized Solution of Hamilton–Jacobi Equations*, Pitman, London.

A. Majda and J. A. Sethian (1984), 'Derivation and numerical solution of the equations of low mach number combustion', *Combustion Science and Technology* **42**, 185–205.

R. Malladi, D. Adalsteinsson and J. A. Sethian (1995a), 'A fast level set algorithm for 3D shape recovery', *IEEE Transactions on PAMI*. Submitted.

R. Malladi and J. A. Sethian (1994), A unified approach for shape segmentation, representation, and recognition, Technical Report 36069, Lawrence Berkeley Laboratory, University of California, Berkeley.

R. Malladi and J. A. Sethian (1995), 'Image processing via level set curvature flow', *Proc. Natl. Acad. of Sci., USA* **92**, 7046–7050.

R. Malladi and J. A. Sethian (1996a), 'Image processing: Flows under min/max curvature and mean curvature', *Graphical Models and Image Processing*. In press.

R. Malladi and J. A. Sethian (1996b), 'A unified approach to noise removal, image enhancement, and shape recovery', *IEEE Image Processing*. In press.

R. Malladi, J. A. Sethian and B. C. Vemuri (1994), Evolutionary fronts for topology-independent shape modeling and recovery, in *Proceedings of Third European Conference on Computer Vision*, LNCS Vol. 800, Stockholm, pp. 3–13.

R. Malladi, J. A. Sethian and B. C. Vemuri (1995b), 'Shape modeling with front propagation: A level set approach', *IEEE Trans. on Pattern Analysis and Machine Intelligence* **17**, 158–175.

J. P. McVittie, J. C. Rey, A. J. Bariya et al. (1991), SPEEDIE: A profile simulator for etching and deposition, in *Proceedings of the SPIE The International Society for Optical Engineering*, Vol. 1392, pp. 126–38.

B. Merriman, J. Bence and S. J. Osher (1994), 'Motion of multiple junctions: A level set approach', *J. Comp. Phys.* **112**, 334–363.

B. Milne and J. A. Sethian (1995), 'Adaptive mesh refinement for level set methods for propagating interfaces', *J. Comp. Phys.* Submitted.

W. Mulder, S. J. Osher and J. A. Sethian (1992), 'Computing interface motion in compressible gas dynamics', *J. Comp. Phys.* **100**, 209–228.

W. W. Mullins and R. F. Sekerka (1963), 'Morphological stability of a particle growing by diffusion or heat flow', *J. Appl. Phys.* **34**, 323–329.

W. Noh and P. Woodward (1976), A simple line interface calculation, in *Proceedings, Fifth International Conference on Fluid Dynamics* (A. I. van de Vooran and P. J. Zandberger, eds), Springer, Berlin.

S. Osher and L. I. Rudin (1990), 'Feature-oriented image enhancement using shock filters', *SIAM J. Num. Anal.* **27**, 919–940.

S. Osher and L. I. Rudin (1992), Rapid convergence of approximate solutions of shape-from-shading. To appear.

S. Osher and J. A. Sethian (1988), 'Fronts propagating with curvature dependent speed: Algorithms based on Hamilton–Jacobi formulation', *J. Comp. Phys.* **79**, 12–49.

S. Osher and C. Shu (1991), 'High-order nonoscillatory schemes for Hamilton–Jacobi equations', *J. Comp. Phys.* **28**, 907–922.

P. Perona and J. Malik (1990), 'Scale-space and edge detection using anisotropic diffusion', *IEEE Trans. Pattern Analysis and Machine Intelligence* **12**, 629–639.

M. Z. Pindera and L. Talbot (1986), Flame-induced vorticity: The effects of stretch, in *Twenty-First Symposium (Int'l) on Combustion*, The Combustion Institute, Pittsburgh, pp. 1357–1366.

W. H. Press, B. P. Flannery, S. A. Teukolsky and W. T. Vetterling (1988), *Numerical Recipes*, Cambridge University Press.

E. G. Puckett (1991), A volume-of-fluid interface tracking algorithm with applications to computing shock wave refraction, in *Proceedings of the 4th International Symposium on Computational Computational Fluid Dynamics, Davis, California*.

J. C. Rey, L.-Y. Cheng, J. P. McVittie and K. C. Saraswat (1991), 'Monte Carlo low pressure deposition profile simulations', *J. Vac. Sci. Tech. A* **9**, 1083–7.

C. Rhee, L. Talbot and J. A. Sethian (1995), 'Dynamical study of a premixed V flame', *J. Fluid Mech.* **300**, 87–115.

E. Rouy and A. Tourin (1992), 'A viscosity solutions approach to shape-from-shading', *SIAM J. Num. Anal.* **29**, 867–884.

L. Rudin, S. Osher and E. Fatemi (1992), Nonlinear total variation-based noise removal algorithms, in *Modelisations Mathématiques pour le Traitement d'Images*, INRIA, pp. 149–179.

G. Sapiro and A. Tannenbaum (1993a), 'Affine invariant scale-space', *Int. J. Comp. Vision* **11**, 25–44.

G. Sapiro and A. Tannenbaum (1993b), Image smoothing based on affine invariant flow, in *Proc. of the Conference on Information Sciences and Systems*, Johns Hopkins University, Baltimore.

G. Sapiro and A. Tannenbaum (1994), Area and length preserving geometric invariant scale-spaces, in *Proc. of Third European Conference on Computer Vision*, Vol. 801 of *LNCS*, Stockholm, pp. 449–458.

E. W. Scheckler (1991), PhD thesis, EECS, University of California, Berkeley.

E. W. Scheckler, K. K. H. Toh, D. M. Hoffstetter and A. R. Neureuther (1991), 3D lithography, etching and deposition simulation, in *Symposium on VLSI Technology*, Oiso, Japan, pp. 97–98.

R. Sedgewick (1988), *Algorithms*, Addison-Wesley, New York.

J. A. Sethian (1982), An Analysis of Flame Propagation, PhD thesis, Department of Mathematics, University of California, Berkeley.

J. A. Sethian (1984), 'Turbulent combustion in open and closed vessels', *J. Comp. Phys.* **54**, 425–456.

J. A. Sethian (1985), 'Curvature and the evolution of fronts', *Comm. Math. Phys.* **101**, 487–499.

J. A. Sethian (1987), Numerical methods for propagating fronts, in *Variational Methods for Free Surface Interfaces* (P. Concus and R. Finn, eds), Springer, New York.

J. A. Sethian (1989), Parallel level set methods for propagating interfaces on the connection machine. Unpublished manuscript.

J. A. Sethian (1990), 'Numerical algorithms for propagating interfaces: Hamilton–Jacobi equations and conservation laws', *J. Diff. Geom.* **31**, 131–161.

J. A. Sethian (1991), A brief overview of vortex methods, in *Vortex Methods and Vortex Motion* (K. Gustafson and J. A. Sethian, eds), SIAM Publications, Philadelphia.

J. A. Sethian (1994), 'Curvature flow and entropy conditions applied to grid generation', *J. Comp. Phys.* **115**, 440–454.

J. A. Sethian (1995a), 'Algorithms for tracking interfaces in CFD and material science', *Annual Review of Computational Fluid Mechanics*.

J. A. Sethian (1995b), Level set techniques for tracking interfaces; fast algorithms, multiple regions, grid generation and shape/character recognition, in *Curvature flow and related topics*, Gakuto Int. Series, Volume 5, Tokyo.

J. A. Sethian (1995c), 'A marching level set method for monotonically advancing fronts', *Proc. Nat. Acad. Sci.* To appear.

J. A. Sethian (1996), *Level Set Methods: Evolving Interfaces in Geometry, Fluid Mechanics, Computer Vision and Material Science*, Cambridge University Press. To appear.

J. A. Sethian, D. Adalsteinsson and R. Malladi (1996), 'Efficient fast marching level set methods'. In progress.

J. A. Sethian and J. D. Strain (1992), 'Crystal growth and dendritic solidification', *J. Comp. Phys.* **98**, 231–253.

P. E. Souganidis (1985), 'Approximation schemes for viscosity solutions of Hamilton–Jacobi equations', *J. Diff. Eqns.* **59**, 1–43.

J. Strain (1988), 'Linear stability of planar solidification fronts', *Physica D* **30**, 297–320.

J. Strain (1989), 'A boundary integral approach to unstable solidification', *J. Comp. Phys.* **85**, 342–389.

J. Strain (1990), 'Velocity effects in unstable solidification', *SIAM J. Appl. Math.* **50**, 1–15.

M. Sussman, P. Smereka and S. J. Osher (1994), 'A level set method for computing solutions to incompressible two-phase flow', *J. Comp. Phys.* **114**, 146–159.

J. E. Taylor, J. W. Cahn and C. A. Handwerker (1992), 'Geometric models of crystal growth', *Acta Metallurgica et Materialia* **40**, 1443–74.

D. Terzopoulos, A. Witkin and M. Kass (1988), 'Constraints on deformable models: Recovering 3d shape and nonrigid motion', *Artificial Intelligence* **36**, 91–123.

K. K. H. Toh (1990), PhD thesis, EECS, University of California, Berkeley.

K. K. H. Toh and A. R. Neureuther (1991), 'Three-dimensional simulation of optical lithography', *Proceedings SPIE, Optical/Laser Microlithography IV* **1463**, 356–367.

M. S. Young, D. Lee, R. Lee. and A. R. Neureuther (1993), 'Extension of the Hopkins theory of partially coherent imaging to include thin-film interference effects', *Proceedings SPIE, Optical/Laser Microlithography VI* **1927**, 452–463.

J. Zhu and P. D. Ronney (1995), 'Simulation of front propagation at large non-dimensional flow disturbance intensities', *Comb. Sci. Tech.* To appear.

J. Zhu and J. A. Sethian (1992), 'Projection methods coupled to level set interface techniques', *J. Comp. Phys.* **102**, 128–138.